CONVERSION FACTORS AND SELECTED CONSTANTS

$\ln x = 2.303 \log_{10} x$

$\ln 10 = 2.303 \ldots$

$e = 2.718 \ldots$

$\pi = 3.1416 \ldots$

1 in. $= 2.54$ cm $= 0.0254$ m

1 lb mass $= 454$ g $= 0.454$ kg

1 Å $= 10^{-8}$ cm

$^\circ C = (^\circ F - 32)/1.8$

$K = {^\circ C} + 273$

$^\circ R = {^\circ F} + 460$

1 poise $= 0.1$ Pa-sec

1 lb force/in.$^2 = 6.9 \times 10^3$ N/m$^2 = 6.9 \times 10^{-3}$ MN/m$^2 = 7.03 \times 10^{-4}$ kg/mm^2

1 cal $= 4.186$ J

1 eV $= 1.6 \times 10^{-19}$ J

1 erg $= 1 \times 10^{-7}$ J

$R =$ gas constant $= 1.987$ cal/mole

$k =$ Boltzmann's constant $= 1.3 \times 10^{-16}$ erg/K

$h =$ Planck's constant $= 6.62 \times 10^{-27}$ erg-sec

Avogadro's number $= 6.02 \times 10^{23}$ atoms/at. wt.

$\qquad\qquad\quad = 6.02 \times 10^{23}$ molecules/mol. wt.

Density of water $= 1$ g/cm$^3 = 62.4$ lb/ft$^3 = 0.0361$ lb/in.^{3}

1 gal water weighs 8.33 lb

Electron charge $=$ electron hole charge $= 1.6 \times 10^{-19}$ coul

1 Bohr magneton $= 9.27 \times 10^{-24}$ amp-m$^2 = 0.927$ erg/gauss

$c =$ velocity of light $= 3 \times 10^{10}$ cm/sec

$$BCC \Rightarrow (4r)^2 = 3a_0^2$$

ENGINEERING MATERIALS AND THEIR APPLICATIONS

ENGINEERING
MATERIALS
AND THEIR
APPLICATIONS

Richard A. Flinn
Professor of Materials and
Metallurgical Engineering
University of Michigan, Ann Arbor

Paul K. Trojan
Professor of Materials and
Metallurgical Engineering
University of Michigan, Dearborn

HOUGHTON MIFFLIN COMPANY / BOSTON
Atlanta   Dallas   Geneva, Ill.   Hopewell, N.J.   Palo Alto   London

To Edwina and Barbara

Library of Congress Catalog Card Number: 74-15399

ISBN: 0-395-18916-0

CONTENTS

6 Steel, Superalloys, Cast Iron, Ductile Iron, Malleable Iron   166

13 Electrical Properties of Materials   452

PREFACE

In the past two decades we have seen several swings of the pendulum in faculty thinking and student response to courses in materials for the general engineering student. In the 1950s most of the elementary texts emphasized the mechanical properties of materials with long, dull lists of chemical specifications and descriptions of processing. Then in the 1960s it became widely recognized that the properties of a material depend not merely on its chemical analysis but also on its structure. By "structure" we mean not only the kinds of atoms present, but how the atoms are arranged in the unit cell, the assembly of unit cells into grains, and the resulting microstructure. The application of the phase diagram and of kinetics of reactions to the control of structure became important. At the present time, with the added input of solid-state physics and chemistry, the mass of theoretical knowledge of the structure of materials is indeed impressive. And so there is a great temptation in a general course on materials to dwell on only the basic concepts of materials science and to assume that students will learn the applications in their professional work.

We have found, however, that just as it would be unfair to expect students to design a gas turbine after taking only a course in theoretical physics, we cannot expect students to select materials intelligently if they have studied only the fundamental concepts, such as the behavior of single crystals.

The objective of this text is to give students a thorough grasp of the structures encountered in the principal families of materials—metals, ceramics, and polymers—and then to show how the properties of the important engineering materials depend on these structures. At first this may appear to be an impossibly broad objective, but it rests basically on the power of the structural approach to materials. Let us illustrate this by an example. The steels constitute over 90 percent of our metallic materials and include thousands of chemical compositions. Despite this complexity, it is possible to give the student a thorough grasp of the entire field as follows.

First we point out that there are only four basic structures in this myriad of compositions, and we analyze the unit cells of austenite, ferrite, iron carbide, and martensite. Next we discuss how the amount, size, shape, and distribution of these phases may be controlled from a combination of a knowledge of the iron–iron carbide phase diagram and the kinetic concepts of hardenability. With this background, the production and control of structures with different mechanical properties, from plain-carbon constructional steels through the hardenable alloy grades, can be readily understood. If the instructor or student wishes to go further, optional sections which build on this foundation are provided; these cover the structures and properties of the high-alloy steels (such as the stainless and tool steel grades), the superalloys, and also cast iron, ductile iron, and malleable iron. The same approach is used for the ceramics and the polymers, although in the case of the polymers the emphasis is on the molecular structure.

We hope our colleagues will understand that we are not advocating the substitution of descriptive applications for basic theory. We find, however, that students obtain a firmer grasp of the more important points of materials science by applying them in a variety of relevant situations to important materials. If we may be forgiven a simple analogy to illustrate the student response, we may recall the various schools of violin teaching. In the old European school, students spend long hours on carefully designed exercises to develop technique; in the other school, students are allowed to apply the technique as they learn it to the beautiful compositions of the musical literature. In a parallel way, we have found that after taking up the simple metallic unit cells, mechanisms of atom movement by deformation and thermal activation, and the phase diagrams, students are ready to apply their knowledge in understanding important alloys of aluminum, copper, and iron, and in specifying selection and heat treatment of these alloys. The same procedure is then followed in studying the ceramics and polymers. In many cases, after understanding the relevance of materials science to professional engineering, students will be motivated to study the science in greater depth in advanced courses.

A few words on the use of this text in teaching and learning are in order. The complete text is generally considered too long for a one-semester course. For this reason the sections that can be omitted without loss of the background needed for later sections are marked with a rule at the right margin of the page. In this way the text is easily covered in one semester. For example, in the rather lengthy chapter on steel, a good deal of the material can be omitted without affecting the understanding of later chapters. We believe, however, that it is helpful to have this material available for students for self-study as the need arises and as source material for the instructor. Also, to facilitate the use of the text early in the curriculum, we have avoided the use of calculus but have employed important relations, as in diffusion, that are derived using calculus.

We have added several features that should be helpful. Most of the microstructures that are illustrated show indentations made with the same load using a Vickers diamond microhardness tester. This gives an indication of the relative hardness of the different phases present and a feel for their individual contributions. Microhardness impressions can only be compared in the same photograph because different percent reductions have been used.

We have also prepared a number of original photomicrographs, using the scanning electron microscope and the electron microprobe, where these techniques gave important additional insight into the structures. Also, instead of supplying repetitive references in each chapter, we have added an annotated list of references at the end of the text.

In addition to providing the basic problems involving typical calculations, we have added a number of problems dealing with materials selection or component failure from our professional experiences. In many cases there is no simple quantitative answer to these problems, and we have prepared a manual for the instructor giving the background of the case and our opinions.

To emphasize the important concepts and their interrelation and to facilitate study, we have added a Summary and Definitions section at the end of each chapter.

Throughout we have used the system of units now in common use in the United States. When this is not the metric system or SI (International System of Units), we have given in parentheses the particular metric or SI equivalent that we believe will be most common in the future. In a few instances we have not included conversion factors because they would be confusing. For example, in the discussion of engineering vs. true strain in a 0.505-in.-diameter standard tensile bar, the metric equivalent of the diameter is not included because it would give the misleading impression that this is the standard bar in European practice. Also, the argument only involves relative differences, so that the system of units is unimportant.

Finally, we wish to acknowledge the many helpful criticisms we received from the staff of Rensselaer Polytechnic Institute as well as from our colleagues at the University of Michigan. In addition, we appreciate the contributions of Professor P. A. Flinn to Chaps. 13, 14, and 15. We are especially grateful to William Talbott for assistance in preparing the final manuscript and the solutions manual, and to Mrs. A. G. Crowe for editorial assistance.

ENGINEERING
MATERIALS
AND THEIR
APPLICATIONS

1

A GENERAL VIEW OF THE PROBLEM OF MATERIALS SELECTION AND DEVELOPMENT

THIS photograph illustrates the three important families of materials available to the engineer—the metallic, such as steel and aluminum; the ceramic, such as glass and china; and the plastics or polymers, such as polyethylene.

In this chapter we will discuss how it is necessary to attain a basic understanding of the structure of each of these groups in order to anticipate how they will perform under service conditions. We will define the meaning of the word "structure" as indicating (1) the nature of the atoms of which a material is composed, (2) the arrangement of the atoms into structural units called "unit cells" (also "molecules" in the case of the polymers), and (3) the grouping of unit cells to form grains or crystals.

A variety of materials are used by the container industry. Metal cans are steel and aluminum. Glass containers still have metal caps. Polymeric materials either make up the entire container or are used to coat paper cartons, such as those commonly used for milk.

1.1 General

Before setting out on a trip into new territory, the seasoned canoeist or backpacker will spend a good deal of time studying the overall maps of the country to be traveled and planning the time and resources carefully for the days ahead. In the same way a good professional engineer will always try to map out the problem as a whole before becoming involved in the details.

In this chapter we will take an overall view of the course we will follow—the examination of the key features of the general problem of selecting and developing materials for any application. We will begin with a simple illustration, the problem of containers for beverages. Next, we will discuss how the *structure* of a material—the atoms, unit cells, and the microstructures—basically determines its properties. Following this we will consider how, on the basis of structure, all materials can be divided into three groups—metals, ceramics, and polymers—and how these structures lead to characteristic properties and service performance.

1.2 A typical illustration

To come to grips quickly with the problem as a whole, consider the containers shown in the frontispiece for Chap. 1. These cans, glass bottles, and plastic containers represent the three groups of materials—metals, ceramics, and plastics. Why are these materials being used instead of others? Although the containers themselves are inexpensive, total consumption has been great enough to call for the expenditure of many millions of dollars in research and development. To illustrate the problem, on page 5 we list in a very simple qualitative way the factors involved in the selection and evaluation of the materials.

The ideal bottle would be strong, tough, reusable, and light, as well as in compliance with the Food and Drug Administration rules. We see that none of the materials meets all these criteria.

The essence of materials engineering is first to understand the application, then to establish the characteristics of an ideal material, and finally to optimize, to make the best choice of available materials to produce the most economical and safest component. Everyone is aware of the competition between aluminum and glass as containers for soda pop and beer, but there is also competition within the metals industry itself between steel and aluminum. The makers of metallic materials point to their shock resistance, low weight, and rapid cooling, while the backers of glass bottles emphasize better corrosion resistance, leading to avoidance of interference with flavor and reusability. The balance is close, and in many cases ecological factors can tip the scales. In the milk container field, plastic has largely replaced glass on the basis of weight, toughness, and cost, where reusability is not a factor. Within the plastics

	Metal cans	Glass bottles	Plastic bottles
Typical material	Aluminum	Glass	Polyethylene
Strength	High	High	Moderate
Toughness	High	Low	High
Hardness	Low	High	Low
Corrosion resistance	High	High	High
Thermal conductivity	High	Low	Low
Ability to be shaped	High	High	High
Reusability	Low	High	Low
Relative weight	Low	Moderate	Low
Cost	All closely competitive		

industry the competition between plastic-coated paper containers and all-plastic bottles is intense.

Finally, an additional factor, recycling, has become important recently. Metal and glass containers are easily recycled, and there are possibilities of using the plastics as fuel or reclaiming them for reuse.

1.3   The structure of materials

At this point we have gone as far as we can in analyzing the container problem. To understand this type of problem better and to make the proper choice of materials, we must develop a deeper knowledge of the materials themselves and the reasons for their strength, toughness, and corrosion resistance.

To understand the structure of a material we need to know (1) which atoms are present and (2) how they are bonded (arranged). We can usually show the arrangement by means of a very simple unit cell, such as a cubic or hexagonal structure, consisting of just a few atoms. We find that millions of these cells are arranged together to form the grains of the material. The grains are visible with a microscope and often are large enough to be seen with the naked eye, as in a brass doorknob.

By understanding these elements of structure—atomic structure, unit cell structure, and microstructure—we can classify and simplify the hundreds of thousands of materials into relatively few typical structures and their combinations.

For simplicity, in the following pages when we use the term "structure" we will be referring to any or all of these aspects. For example, if we talk about the effect of nickel on the structure of iron, we mean the effect of atomic interaction, the effect on the dimensions of the unit cell, and the changes in microstructure and macrostructure. Naturally we will discuss each effect in detail where necessary.

After describing these structures and how they respond to stress, temperature, electrical potential, and magnetic fields, we will be able not only to comprehend the specific properties of these materials, but also to predict how they will behave under unusual circumstances.

Let us digress a moment to illustrate the basic differences between understanding the structure and simply knowing the chemical analysis of a material. Consider a simple coat hanger of steel wire, as shown in Fig. 1.1, which performs adequately supporting a heavy pair of pants. Now suppose we heat the hanger to a red color (about 1700°F or 910°C) and let it cool slowly. The chemical analysis is unchanged, but the change in structure, as seen in the photomicrograph, leads to the hanger's failure to perform. In Chap. 3 we will take up in detail the structural changes occurring on heating. In Chap. 6 we shall see that it is possible to understand the properties of thousands of different steels, processed in many ways, by following the changes in only seven simple microstructures.

The greater simplicity of this new structural approach is only one of its advantages. With the old approach it is possible to look up mechanical properties in a handbook. However, if the steel is subsequently to be processed in fabricating a component—by welding or rolling, for example—it is far easier to anticipate how the properties will be affected from a basic knowledge of how the microstructures will be changed rather than from handbook data, if these can be found at all. In addition, in analyzing the success or failure of the component under complex service conditions involving wear, impact, fatigue, and other problems, valuable information can be found by observing the changes in the structure in service. Now let us see how the structures of engineering materials vary.

1.4   The basic divisions of engineering materials

If we survey the entire field of engineering materials, we find that there are three principal groups:

1. Metals: steel, copper, cast iron, etc.
2. Ceramics: glass, brick, cement, etc.
3. Plastics and other high polymers: polyethylene, rubbers, etc.

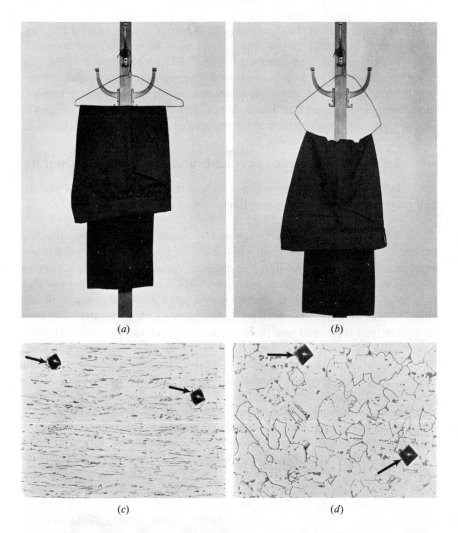

Fig. 1.1 *Effect of annealing and cold work on the microstructure and service performance of a clothes hanger. (a) Cold-worked hanger, normal form. The grains are elongated in the direction in which the wire was drawn through a die. (b) An annealed clothes hanger. The larger equiaxed grains make the wire hanger too soft and weak. (c) Normal hanger, Vickers hardness number (VHN) 173 to 183, 500X, 2 percent nital etch. (d) Hanger annealed at 1700°F (910°C) for 0.5 hr, VHN 108 to 118, 500X, 2 percent nital etch.*

The grain structures shown were from samples which were longitudinally cut, then polished and etched with 2 percent nitric acid in alcohol (nital). The "normal" hanger was work-hardened by drawing through a die. When heated, new softer grains were formed.

The VHN hardness was obtained by pressing a light 25-g load into individual grains with a pyramid-shaped diamond indenter (see arrows).

The differences in the characteristics of each group can be traced to basic differences in bonding between atoms and groups of atoms. For example, in the metals the metallic bond leads to ductility and high electrical conductivity; in the ceramics the ionic and covalent bonds give rise to generally high hardness, brittle fracture, and high electrical resistance; and in the plastics the covalent and intermolecular van der Waals forces lead to the variety of properties encountered in thermoplastic and thermosetting resins.

1.5   Metallic, ionic, covalent, and van der Waals bonds

Let us examine now in simplified fashion these different types of bonding and observe how they lead to the differences in properties.

The Metallic Bond. We know from elementary chemistry that the atomic structure of any element is made up of a positively charged nucleus surrounded by electrons revolving at different distances. The atomic number is equal to the number of these "planetary" electrons. We know further that it is helpful to consider these electrons as present at different energy levels, called "shells" or "rings." For example, magnesium, which has an atomic number of 12, contains two electrons in the first shell (nearest the nucleus), eight electrons in the second shell, and two electrons in the outer shell (Fig. 1.2a).

A common characteristic of the metallic elements is that they contain only one, two, or three electrons in the outer shell. This is the key to the formation of the metallic bond. When an element has only one, two, or three electrons in the outer shell, these are bonded relatively loosely to the nucleus. Therefore, when we put a number of magnesium atoms together in a block of magnesium, the outer electrons leave individual atoms to enter a common "electron gas" (Fig. 1.2b). The atoms are therefore changed to Mg^{2+} ions. These

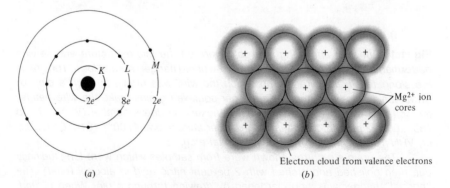

(a) (b)

Fig. 1.2 (a) *Simplified sketch of a magnesium atom with its electron shells.* (b) *Atoms (ions) in solid magnesium surrounded by "electron gas."*

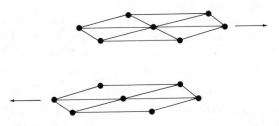

Fig. 1.3 *Slip between close-packed atom planes in magnesium.*

repel each other but are still held together in the block because of the attraction of the negative electrons to the positive ions. The end result is that the Mg^{2+} ions arrange themselves in the regular pattern shown. This is the arrangement in one plane cut through a crystal of magnesium. It represents the closest possible packing for a group of equally sized spheres and is the arrangement we would encounter if we shook a box of tennis balls. Many other metals show this same structure. Let us stop our discussion of metallic structure at this point and see how even this very simple model helps us to understand a few metallic properties.

First the high electrical conductivity of metals can be explained. If we apply a voltage across the crystal, the electrons in the electron gas (which are loosely bonded) will move readily, giving a current. Next the ductility or ability to deform without fracturing can be understood. We note that the atoms are closely packed in planes. If we apply a shearing stress, as in Fig. 1.3, one plane will slide over the other and there will be no fracture, because after a movement of one atomic diameter the same forces between atoms are still operative. We will discuss many other features of the metallic structure in the chapters that follow, but this brief illustration will serve for the present.

The ceramics and plastics contain by contrast ionic, covalent, and van der Waals bonds.

The Ionic Bond. This bond occurs between metallic and nonmetallic elements (Fig. 1.4a). We recall from chemistry that in a nonmetallic element the outer shell contains a larger number of electrons than it does in a metal. For example, the outer shell of a chlorine atom contains seven electrons and there is a strong driving force to attract an electron to make a stable group of eight. When sodium with one valence electron and chlorine with seven react, the sodium gives up its electron to form a stable outer shell for the chlorine. This is called an ionic bond because Na^+ and Cl^- ions are formed and are naturally attracted because of their opposite charges. If the sodium chloride is dissolved in water, the ions exist independently in solution. In a *crystal* of sodium chloride, however, the ions arrange themselves in an electrically balanced structure, as shown by a section through the crystal (Fig. 1.4b).

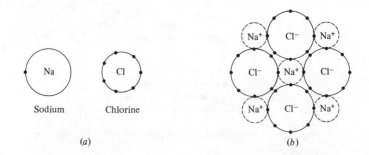

(a)   (b)

Fig. 1.4 *Greatly simplified sketch of the formation of the ionic bond in sodium chloride. (a) Un-ionized sodium and chlorine atoms. (b) Ionized sodium and chloride ions. (The eight outer electrons in the chloride ion are at different energy levels and not at fixed distances.) The valence electron of sodium is bonded to the chloride ion. For simplicity, only the outer electron shells are shown.*

Let us compare the properties of this structure with those of the metallic structure. The electrical conductivity of the solid is many orders of magnitude lower than that of the metals because the electrons are tightly held in place instead of being free in the electron gas. The crystal fractures in brittle fashion because when we attempt to slide one plane of ions across another, the electrical fields of different ions are opposed. Furthermore, the actual fracture follows cleavage planes which have certain ionic arrangements, such as that shown in Fig. 1.4b.

The Covalent Bond. In many ceramic materials the covalent bond is also encountered. The feature of this bond is that electrons are tightly held and equally shared by the participating atoms. An outstanding example is found in the diamond crystal (Fig. 1.5a), where each carbon atom is in the center of a tetrahedron, sharing each of its four electrons with the adjoining atoms. In many cases, such as silica (SiO_2), only a portion of the bonding is covalent. This takes place when there is a compromise between perfectly balanced attraction for the shared electrons, as in diamond, and strong ionic attraction, as in sodium chloride. In the case of silica, the four outer electrons of silicon can be shared with the outer shell electron from each of four oxygen atoms to make a stable shell of eight. Thus we have each silicon atom surrounded by a tetrahedron of four oxygen atoms. Each oxygen atom is placed between two silicon atoms and shares an electron on each side (Fig. 1.5b). The bond is also partly ionic because the oxygen is somewhat more electronegative and tends to attract electrons more strongly. In one form of silica, quartz, this bonding results in high hardness and low electrical conductivity.

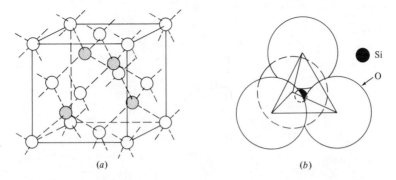

Fig. 1.5 (a) *Covalent bonding as seen in the unit cell of diamond. Atoms at the centers of the tetrahedra have been shaded.* (b) *Tetrahedral structure of an* SiO_4^{4-} *ion. Compare the Si—O bonds with those of diamond.*

EXAMPLE 1.1 Sketch the valence electron configuration in an SO_4^{2-} ion.

ANSWER We will assume that the valence electrons of sulfur and oxygen interact and are shared. In the figure the electrons for atoms are denoted by • or × to indicate the parent element, although it must be appreciated that the identity is lost in the electron-sharing process. (Sulfur and oxygen both have six electrons in their outer shells in the un-ionized state.)

Two electrons from some outside source to satisfy the desired configuration of eight. (Such a source might be one each from Na atoms which then become Na^+ and form the compound Na_2SO_4.)

The van der Waals Bond. This type of bond occurs to some extent in all materials but is particularly important in plastics or high polymers. The molecules are made up of a backbone or skeleton of covalently bonded carbon atoms with other atoms—such as hydrogen, nitrogen, fluorine, chlorine, sulfur, and oxygen—attached to the other carbon bonds at the sides (Fig. 1.6).

The covalent bonds within the molecule are very strong and rupture only under extreme conditions. It is the bonds between the molecules that follow the sliding and finally rupture which are of great importance. These bonds are called "van der Waals forces" and require separate discussion.

Even when the ionic and covalent bonds are present, there is still some imbalance in the electrical charge in a molecule. For example, it can be shown

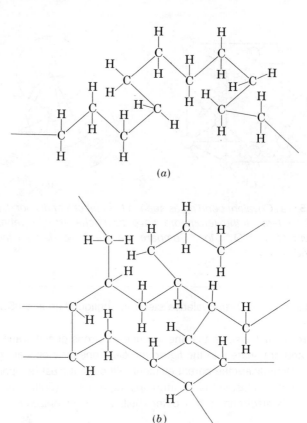

Fig. 1.6 (a) *Carbon backbone in a linear polymer (thermoplastic), and (b) carbon network in a thermosetting polymer.*

that in a simple water molecule the hydrogen atoms are connected to the oxygen by bonds at an angle of 109°. This gives a positive polarity at the hydrogen-rich end of the molecule and a negative polarity at the other end (Fig. 1.7). As a result, the molecules attract one another. This is the force holding water molecules together in the liquid state. Let us see how these concepts explain some properties of the plastics.

As molecules become larger, the total of the van der Waals forces between the molecules increases. In polyethylene molecules, for example, we can have the same ratio of hydrogen to carbon atoms as in simple ethylene gas, but as the number of atoms in the molecule increases, we pass from a gas to a liquid and finally to a solid plastic.

Later we shall discuss two important families of high polymers, the thermoplastic and thermosetting materials. We can mention here, however, that

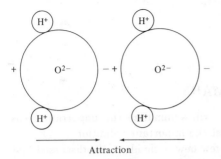

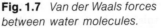
Attraction

Fig. 1.7 *Van der Waals forces between water molecules.*

the thermoplastic materials are converted to liquids by heating and then harden on cooling. The effect of the heat is to overcome the van der Waals forces between the molecules by thermal agitation, making flow possible. By contrast, in the thermosetting materials, instead of the long chainlike molecules of the thermoplastics, we have three-dimensional networks of covalent bonds. These are not easily broken by heating, and we therefore have harder, stronger materials; on the other hand, these materials cannot be formed into shapes as easily as the thermoplastics (Fig. 1.6).

These are just a few examples of the correlation between structure and properties. Later we shall see another example of this, when we show how the extraordinary elastic extension of rubber is due to the shape and distribution of its molecules.

1.6 Outline of the text

From the preceding discussion we see that we have three types of materials to understand and that these materials exhibit a wide range of properties. In the chapters ahead we shall follow the following program:

1. We first study separately the structures of metals, ceramics, and plastics (Chaps. 1 to 10). In each case the structures are discussed, and then the analyses and some mechanical properties and applications of important commercial materials are taken up.
2. With this knowledge of the different structures in mind, it is then important to compare the properties of different materials. For example, in choosing a magnet for a television set, the engineer needs to consider not only magnets of metallic materials but also the newly developed ceramic magnets. For the comparison of properties the following sections are provided:

Mechanical properties of materials, Chap. 11, in which fatigue, impact, and creep resistance are discussed, in addition to tensile, compressive, and hardness levels. Composite materials and internal stresses are also considered.
Corrosion resistance, Chap. 12.

Electrical properties, Chap. 13.
Magnetic properties, Chap. 14.
Optical and thermal properties, Chap. 15.

SUMMARY

At the end of each chapter we will summarize the important ideas introduced in that chapter and then list the important definitions.

In this chapter there are only a few new basic ideas. We discussed first the problems of a large industry in the selection of materials for containers. However, it was apparent that at this point in the course we could only treat the three classes of materials—metals, ceramics, and polymers—in a superficial way. To understand and predict properties, it is necessary to understand the structures of the different types of materials. With the illustration of the cold-worked and then heat-treated coat hangers, we saw that structure is not just a matter of which atoms are present but concerns how they are organized in the microstructure of the material. Next we discussed the type of bonding in each group of materials and how it affects the properties. In metals the metallic bond leads to an electron gas which provides excellent conductivity. In the ceramics the combination of ionic and covalent bonds leads to high strength, brittle fracture, and cleavage. In the polymers the covalent bonds within the molecules and van der Waals forces between molecules lead to phenomenal elongation under stress in linear (thermoplastic) polymers. On the other hand, the network structure in thermosetting polymers leads to high strength and low elongation.

The contents of the text were then outlined.

DEFINITIONS

Structure The organization of a material, defined by specifying which atoms are present and their amounts, how they are arranged in unit cells, and the grains resulting from the unit cells. Also the presence of several different unit cells leads to grains or phases with different properties.

Bonding

    Metallic Bonding in which metal atoms give up their electrons to an electron gas and take up a regular arrangement.

    Ionic Bonding in which metal and nonmetal ions bond by the metal giving up an electron or electrons, which are taken up by the outer shell of the nonmetal. Therefore, positive ions of the metal and negative ions of the nonmetal are produced and attract each other.

    Covalent The formation of bonds in which atoms of the same element (such as carbon) or different elements share electrons, resulting in a strong bond.

Bonding

Van der Waals A secondary bonding between molecules, primarily due to charge attraction. The attraction results from unsymmetrical charge distribution rather than primary bonds. Note that molecules rather than ions (as in the ionic bond) are attracted.

Metallic materials Materials characterized by metallic bonds.

Ceramic materials Materials exhibiting ionic or covalent bonds or both.

Plastics (high polymers) Materials exhibiting principally covalent bonding. The residual bonding forces, van der Waals bonds, are also important.

PROBLEMS

1.1 List a number of important components in an automobile and catalog each as metallic, ceramic, plastic, or composite. Examples: engine block, spark plug, body, windows.

1.2 Review the Periodic Table of the elements and classify the elements as

   a. Metals
   b. Strongly nonmetallic
   c. Between metals and nonmetals

   (See Table 2.2.)

1.3 Graphite and diamond are both pure carbon. The bonding in one structure is perfectly covalent, while that of the other is partly metallic. One material is a fairly good conductor of electric current, and the other is an insulator. Identify each type of bonding with its properties.

1.4 Despite the strength of the ionic bond, many materials with ionic bonds are not considered "good engineering materials." Why not? (*Hint:* Consider how ionics react to different environments.)

1.5 Sketch the valence electron configuration for a $CO_3{}^{2-}$ ion (see Example 1.1).

1.6 Sketch the valence electron configuration for water (H_2O) and formaldehyde (CH_2O).

1.7 In a given type of bonding the different bond strengths manifest themselves macroscopically as variations in melting or boiling points. The materials listed below are covalently bonded with secondary van der Waals bonds. Arrange them in the anticipated order of lowest to highest boiling points. (If necessary use sketches to support your conclusions.)

   a. C_2H_4, a single ethylene molecule
   b. CH_4, a single methane molecule
   c. C_2H_3Cl, a single vinylchloride molecule
   d. $(C_2H_4)_n$, multiple joined ethylene molecules (polyethylene)

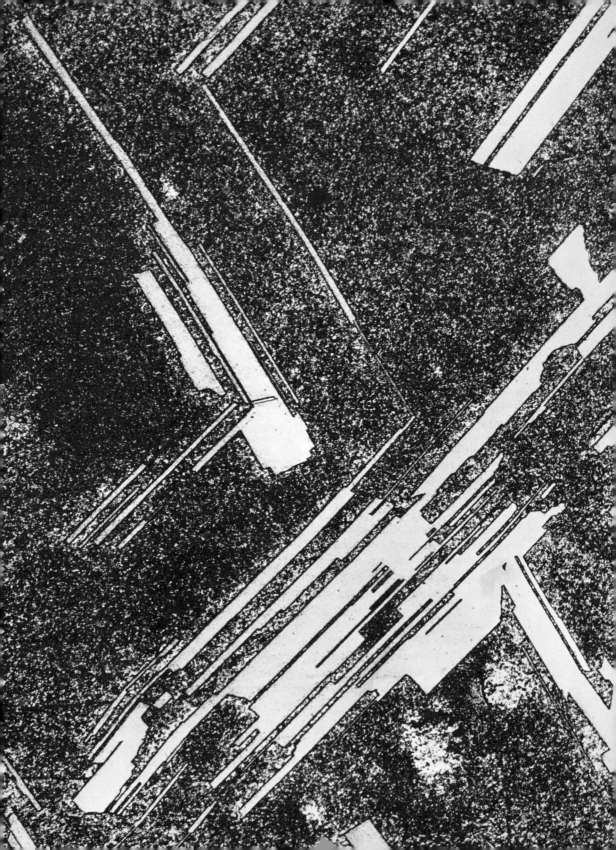

2

THE METALLIC STRUCTURES;
THE UNIT CELL

THIS is a photograph of a specimen of native copper collected
in the upper peninsula of Michigan. The sample was sectioned,
polished, and etched with a weak acid solution to disclose the
crystalline structure. The picture is called a photomicrograph
because it is taken through a microscope (magnification 100X).

The crystals or grains show a cubic structure, and in this
chapter we shall discuss how this is due to the arrangement
of atoms in cubic unit cells. From an understanding of these
elementary building blocks we will be able to show how metals
and materials in general behave under stress and other service
conditions.

2.1 General

In this chapter we begin our study of metallic materials. We will find that the structures of these materials are simpler than those of the ceramics or plastics, but what we learn about the basic features of metallic structures will be valuable in our understanding of the other materials. To obtain a basic knowledge of any material, it is not enough merely to look at it under the microscope. After all, the fundamental building blocks are the *atoms* themselves, so we need to start with some atomic properties. Next we find that the atoms are grouped in *unit cells,* and that by assembling millions of unit cells we obtain the *grains* we see under the microscope. It is important to understand the grain structure because a metal part does not stretch uniformly like a piece of taffy; it stretches by deformation of individual grains. We will find that the properties of a metal are different in different directions in the grain. We will begin, therefore, with a short review of the properties of metallic atoms. Then, after discussing the geometry of unit cells, we will examine metallic unit cells and their role in the structure of grains. We will also begin our study of alloys, which are simply the materials produced when one or more elements are added to a metallic element (without forming a covalent or ionic compound); for example, copper plus 30 percent zinc gives an alloy we know commonly as brass.

2.2 Metal atoms

The properties of a metal atom are largely determined by the number of electrons around the nucleus and their arrangement. We recall from chemistry that the atomic number tells us the number of electrons in orbit around the nucleus and that the nucleus has a positive charge balancing the negative charge of the electrons.

Even the early investigators realized that the electrons had different energies and rotated at different distances from the nucleus. From this early concept simple sketches of the atom were made showing the electrons in rings or shells of different energy levels, as illustrated for magnesium (atomic number = 12) in Fig. 2.1. It is important to review this early concept because some of the nomenclature is still used. For example, we still call the shell or ring closest to the nucleus the K shell, the next the L shell, and so forth (Fig. 1.2a).

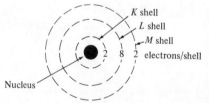

Fig. 2.1 *Early concept of the atomic structure of magnesium (atomic number = 12). Numerals on shells indicate the number of electrons per shell.*

Later investigators, however, found differences in energy and in spin direction among electrons in the same shell, and it became necessary to assign four quantum numbers to each electron rather than merely to state the shell in which it was found.†

In simple terms we can say that if we are to locate an object, such as an electron, in space, we need three coordinates and a fourth if we want to define the direction of spin. Of course the basic explanation is found in quantum theory and is much more complex. The symbols for these four numbers are:

n The principal quantum number gives the shell or ring. The numbering begins with the inner shell. Thus for the electrons in the K shell $n = 1$, for the L shell $n = 2$, etc.

l The second quantum number is limited by the value of n and must be a positive integer equal to or less than $n - 1$. Therefore, in the second shell of magnesium $n = 2$ and l can have values of 1 or 0.

m_l The third quantum number is equal to or less than l but can be positive or negative. Therefore, in the second shell of magnesium m_l can have values of $+1$, 0, or -1.

m_s The fourth quantum number can have values of only $+\frac{1}{2}$ or $-\frac{1}{2}$. This number tells the direction of spin of the electron. We shall see in Chap. 14 that the magnetic properties are basically related to spin.

With one other concept we can proceed to develop the quantum numbers of the electrons in magnesium as an example. The Pauli postulate states that no two electrons can have the same four quantum numbers. With this in mind and starting with the inner ring $n = 1$, we can write the quantum numbers for the 12 electrons of magnesium as follows:

Electron Number	n	l	m_l	m_s	Abbreviated Designation
1	1	0	0	$+\frac{1}{2}$	$1s^2$
2	1	0	0	$-\frac{1}{2}$	
3	2	0	0	$+\frac{1}{2}$	$2s^2$
4	2	0	0	$-\frac{1}{2}$	
5	2	1	-1	$+\frac{1}{2}$	
6	2	1	0	$+\frac{1}{2}$	
7	2	1	$+1$	$+\frac{1}{2}$	$2p^6$
8	2	1	-1	$-\frac{1}{2}$	
9	2	1	0	$-\frac{1}{2}$	
10	2	1	$+1$	$-\frac{1}{2}$	
11	3	0	0	$+\frac{1}{2}$	$3s^2$
12	3	0	0	$-\frac{1}{2}$	

†The vertical line at the right indicates optional material, as explained in the Preface and the Instructor's Manual.

Note that we have also given an abbreviated designation, which is often used as a short-hand notation for the more cumbersome quantum numbers. The first number is the first or principal quantum number, while the letters s, p, etc., designate the value of l as follows:

$$l = 0 \; 1 \; 2 \; 3 \; 4 \; 5$$
$$\text{Letter} = s \; p \; d \; f \; g \; h$$
$$\text{Maximum electrons} = 2 \; 6 \; 10 \; 14 \; 20 \; 24$$

The exponent tells how many electrons of this type are present.

Thus we use the abbreviations $1s^2$, $2s^2$, $2p^6$, $3s^2$ to mean two electrons $n = 1$, $l = 0$; two electrons $n = 2$, $l = 0$; six electrons $n = 2$, $l = 1$; two electrons $n = 3$, $l = 0$.

One other fact is that in p, d, and f electrons we find first the maximum number of positive spins, then of negative spins, instead of alternating directions. For example, electrons 5, 6, and 7 all have $s = +\frac{1}{2}$. (This is called "Hund's rule.") The electronic structures for all the elements are given in Table 2.1.

Table 2.1 ELECTRONIC CONFIGURATION OF THE ELEMENTS*

Atomic No.	Ele- ment	K — 1 — s	L — 2 — s p	M — 3 — s p d	N — 4 — s p d f	O — 5 — s p d f	P — 6 — s p d f	Q — 7 — s
1	H	1						
2	He	2						
3	Li	2	1					
4	Be	2	2					
5	B	2	2 1					
6	C	2	2 2					
7	N	2	2 3					
8	O	2	2 4					
9	F	2	2 5					
10	Ne	2	2 6					
11	Na	2	2 6	1				
12	Mg	2	2 6	2				
13	Al	2	2 6	2 1				
14	Si	2	2 6	2 2				
15	P	2	2 6	2 3				
16	S	2	2 6	2 4				
17	Cl	2	2 6	2 5				
18	Ar	2	2 6	2 6				
19	K	2	2 6	2 6 ..	1			
20	Ca	2	2 6	2 6 ..	2			

Table 2.1 ELECTRONIC CONFIGURATION OF THE ELEMENTS* (*Continued*)

Atomic No.	Ele- ment	K 1 s	L 2 s p	M 3 s p d	N 4 s p d f	O 5 s p d f	P 6 s p d f	Q 7 s
21	Sc	2	2 6	2 6 1	2			
22	Ti	2	2 6	2 6 2	2			
23	V	2	2 6	2 6 3	2			
24	Cr	2	2 6	2 6 5†	1			
25	Mn	2	2 6	2 6 5	2			
26	Fe	2	2 6	2 6 6	2			
27	Co	2	2 6	2 6 7	2			
28	Ni	2	2 6	2 6 8	2			
29	Cu	2	2 6	2 6 10†	1			
30	Zn	2	2 6	2 6 10	2			
31	Ga	2	2 6	2 6 10	2 1			
32	Ge	2	2 6	2 6 10	2 2			
33	As	2	2 6	2 6 10	2 3			
34	Se	2	2 6	2 6 10	2 4			
35	Br	2	2 6	2 6 10	2 5			
36	Kr	2	2 6	2 6 10	2 6			
37	Rb	2	2 6	2 6 10	2 6 ..	1		
38	Sr	2	2 6	2 6 10	2 6 ..	2		
39	Y	2	2 6	2 6 10	2 6 1	2		
40	Zr	2	2 6	2 6 10	2 6 2 ..	2		
41	Nb	2	2 6	2 6 10	2 6 4† ..	1		
42	Mo	2	2 6	2 6 10	2 6 5 ..	1		
43	Tc	2	2 6	2 6 10	2 6 6 ..	1		
44	Ru	2	2 6	2 6 10	2 6 7 ..	1		
45	Rh	2	2 6	2 6 10	2 6 8 ..	1		
46	Pd	2	2 6	2 6 10	2 6 10† ..	..		
47	Ag	2	2 6	2 6 10	2 6 10 ..	1		
48	Cd	2	2 6	2 6 10	2 6 10 ..	2		
49	In	2	2 6	2 6 10	2 6 10 ..	2 1		
50	Sn	2	2 6	2 6 10	2 6 10 ..	2 2		
51	Sb	2	2 6	2 6 10	2 6 10 ..	2 3		
52	Te	2	2 6	2 6 10	2 6 10 ..	2 4		
53	I	2	2 6	2 6 10	2 6 10 ..	2 5		
54	Xe	2	2 6	2 6 10	2 6 10	2 6		
55	Cs	2	2 6	2 6 10	2 6 10 ..	2 6	1	
56	Ba	2	2 6	2 6 10	2 6 10 ..	2 6	2	
57	La	2	2 6	2 6 10	2 6 10 ..	2 6 1 ..	2	
58	Ce	2	2 6	2 6 10	2 6 10 2†	2 6	2	
59	Pr	2	2 6	2 6 10	2 6 10 3	2 6	2	
60	Nd	2	2 6	2 6 10	2 6 10 4	2 6	2	
61	Pm	2	2 6	2 6 10	2 6 10 5	2 6	2	
62	Sm	2	2 6	2 6 10	2 6 10 6	2 6	2	

(*Continued*)

Table 2.1 ELECTRONIC CONFIGURATION OF THE ELEMENTS* (*Continued*)

Atomic No.	Element	K 1 s	L 2 s p	M 3 s p d	N 4 s p d f	O 5 s p d f	P 6 s p d f	Q 7 s
63	Eu	2	2 6	2 6 10	2 6 10 7	2 6	2	
64	Gd	2	2 6	2 6 10	2 6 10 7	2 6 1 ..	2	
65	Tb	2	2 6	2 6 10	2 6 10 9*	2 6	2	
66	Dy	2	2 6	2 6 10	2 6 10 10	2 6	2	
67	Ho	2	2 6	2 6 10	2 6 10 11	2 6	2	
68	Er	2	2 6	2 6 10	2 6 10 12	2 6	2	
69	Tm	2	2 6	2 6 10	2 6 10 13	2 6	2	
70	Yb	2	2 6	2 6 10	2 6 10 14	2 6	2	
71	Lu	2	2 6	2 6 10	2 6 10 14	2 6 1 ..	2	
72	Hf	2	2 6	2 6 10	2 6 10 14	2 6 2 ..	2	
73	Ta	2	2 6	2 6 10	2 6 10 14	2 6 3 ..	2	
74	W	2	2 6	2 6 10	2 6 10 14	2 6 4 ..	2	
75	Re	2	2 6	2 6 10	2 6 10 14	2 6 5 ..	2	
76	Os	2	2 6	2 6 10	2 6 10 14	2 6 6 ..	2	
77	Ir	2	2 6	2 6 10	2 6 10 14	2 6 9† ..	0	
78	Pt	2	2 6	2 6 10	2 6 10 14	2 6 9 ..	1	
79	Au	2	2 6	2 6 10	2 6 10 14	2 6 10 ..	1	
80	Hg	2	2 6	2 6 10	2 6 10 14	2 6 10 ..	2	
81	Tl	2	2 6	2 6 10	2 6 10 14	2 6 10 ..	2 1	
82	Pb	2	2 6	2 6 10	2 6 10 14	2 6 10 ..	2 2	
83	Bi	2	2 6	2 6 10	2 6 10 14	2 6 10 ..	2 3	
84	Po	2	2 6	2 6 10	2 6 10 14	2 6 10 ..	2 4	
85	At	2	2 6	2 6 10	2 6 10 14	2 6 10 ..	2 5	
86	Rn	2	2 6	2 6 10	2 6 10 14	2 6 10 ..	2 6	
87	Fr	2	2 6	2 6 10	2 6 10 14	2 6 10 ..	2 6	1
88	Ra	2	2 6	2 6 10	2 6 10 14	2 6 10 ..	2 6	2
89	Ac	2	2 6	2 6 10	2 6 10 14	2 6 10 ..	2 6 1 ..	2
90	Th	2	2 6	2 6 10	2 6 10 14	2 6 10 ..	2 6 2 ..	2
91	Pa	2	2 6	2 6 10	2 6 10 14	2 6 10 2*	2 6 1 ..	2
92	U	2	2 6	2 6 10	2 6 10 14	2 6 10 3	2 6 1 ..	2
93	Np	2	2 6	2 6 10	2 6 10 14	2 6 10 4	2 6 1 ..	2
94	Pu	2	2 6	2 6 10	2 6 10 14	2 6 10 6	2 6	2
95	Am	2	2 6	2 6 10	2 6 10 14	2 6 10 7	2 6	2
96	Cm	2	2 6	2 6 10	2 6 10 14	2 6 10 7	2 6 1 ..	2
97	Bk	2	2 6	2 6 10	2 6 10 14	2 6 10 9	2 6	2
98	Cf	2	2 6	2 6 10	2 6 10 14	2 6 10 10	2 6	2
99	Es	2	2 6	2 6 10	2 6 10 14	2 6 10 11	2 6	2
100	Fm	2	2 6	2 6 10	2 6 10 14	2 6 10 12	2 6	2
101	Md	2	2 6	2 6 10	2 6 10 14	2 6 10 13	2 6	2

*Devised by Laurence S. Foster.   †Note irregularity.

References: Therald Moeller, "Inorganic Chemistry," John Wiley & Sons, New York, 1952, pp. 98–101. Joseph J. Katz and Glenn T. Seaborg, "The Chemistry of the Actinide Elements," Metheun & Co., Ltd., London, 1957; John Wiley & Sons, New York, 1957, p. 464.

2.3 The periodic table

This brief review of the quantum numbers of the electrons is important at this point because it is part of the background of the Periodic Table, Table 2.2. This arrangement is obtained by placing the elements with $n = 1$ (hydrogen and helium) in the top row, those with $n = 2$ in the next row, and so forth.

The table is extremely valuable in the practical development of metallic materials because of the interrelation of elements in the same column and in some cases in the same row. Let us consider some examples.

The first vertical column, group IA, omitting hydrogen which is a special case, is composed of the very active metals lithium through francium. This reactivity is due to the fact that all have one loosely held s electron in the outer shell. All will react rapidly with oxygen (which has a high affinity for electrons to attain an outer shell of eight) and for this reason are used to eliminate undesired oxygen from liquid melts of other metals. Lithium, for example, is commonly added to liquid copper to remove oxygen as an impurity and thereby produce a pure copper structure with high electrical conductivity.

The relative tendency of an element to gain an electron is expressed by a factor called "electronegativity." As we would expect, the active metals just discussed have the lowest values, as shown in Table 2.3. We can also say that these metals are strongly electropositive. In general, electronegativity *decreases* as the total number of electrons in a vertical column increases because the outer electrons are farther from the positive nucleus. Also, electronegativity *increases* as we go across the table because the higher the number of electrons in the outer shell, the greater the attraction for other electrons until a stable group of eight is reached (two for helium).

The second column, group IIA, contains elements with two loosely held electrons in the outer shell, and these are only slightly less active than the elements in group IA. However, magnesium can be protected with a suitable coating or by the addition of other elements and is used extensively even though it may react in pure form. For example, the engine block of a Volkswagen automobile is principally magnesium.

From group IIIB to group VIII we have the first group of transition metals (atomic numbers 21 through 28). In this range the changes in electrons are in the inner rings. In the usual case (potassium and calcium) the $4s$ level fills before the $3d$ level.

EXAMPLE 2.1 What is the electron configuration (short-hand notation) for the three transition elements vanadium (atomic number = 23), iron (atomic number = 26), and nickel (atomic number = 28)?

ANSWER By definition the transition elements are those with partially filled $3d$ shells, but the $3d$ shells are not the outer shells. These elements have similar properties.

V	$1s^2, 2s^2, 2p^6, 3s^2, 3p^6, 3d^3, 4s^2$
Fe	$1s^2, 2s^2, 2p^6, 3s^2, 3p^6, 3d^6, 4s^2$
Ni	$1s^2, 2s^2, 2p^6, 3s^2, 3p^6, 3d^8, 4s^2$

The so-called "lanthanide series" of rare earths, from cerium (atomic number = 58) to lutetium (atomic number = 71), are even more alike than the transition metals, because the only changes in electrons are in the second shell below the outer shell. That is, the $5p$ and $6s$ levels are filled first, then the $4f$ levels. Since the $4f$ level can contain 14 electrons, there are 14 rare earths. All exhibit typical metallic properties and are finding increasing use. Cerium, for instance, is used in lighter "flints" and in producing a new casting alloy called "ductile iron." Europium has unique emission properties which lead to its use in the red phosphor in television tubes. Samarium is alloyed with cobalt to form the strongest known permanent magnets.

Group IB contains the noble metals copper, silver, and gold. Although the outer shell of these metals contains only one electron, this electron is tightly held because it is close to the energy level of the $3d$ electrons and tends to associate with the other electrons. For example, the electron structure of copper is $1s^2, 2s^2, 2p^6, 3s^2, 3p^6, 3d^{10}, 4s^1$, and the $4s$ electron is close to the $3d$ electron. By contrast, in the alkali metals, in group IA, the shell beneath the outer s electron is a very stable group of either two or eight electrons and the outer electron is loosely held.

For the same reason the elements of group IIB—zinc, cadmium, and mercury—are stabler than those of IIA, such as magnesium and calcium, which have a shell of eight beneath the outer electrons. Zinc, for example, is used widely in die castings for automobiles.

The most important element in group IIIA is aluminum, and despite the fact that its three outer electrons are loosely held (there is a shell of eight beneath), making it quite reactive, it is used widely as a building material. We shall see in Chap. 12 that this successful use of aluminum depends on the formation of an adherent film of aluminum oxide which protects it from further reaction with the air.

Finally in group IVA we reach the end of the metallic elements. The lower members of the group, such as lead and tin, have metallic characteristics, while the upper members, such as germanium and silicon, are semiconductors, and the diamond form of carbon shows only covalent bonding and is an insulator. We will discuss this behavior in detail in Chap. 13 when we consider transistor materials.

We will use this table often in discussing the development of different alloys.

The rest of the elements in the Periodic Table are generally considered to be nonmetallic although a few, such as antimony and bismuth, possess semimetallic characteristics. We will, of course, return to a number of these nonmetallic elements in our discussion of polymers, where carbon, nitrogen, and oxygen are important, and in ceramics, which are defined as combinations of metallic and nonmetallic elements.

Now let us go on to the discussion of the unit cell and grains.

2.4 Crystals and grains

Instead of describing the arrangement of many millions of atoms (10^{10} millions) in a grain, it is much easier to describe the unit cell. Just as we can understand the construction of a large building by analyzing the modules or units of which it is made, so we can visualize a grain as made up of identical unit cells. It is basic to later problems to understand this concept of grain structure, so let us spend a few more moments on this point.

In cavities in rocks we encounter cubic crystals of copper. The relation between a *crystal* of copper and a *grain* is quite simple. The grain is merely a crystal without smooth faces because its growth was impeded by contact with another grain or a restraining surface. Within the grain the arrangement of unit cells is just as perfect as it is within a crystal with smooth faces.

To indicate the planes formed by alignment of the unit cells of copper, let us perform a simple experiment. First we cut a small bar of copper 1 in. long and 0.030 by 0.030 in. square. We polish one surface and find that we can observe grains (Fig. 2.2*a*).

Now let us bend the specimens as shown in Fig. 2.2*b*. We see that straight dark lines appear in the grains near both edges of the sample. If we watch carefully while the sample is being bent, we see that the metal is being displaced on either side of the black lines by a shearing, slipping type of action. In Chap. 3 we will show that the black lines or slip bands are simply the intersections of slip planes with the surface. Furthermore, we will see that the slip plane is made up of a very closely packed mass of atoms on a common plane. Finally, we should realize that although the grains at the upper edge of the sample are in tension and those at the lower edge are in compression, the mechanism of deformation is by shearing in both cases.

2.5 The unit cell

Let us now proceed to a description of the unit cell, which will allow us to describe these movements more accurately. To describe the unit cell and later to describe the movement of an atom in the cell, we need a system for specifying:

1. Atom positions or coordinates
2. Directions in the cell
3. Planes in the cell

Table 2.2 PERIODIC TABLE OF THE ELEMENTS

1a	2a	3b	4b	5b	6b	7b	8	8	8	1b	2b	3a	4a	5a	6a	7a	0	Orbit
1 +1 −1 H 1.0079 1		KEY TO CHART Atomic Number → **50** +2 ← Oxidation States Symbol → Sn +4 Atomic Weight → 118.69 18 18 4 ← Electron Configuration															**2** 0 He 4.00260 2	K
3 +1 Li 6.94, 2-1	**4** +2 Be 9.01218 2-2											**5** +3 B 10.81 2-3	**6** +2 +4 −4 C 12.011 2-4	**7** +1 +2 +3 +4 +5 −1 −2 −3 N 14.0067 2-5	**8** −2 O 15.9994 2-6	**9** −1 F 18.99840 2-7	**10** 0 Ne 20.17_9 2-8	K-L
11 +1 Na 22.98977 2-8-1	**12** +2 Mg 24.305 2-8-2	Transition Elements						Group 8			Transition Elements	**13** +3 Al 26.98154 2-8-3	**14** +2 +4 −4 Si 28.086 2-8-4	**15** +3 +5 −3 P 30.97376 2-8-5	**16** +4 +6 −2 S 32.06 2-8-6	**17** +1 +5 +7 −1 Cl 35.453 2-8-7	**18** 0 Ar 39.948 2-8-8	K-L-M
19 +1 K 39.09_8 -8-8-1	**20** +2 Ca 40.08 -8-8-2	**21** +3 Sc 44.9559 -8-9-2	**22** +2 +3 +4 Ti 47.90 -8-10-2	**23** +2 +3 +4 +5 V 50.941_4 -8-11-2	**24** +2 +3 +6 Cr 51.996 -8-13-1	**25** +2 +3 +4 +7 Mn 54.9380 -8-13-2	**26** +2 +3 Fe 55.847 -8-14-2	**27** +2 +3 Co 58.9332 -8-15-2	**28** +2 +3 Ni 58.71 -8-16-2	**29** +1 +2 Cu 63.546 -8-18-1	**30** +2 Zn 65.38 -8-18-2	**31** +3 Ga 69.72 -8-18-3	**32** +2 +4 Ge 72.59 -8-18-4	**33** +3 +5 −3 As 74.9216 -8-18-5	**34** +4 +6 −2 Se 78.96 -8-18-6	**35** +1 +5 −1 Br 79.904 -8-18-7	**36** 0 Kr 83.80 -8-18-8	-L-M-N
37 +1 Rb 85.467_8 -18-8-1	**38** +2 Sr 87.62 -18-8-2	**39** +3 Y 88.9059 -18-9-2	**40** +4 Zr 91.22 -18-10-2	**41** +3 +5 Nb 92.9064 -18-12-1	**42** +6 Mo 95.94 -18-13-1	**43** +4 +6 +7 Tc 98.9062 -18-13-2	**44** +3 Ru 101.07 -18-15-1	**45** +3 Rh 102.9055 -18-16-1	**46** +2 +4 Pd 106.4 -18-18-0	**47** +1 Ag 107.868 -18-18-1	**48** +2 Cd 112.40 -18-18-2	**49** +3 In 114.82 -18-18-3	**50** +2 +4 Sn 118.69 -18-18-4	**51** +3 +5 −3 Sb 121.75 -18-18-5	**52** +4 +6 −2 Te 127.60 -18-18-6	**53** +1 +5 +7 −1 I 126.9045 -18-18-7	**54** 0 Xe 131.30 -18-18-8	-M-N-O
55 +1 Cs 132.9054 -18-8-1	**56** +2 Ba 137.34 -18-8-2	**57*** +3 La 138.9055 -18-9-2	**72** +4 Hf 178.49 -32-10-2	**73** +5 Ta 180.947_9 -32-11-2	**74** +6 W 183.85 -32-12-2	**75** +4 +6 +7 Re 186.2 -32-13-2	**76** +3 +4 Os 190.2 -32-14-2	**77** +3 +4 Ir 192.22 -32-15-2	**78** +2 +4 Pt 195.09 -32-16-2	**79** +1 +3 Au 196.9665 -32-18-1	**80** +1 +2 Hg 200.59 -32-18-2	**81** +1 +3 Tl 204.37 -32-18-3	**82** +2 +4 Pb 207.2 -32-18-4	**83** +3 +5 Bi 208.9808 -32-18-5	**84** +2 +4 Po (209) -32-18-6	**85** At (210) -32-18-7	**86** 0 Rn (222) -32-18-8	-N-O-P
87 +1 Fr (223) -18-8-1	**88** +2 Ra 226.0254 -18-8-2	**89**** +3 Ac (227) -18-9-2	**104** — -32-10-2	**105** —														O P Q

*Lanthanides	**58** +3 +4 Ce 140.12 -20-8-2	**59** +3 Pr 140.9077 -21-8-2	**60** +3 Nd 144.24 -22-8-2	**61** +3 Pm (145) -23-8-2	**62** +2 +3 Sm 150.4 -24-8-2	**63** +2 +3 Eu 151.96 -25-8-2	**64** +3 Gd 157.25 -25-9-2	**65** +3 Tb 158.9254 -27-8-2	**66** +3 Dy 162.50 -28-8-2	**67** +3 Ho 164.9304 -29-8-2	**68** +3 Er 167.26 -30-8-2	**69** +3 Tm 168.9342 -31-8-2	**70** +2 +3 Yb 173.04 -32-8-2	**71** +3 Lu 174.97 -32-9-2		N O P
**Actinides	**90** +4 Th 232.0381 -18-10-2	**91** +5 +4 Pa 231.0359 -20-9-2	**92** +3 +4 +5 +6 U 238.029 -21-9-2	**93** +3 +4 +5 +6 Np 237.0482 -22-9-2	**94** +3 +4 +5 +6 Pu (244) -24-8-2	**95** +3 +4 +5 +6 Am (243) -25-8-2	**96** +3 Cm (247) -25-9-2	**97** +3 +4 Bk (247) -27-8-2	**98** +3 Cf (251) -28-8-2	**99** Es (254) -29-8-2	**100** Fm (257) -30-8-2	**101** Md (256) -31-8-2	**102** No (254) -32-8-2	**103** Lr -32-9-2	**104** (257)	O P Q

Numbers in parentheses are mass numbers of most stable isotope of that element.

(From "Handbook of Chemistry and Physics," © The Chemical Rubber Co., 1966. Used by permission of The Chemical Rubber Co.)

Table 2.3 ELECTRONEGATIVITIES OF THE ELEMENTS*

IA	IIA	IIIB	IVB	VB	VIB	VIIB	VIIIB	VIIIB	VIIIB	IB	IIB	IIIA	IVA	VA	VIA	VIIA	O
1 H 2.1																	2 He —
3 Li 1.0	4 Be 1.5											5 B 2.0	6 C 2.5	7 N 3.0	8 O 3.5	9 F 4.0	10 Ne —
11 Na 0.9	12 Mg 1.2											13 Al 1.5	14 Si 1.8	15 P 2.1	16 S 2.5	17 Cl 3.0	18 Ar —
19 K 0.8	20 Ca 1.0	21 Sc 1.3	22 Ti 1.5	23 V 1.6	24 Cr 1.6	25 Mn 1.5	26 Fe 1.8	27 Co 1.8	28 Ni 1.8	29 Cu 1.9	30 Zn 1.6	31 Ga 1.6	32 Ge 1.8	33 As 2.0	34 Se 2.4	35 Br 2.8	36 Kr —
37 Rb 0.8	38 Sr 1.0	39 Y 1.2	40 Zr 1.4	41 Nb 1.6	42 Mo 1.8	43 Tc 1.9	44 Ru 2.2	45 Rh 2.2	46 Pd 2.2	47 Ag 1.9	48 Cd 1.7	49 In 1.7	50 Sn 1.8	51 Sb 1.9	52 Te 2.1	53 I 2.5	54 Xe —
55 Cs 0.7	56 Ba 0.9	57–71 La–Lu 1.1–1.2	72 Hf 1.3	73 Ta 1.5	74 W 1.7	75 Re 1.9	76 Os 2.2	77 Ir 2.2	78 Pt 2.2	79 Au 2.4	80 Hg 1.9	81 Tl 1.8	82 Pb 1.8	83 Bi 1.9	84 Po 2.0	85 At 2.2	86 Rn —
87 Fr 0.7	88 Ra 0.9	89— Ac— 1.1–1.7															

*From Linus Pauling, "The Nature of the Chemical Bond," 3d ed. Copyright 1939 and 1940, 3d ed. © 1960, by Cornell University Press.

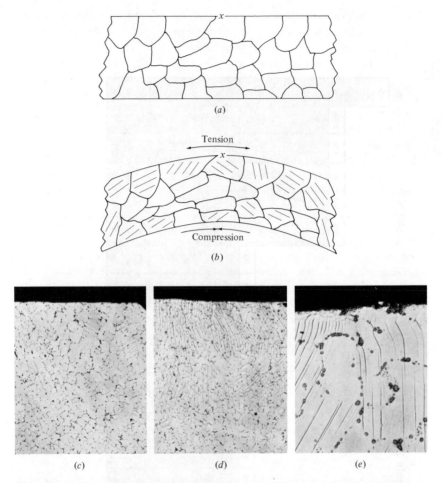

Fig. 2.2 (a) *Grains in a polished bar of copper.* (b) *Slip occurring during bending.* (c) *Photomicrograph of region x in* (a), *100X.* (d) *Photomicrograph of region x in* (b), *100X.* (e) *Photomicrograph of region x in* (b), *500X.*

Position. An atom position is described with reference to the axes of the unit cell and the unit dimensions of the cell. To take the general case, suppose that the grain or crystal is built of unit cells of dimensions a_0,† b_0, c_0 angstroms ($\text{Å} = 10^{-8}$ cm), as shown in Fig. 2.3. In this case the axes are at right angles to each other; to construct the cell, we merely lay out a_0, the lattice parameter, in the x direction, b_0 in the y direction, and c_0 in the z

†The subnumeral 0 specifies that this dimension is measured at a standard temperature, usually 68°F (20°C). Naturally the unit cell expands with temperature and a, b, and c change.

direction. The coordinates of the corners of the cell are shown in the figure. An atom at the center would have coordinates $\frac{1}{2}$, $\frac{1}{2}$, $\frac{1}{2}$, while an atom in the center of the face in the xy plane would have coordinates $\frac{1}{2}$, $\frac{1}{2}$, 0.

Up to this point we have used relatively simple cells as examples. In nature 14 different types of crystal lattices are found; these are illustrated in Fig. 2.4. These lattices cover the variations in length of a, b, and c and in the angles between the axes.

Fortunately, in the metals we need to concentrate mostly on the three simple types of cells shown in Fig. 2.5: body-centered cubic (BCC), face-centered cubic (FCC), and hexagonal close-packed (HCP).

Some of the other types will be encountered occasionally in metals and ceramics.

Direction. To specify a direction in the unit cell, we merely place the base of the arrow of the direction vector at the origin and follow the shaft until we encounter integral coordinates (Fig. 2.6). As an alternative to constructing other cells, we can use a point that has fractional intercepts in the unit cell and multiply by the least common denominator. Thus direction A is obviously [111], but B has coordinates 1, $\frac{1}{2}$, 0 at the edge of the cell. These become [210]. Note that in specifying a direction we put square brackets around the numbers to distinguish direction from the notation for coordinates (and for a plane, described later). Note also that we can have negative indices, as shown by C, and that we indicate these with an overbar.

If we wish to find the indices of a direction that does not pass through

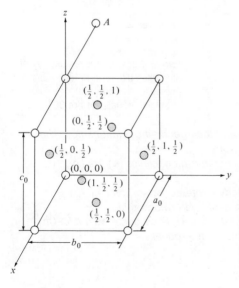

Fig. 2.3 *Coordinates of atoms in the face-centered positions in a unit cell. The atom A shown in the next unit cell would have coordinates ($\bar{1}$, 0, 1). Note: $\bar{1}$ is equivalent to -1.*

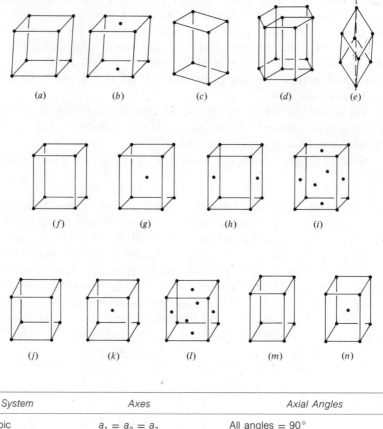

System	Axes	Axial Angles
Cubic	$a_1 = a_2 = a_3$	All angles = 90°
Tetragonal	$a_1 = a_2 \neq c$	All angles = 90°
Orthorhombic	$a \neq b \neq c$	All angles = 90°
Monoclinic	$a \neq b \neq c$	Two angles = 90°; one angle $\neq$ 90°
Triclinic	$a \neq b \neq c$	All angles different; none equals 90°
Hexagonal	$a_1 = a_2 = a_3 \neq c$	Angles = 90° and 120°
Rhombohedral	$a_1 = a_2 = a_3$	All angles equal, but not 90°

Fig. 2.4 *The 14 crystal lattices and their geometric relationships. The lattices continue in three dimensions.*
(a) Simple monoclinic. (b) End-centered monoclinic. (c) Triclinic. (d) Hexagonal. (e) Rhombohedral.
(f) Simple orthorhombic. (g) Body-centered orthorhombic. (h) End-centered orthorhombic. (i) Face-centered orthorhombic.
(j) Simple cubic. (k) Body-centered cubic. (l) Face-centered cubic. (m) Simple tetragonal. (n) Body-centered tetragonal.

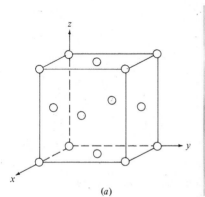

(a)

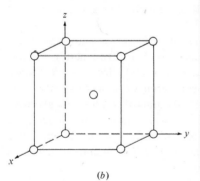

(b)

12

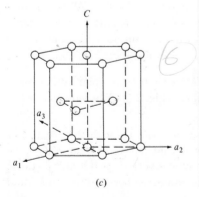

(c)

$\frac{1}{2}$

$2\frac{1}{2}$

$\frac{7}{2} = 12 X$

$x = \frac{7}{14}$

Fig. 2.5 *The principal crystal structures of the metals. (a) Face-centered cubic. (b) Body-centered cubic. (c) Close-packed hexagonal. A sketch and a photograph of the hard sphere model of each case are shown.*

$(4 r)^3 = 3(a d)^3$

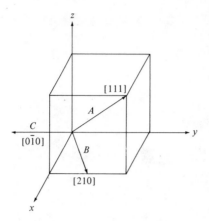

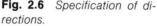

Fig. 2.6 *Specification of directions.*

the origin, we merely pass a parallel direction through the origin and proceed as before.

In the cubic system the dimensions of the unit cell are the same, $a_0 = b_0 = c_0$. The order in a given set of indices, such as [110], merely depends on which directions we happened to choose for x, y, and z in the crystal, since in the cubic system the indices are identical. Often we wish to signify a set of directions that are related, such as [110], [101], and [011]. These are all directions across the face diagonal of the cube. To specify these similar directions, we use pointed brackets and the indices of only one direction. Thus $\langle 110 \rangle = $ [110], [101], and [011]. Directions with negative indices, such as [$\bar{1}$10], are also considered part of the same set.

It is important to realize that unit distances in the three directions x, y, and z are a_0, b_0, and c_0, respectively. Therefore, the [111] direction in a noncubic system [orthorhombic (three side lengths unequal); Fig. 2.4] means a direction starting at coordinates 0, 0, 0 with the tip at a_0 distance in the x direction, b_0 distance in the y direction, and c_0 distance in the z direction.

Plane. The method for finding the Miller indices of a plane differs from that for describing coordinates and directions (Fig. 2.7). The sequence below should be followed carefully.

1. Select a plane in the unit cell that does not pass through the origin. For example, to describe the plane at face A that passes through (0,0,0), we use plane B, which is in the same set of planes because it is a unit distance away from A. (That is, B would have the same number of atoms and would be parallel to the original plane.)
2. Record the intercepts of the plane as multiples of a_0, b_0, and c_0 on the x, y, and z axes in that order.
3. Take reciprocals and clear fractions. This gives (001) for plane B.† Note that

---

† $\dfrac{1}{\infty}, \dfrac{1}{\infty}, \dfrac{1}{1}$

parentheses are used in this case to show that we are describing a plane rather than a direction. Also commas do not separate the indices, as they do in atom positions. The plane should be read as the "zero-zero-one" plane.

Following the same rules, we find the indices for plane C of intercepts $1, \frac{2}{3}, \frac{1}{3}$:

Reciprocals: $1, \frac{3}{2}, 3$
Clearing fractions: 2, 3, 6 or (236)

As in the case of directions, often we want to specify a family of planes such as the cube faces. Here we use braces; thus $\{100\} = (100), (010), (001)$. We can also have the corresponding negative values, such as $(\bar{1}00)$, depending on our choice of position for the origin (0,0,0).

2.6 Correlation of unit cell data with density measurements

It is interesting and important to see whether we can check the unit cell dimensions with engineering data such as density. After all, we are not going to use the metals in unit cell size pieces, and we should satisfy ourselves that the characteristics of the unit cell relate to engineering-size components.

For example, the density of copper is 8.96 g/cm³ (8.96 $\times$ 10³ kg/m³) at 20°C (68°F). If our data are correct, we should find the same value using the mass of copper in the volume of a unit cell.

The mass (M) and volume (V) can be found from previous definitions. We know the weight of an atom of copper, since each gram atomic weight of copper (63.5 g/at. wt.) has 6.02 $\times$ 10²³ atoms (Avogadro's number). The volume of a unit cell is merely the lattice dimension cubed. Before we can complete the calculation, however, we must determine the number of atoms per unit cell.

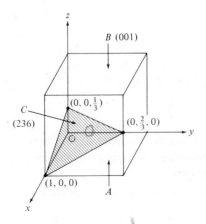

Fig. 2.7 *Calculation of Miller indices of a plane.*

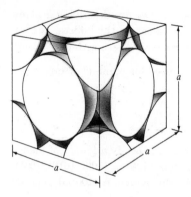

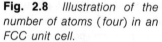

Fig. 2.8 *Illustration of the number of atoms (four) in an FCC unit cell.*

Figure 2.8 illustrates the technique for determining the number of atoms in a single cell. If we were to set up thin glass walls as faces of the FCC, these would pass through the centers of the atoms in the faces and extend only to the centers of the corner atoms. Counting the portions of the atoms that are within the cell, we have

6 face atoms of which $\frac{1}{2}$ of each is inside the glass $= 3$ atoms

8 corner atoms of which $\frac{1}{8}$ of each is inside the glass $= \underline{1 \text{ atom}}$

Total atoms per unit cell $= 4$ atoms

In a similar manner we find that the number of atoms per unit cell in a BCC is two, while that in an HCP is six.

EXAMPLE 2.2 We know from x-ray data discussed later in this chapter that $a_0 = 3.61$ Å at 20°C (68°F) for copper. Calculate the theoretical density of copper, using the knowledge that the cell of this element is an FCC.

ANSWER

$$\text{Density} = \frac{M}{V} = \frac{4 \text{ atoms/unit cell} \dfrac{63.5 \text{ g/at. wt.}}{6.02 \times 10^{23} \text{ atoms/at.wt.}}}{(3.61)^3 \times 10^{-24} \text{ cm}^3/\text{unit cell}}$$

$$= 8.98 \ \frac{\text{g}}{\text{cm}^3} \ \left(8.98 \times 10^3 \frac{\text{kg}}{\text{m}^3}\right)$$

This is quite close to the density of a block of copper (8.96 g/cm), and in the next chapter we shall see that there are missing atoms (vacancies and dislocations) which explain the difference.

2.7 Other unit cell calculations: atomic radius, planar density, and linear density

Atomic Radius. Soon we will be discussing the design of alloys, and we will see that when a second element is added, its atoms can substitute for some of the atoms of the major element. The extent to which this substitution can take place is governed by the similarity of the two metals. One important measure of similarity is the atomic radius.

We can easily calculate the atomic radius once we know the unit cell dimensions. First we consider the atoms as spheres and find some dimension of the unit cell along which the spheres are in contact. We can calculate any dimension of the cube, such as a body or face diagonal, by geometry and then merely divide the figure by the number of atomic radii present (Fig. 2.9).

Planar Density. We will see that when slip occurs under stress (plastic deformation), it takes place on the planes on which the atoms are most densely packed. To calculate planar density, we use the following convention.

If an atom belongs entirely to a given area, such as the atom in the center of a face in an FCC structure, we note that the trace of the atom on the plane is a circle (Fig. 2.10). Therefore, in the area a_0^2 we count one atom for the center but one-quarter atom for each of the corners because each has a trace of only one-quarter circle on the area a_0^2. The planar density is $2/a_0^2$ (atoms/Å^2). It should be added that in all these density calculations one of the ground rules is that a plane or a line must pass through the center of an atom or the atom is not counted in the calculations.

Linear Density. This is an important concept because when planes slip over each other, the slip takes place in the direction of the closest packing of atoms on the planes. We calculate this by the following convention.

If a line passes completely through an atom, the trace of the atom on the line is one diameter (Fig. 2.11). In an FCC face the center atom counts for one atom on the face diagonal. The corner atoms make traces equal to only one-half diameter each on the line of length AB. Therefore, the linear density in the AB direction [110] in an FCC structure is

$$\frac{1 + \frac{1}{2} + \frac{1}{2}}{a_0 \sqrt{2}} \frac{\text{atoms}}{\text{Å}} = \frac{2}{a_0 \sqrt{2}} \frac{\text{atoms}}{\text{Å}}$$

By the same reasoning the linear density in a BCC in the [111] direction is

$$\frac{2}{a_0 \sqrt{3}} \frac{\text{atoms}}{\text{Å}}$$

(Recall that the atoms touch along the body diagonal of a BCC.)

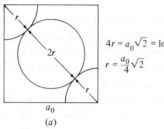

$4r = a_0\sqrt{2}$ = length of face diagonal

$r = \dfrac{a_0}{4}\sqrt{2}$

(a)

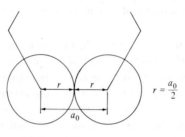

$r = \dfrac{a_0}{2}$

(b)

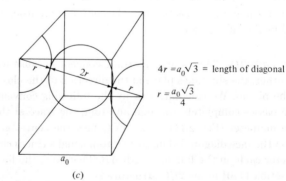

$4r = a_0\sqrt{3}$ = length of diagonal

$r = \dfrac{a_0\sqrt{3}}{4}$

(c)

Fig. 2.9 *Calculation of atomic radius. (a) FCC unit cell; spheres (atoms) in contact along the face diagonal. (b) HCP unit cell; spheres (atoms) in contact on the edge. (c) BCC unit cell; spheres (atoms) in contact along the body diagonal.*

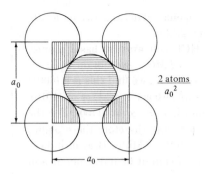

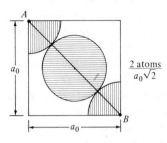

Fig. 2.10 *Calculation of* planar density *on the face plane of an FCC unit cell.*

Fig. 2.11 *Calculation of* linear density *on the face diagonal of an FCC unit cell.*

2.8 Close-packed hexagonal metals

There are a few simple modifications of the Miller indices of directions and planes that are needed to describe hexagonal structures. These are called "Miller-Bravais indices" (Fig. 2.12).

Instead of three axes x, y, z, we use four axes—three in the horizontal xy plane called a_1, a_2, a_3 at 120° to each other and the fourth c in the z direction. The inclusion of the extra axis makes it much easier to see the

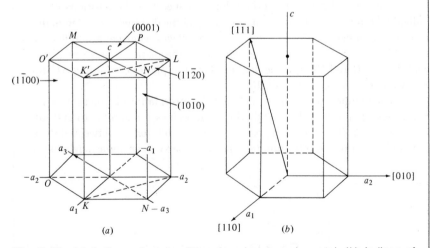

Fig. 2.12 *(a) Indices of some planes in a hexagonal crystal. (b) Indices of some directions in a hexagonal crystal.*

[Part (a) from C. S. Barrett and T. B. Massalski, "The Structure of Metals," 3d ed., McGraw-Hill Book Company, New York, 1973, Fig. 1.7, p. 12. Used with permission of McGraw-Hill Book Company.]

relations between similar planes in the hexagonal structure. Let us locate some planes using this reference frame (Fig. 2.12a).

The most important planes in an HCP cell are the basal planes. Finding the intercepts of one of these ($K'N'LPMO'$), we obtain $a_1 = \infty$, $a_2 = \infty$, $a_3 = \infty$, $c = 1$. Taking reciprocals as before, we find this is a (0001) plane. Now taking the intercepts of K, K', N', N, we get 1, ∞, $\bar{1}$, ∞ or ($10\bar{1}0$). Similarly, for K, O, O', K', we find intercepts 1, $\bar{1}$, ∞, ∞ or ($1\bar{1}00$). For the other faces we find other combinations of $1\bar{1}00$, and therefore these planes are of the family $\{10\bar{1}0\}$. In general terms we call these Miller-Bravais indices h, k, i, and l, and if we select h, k, and l, then i cannot be independent and always equals $-(h + k)$ by vector geometry.

The notation for a direction is developed by using either only three axes, a_1, a_2, and c, or four axes. Note that a_1 and a_2 are at 120°, as mentioned previously. We will use the simple three-axes notation, and Fig. 2.12b shows that this is similar to the method explained for the cubic system.

2.9 Determination of crystal structure; x-ray diffraction

Although our principal interest is in the use of the unit cell dimensions rather than in the technique of their determination, it is helpful to know a few basic features of x-ray methods. For example, in addition to their use in structure determination, x-rays are used in the microprobe to determine the analysis of different microstructures.

The x-ray diffraction method does not measure the positions of the individual atoms directly but measures instead the distances between planes of atoms. To illustrate this in a two-dimensional way, consider a simple cubic unit cell $ABCD$ (Fig. 2.13). If we determine the distance between planes containing AB and CD and the distance between planes containing AC and BD, we have determined the cell.

The manner in which we measure the interplanar distances by x-ray diffraction is shown in Fig. 2.14, where the planes are similar to AB and CD in Fig. 2.13. It is important to point out the differences between diffraction and reflection. In reflection (as with a mirror) we can shine a light beam

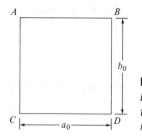

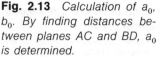

Fig. 2.13 *Calculation of a_0, b_0. By finding distances between planes AC and BD, a_0 is determined.*

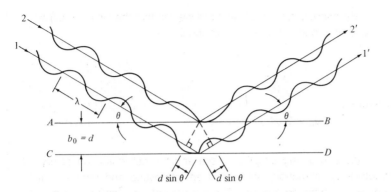

Fig. 2.14 *Diffraction of x-rays from planes of atoms.*

at a surface at any angle and it will be reflected at the same angle. By contrast, in diffraction the x-ray beam penetrates beneath the surface. When it strikes the atoms, they re-emit the radiation at the same wavelength. However, if the radiations being emitted are not in phase, they will cancel out each other and there will be no diffracted beam.

If the radiations are in phase, there will be a beam, as shown in Fig. 2.14. To obtain this beam we must have a certain relationship between the angle of incidence, the wavelength of the radiation λ, and the interplanar distance d. For example, we see that the path of the x-ray beam (1-1′) which goes one planar distance beneath the surface is longer than that of the beam diffracted by the surface atoms (2-2′) by the distance $2d \sin \theta$. To be in phase, $2d \sin \theta$ must equal the wavelength λ, or some multiple, $n\lambda$. This is known as "Bragg's law" and is written

$$n\lambda = 2d \sin \theta$$

Therefore, if we know θ and λ, we can find d. We obtain a beam of x-rays of constant wavelength λ from an x-ray tube and observe the angle θ at which diffraction is found.

EXAMPLE 2.3 A sample of BCC chromium was placed in an x-ray beam of $\lambda = 1.54$ Å. Diffraction from 110 planes was obtained at $\theta = 22.2°$. Calculate a_0. Assume n (the order of diffraction) $= 1$.

ANSWER

$$\lambda = 2d \sin \theta$$

$$d = \frac{1.54 \text{ Å}}{2(0.378)} = 2.04 \text{ Å}$$

By inspecting the BCC unit cell we see that the distance between 110 planes is one-half the face diagonal or

$$d_{110} = \frac{a_0 \sqrt{2}}{2}$$

Therefore,

$$a_0 = 2.04 \sqrt{2} = 2.88 \text{ Å}$$

(This is a very simplified case and the determination of crystal structure by diffraction methods is an interesting and important field covered in advanced courses. The question of x-ray and light emission is discussed in Chap. 15.)

2.10 Definition of a phase; solid phase transformations in metals

At this point it is important to explain what we mean by a *phase* and a *phase transformation*. Everyone is familiar with the fact that H_2O can exist as a gas, a liquid, and a solid. These are called three different phases. The general definition of a phase is a homogeneous aggregation of matter.

It is found that under different conditions of temperature and pressure, different crystal structures, that is, different unit cells, for ice are formed, called ice I, II, III, etc. These are different solid phases of the same chemical composition.

In a similar way we encounter different crystal structures in the same metal. In a bar of iron at room temperature we find a BCC structure. However, if we heat the bar above 1670°F (910°C), we find that the structure changes to FCC. Instead of calling these iron I and iron II, as in the case of ice, the BCC iron is called α (alpha) and the FCC γ (gamma). In general, different Greek letters are used to designate the different phases in solid metals.

In Table 2.4 the crystal structures of the elements are summarized. We see that many of the metallic elements can have different crystal structures. We also note that most of the metallic structures are BCC, FCC, or HCP. When a material has more than one crystal structure, we say it exhibits *allotropy* and has different *allotropic* forms.

Now let us use some lattice parameter data to estimate volume changes resulting from transformation of a metal from one solid phase to another.

EXAMPLE 2.4 On cooling through 880°C (1615°F), titanium goes through a phase change analogous to iron except that in this case the crystal structure changes from BCC to HCP:

BCC: $\qquad$ $a = 3.32$ Å

to HCP: $\qquad$ $a = 2.956$ Å $\qquad$ $c = 4.683$ Å

What is the volume change?

ANSWER The volume of the BCC is $(3.32 \text{ Å})^3 = 36.6 \text{ Å}^3$ for the two atoms per unit cell.

The volume of the HCP is determined from the observation that there are six equilateral triangles in the basal plane:

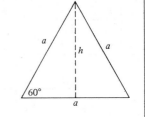

$$\text{Area} = \tfrac{1}{2}ah$$

$$\sin 60° = \frac{\sqrt{3}}{2} = \frac{h}{a}$$

or $\quad$ Area of the triangle $= \dfrac{\sqrt{3}}{4}a^2$

Therefore, the area of the hexagonal base is $\dfrac{6\sqrt{3}}{4}a^2$ and the volume

of the HCP equals

$$\left(6\frac{\sqrt{3}}{4}a^2\right)(c) = (\tfrac{3}{2}\sqrt{3})\,(2.956)^2(4.683) = 106.3 \text{ Å}^3$$

However, the HCP structure just calculated has six atoms per unit cell, and to make the volume comparison we must consider the same number of atoms in each structure. We therefore take one-third of the HCP volume (two atoms) to compare with the two atoms in the BCC, or

$$36.6 \text{ Å}^3 \longrightarrow \frac{106.3}{3} = 35.4 \text{ Å}^3$$

Therefore, there is a 3.3 percent contraction when BCC changes to HCP.

2.11 Effects of the addition of other elements on the structure of pure metals

Only a few elements are widely used commercially in their pure form; pure copper in electrical conductors is one example. Generally, as in the case of iron, other elements are present to produce greater strength, improve corrosion resistance, or simply as impurities because of the cost of refining. In any event, we need to know the effect these elements have on structure, because structure determines properties.

When a second element is added, two basically different structural changes are possible:

Table 2.4 PERIODIC TABLE OF THE ELEMENTS AND THEIR CRYSTAL STRUCTURES*

IA	IIA	IIIA†	IVA	VA	VIA	VIIA	VIII			IB	IIB	IIIB†	IVB	VB	VIB	VIIB	O
1 H A3, A1																	2 He (A3), (A2)
3 Li A2, A1, A3	4 Be (A2), A3											5 B H, T, R	6 C R, H, A4	7 N H, C	8 O C, (R)	9 F C	10 Ne A1
11 Na A2, A3	12 Mg A3											13 Al A1	14 Si A4	15 P C, C	16 S O, M, R	17 Cl A1, A3, A2	18 Ar A1
19 K A2	20 Ca A2, A1, A3	21 Sc (A2), A3	22 Ti A2, A3	23 V A2	24 Cr (A1), A2	25 Mn A2, A1, C	26 Fe A2, A1, A2	27 Co A1, A3	28 Ni A1	29 Cu A1	30 Zn A3	31 Ga O	32 Ge A4	33 As A7	34 Se A8, M	35 Br O	36 Kr A1
37 Rb A2	38 Sr A2, A3, A1	39 Y A2, A3	40 Zr A2, A3	41 Nb A2	42 Mo A2	43 Tc A3	44 Ru A3	45 Rh A1	46 Pd A1	47 Ag A1	48 Cd A3	49 In A6	50 Sn A5, A4	51 Sb A7	52 Te A8	53 I O	54 Xe A1
55 Cs A2	56 Ba A2, (T), (H)	57 La A2, A1, H	72 Hf A2, A3	73 Ta A2	74 W A2	75 Re A3	76 Os A3	77 Ir A1	78 Pt A1	79 Au A1	80 Hg R, T	81 Tl A2, A3	82 Pb A1	83 Bi A7	84 Po R, C	85 At	86 Rn
87 Fr	88 Ra A2	89 Ac A1															

6th (continued) — Lanthanides:

58 Ce A2, A1, H	59 Pr A2, H	60 Nd A2, H	61 Pm	62 Sm (A2), R	63 Eu A2	64 Gd (A2), A3	65 Tb (A2), A3	66 Dy (?), A3	67 Ho (?), A3	68 Er A3	69 Tm A3	70 Yb A2, A1, A3	71 Lu (?), A3

7th (continued) — Actinides:

90 Th A2', A1	91 Pa T	92 U A2, T, O	93 Np A2, T, O	94 Pu A2, T, A1, O, M	95 Am H	96 Cm	97 Bk	98 Cf	99 Es	100 Fm	101 Md	102 No	103 Lw

*These other designations are used to denote the structures of the allotropic forms which are given in the order of their appearance. A1 = FCC; A2 = BCC; A3 = HCP; A4 = diamond cubic; A5 = body-centered tetragonal; A6 = face-centered tetragonal; A7 = rhombohedral; A8 = trigonal; H = hexagonal (usually ABAC . . . close-packed); R = rhombohedral; O = orthorhombic; C = complex cubic; T = tetragonal; M = monoclinic; ( ) = uncertain.

†Note that the authors of this table have used a different convention for the A and B subgroups which has not been followed in Tables 2.2 and 2.3.

(From C. S. Barrett and T. B. Massalski, "The Structure of Metals," 3d ed., McGraw-Hill Book Company, New York, 1973, Table 10.1, p. 227. Used with permission of McGraw-Hill Book Company.)

1. The atoms of the new element form a solid solution with the original element, but there is still only one phase, such as an FCC or liquid.
2. The atoms of the new element form a new second phase, usually containing some atoms of the original element. The entire microstructure may change to this new phase or two phases may be present together.

Let us examine each possibility separately.

2.12 Solid solutions

If we take a crucible containing 70 g of liquid copper and add 30 g of nickel while heating to maintain a completely liquid melt, we obtain a liquid solution of copper and nickel. If we cool slowly and examine the grains, we find that we have an FCC structure and that a_0, the edge of the unit cell, is between the values for copper and nickel. This is called a "substitutional" solid solution because nickel atoms have substituted for copper atoms in the FCC structure, as shown in Fig. 2.15. The advantage of this structure is that it is stronger than either pure copper or pure nickel because the interaction between the atoms gives greater resistance to slip. Another type of solid solution is called "interstitial" because the new atoms are located in the interstices or spaces between the atoms in the matrix. Naturally, the atoms forming interstitial solid solutions are those with small radii, such as hydrogen, carbon, boron, and nitrogen. Carbon, for example, forms an interstitial solid solution in iron, giving steel. Position C $(\frac{1}{2}, \frac{1}{2}, \frac{1}{2})$ in Fig. 2.16 is an interstitial position in the FCC allotropic form of iron which can be filled with carbon.

It should be pointed out that when a substitutional or interstitial solid solution is formed, the Greek letter that was used for the pure metal is retained.

2.13 New phases, intermediate phases

If in place of nickel we added 30 percent lead to the copper melt, we would again obtain a liquid solution. However, when we examine the frozen melt we would find two different phases. There would be grains of copper with

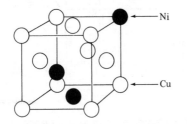

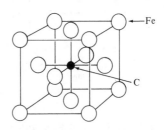

Fig. 2.15 *Substitutional solid solution of nickel in an FCC copper unit cell.*

Fig. 2.16 *Interstitial solid solution of carbon in the FCC form of iron.*

a very small amount of lead in solid solution, and the balance of the lead would be present in small spherical grains of practically pure lead. Both structures are FCC, but the unit cell of lead is much larger, which means that the atomic radius is greater. Therefore, we find two phases α and β listed. Although the lead addition actually lowers the strength, a material of this type is quite useful in bearings. When a load is applied, the lead from the spherical grains tends to flow out and coat the entire surface of the bearing with a film with a low coefficient of friction.

In the case just discussed the new phase that developed had the same structure as the element being introduced. Frequently a new phase is formed with a structure different from either element, and this is called an "intermediate phase." For example, if we add 15 percent tin to copper, some of the tin will go into solution in the FCC copper phase, but we will find another phase with the approximate formula $Cu_{31}Sn_8$. This is a hard phase completely different from copper or tin.

We may ask at this point if there are any general principles that determine whether a given element B will form a separate phase instead of a solid solution when added to another element A. This is an important consideration in alloy development because, in general, a solid solution will be more ductile and more easily shaped, while a two-phase material will usually exhibit greater hardness and strength.

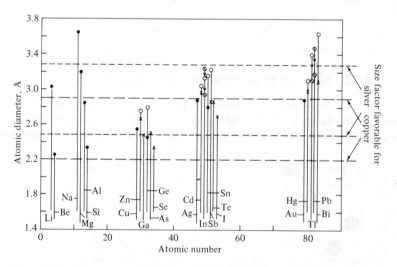

Fig. 2.17 *Favorable atomic diameter for elements soluble in copper and silver. When the size is within ± 15 percent, there is extensive solid solution.*

(Courtesy of W. Hume-Rothery)

The most important considerations in forming solid solutions are related to the atomic radii and chemical properties. To form an extensive solid solution (greater than 10 atomic percent soluble),† we should be aware of the following general rules of W. Hume-Rothery:

1. The difference in atomic radii should be less than 15 percent.
2. Proximity within the Periodic Table is important.
3. For a complete series of solid solutions the metals must have the same crystal structure.

Elements close to one another in the Periodic Table have similar electronegativities (Table 2.3) and are more compatible.

As an example, in Fig. 2.17 we have plotted the atomic diameters ($2r$) of a number of elements. The atomic diameter of copper ± 15 percent is shown by the dashed heavy lines, that of silver ± 15 percent by the dashed light lines. If we include elements with an FCC or HCP structure, we find that zinc, which is in the favorable zone for both copper and silver, forms extensive solid solutions with both elements. Cadmium, however, forms extensive solid solutions with silver but not with copper because it is outside the favorable zone.

SUMMARY

We began this chapter by studying the nature of metal atoms arising from their outer electrons and relative position in the Periodic Table. Next we considered the arrangement of atoms in the unit cell and methods for describing the location of the atoms. We also discussed directions and planes in the unit cell. We found that we can calculate density by dividing the mass of the atoms in the unit cell by the cell volume. Other important calculations included linear and planar density of atoms and atomic radius. The determination of the unit cell by x-ray diffraction and the calculation of volume change resulting from phase transformation were described. The concepts of solid solution of alloying elements in a phase and the formation of a new phase such as an intermetallic compound were considered.

DEFINITIONS

Quantum numbers   A number describing the average position of an electron in the electronic structure.

Short-hand notation   A system for describing the number of electrons in each shell of an atom and their subgrouping.

---

† To find atomic percent, calculate the percentage atoms of element B for the total number of atoms present.

Periodic Table A tabular grouping of the elements in order of atomic number. The metallic elements are on the left-hand side, since in general these elements contain one, two, or three electrons in the outer shell.

Unit cell A geometric figure illustrating the grouping of the atoms in the solid. This group or module is repeated many times in space within a grain or crystal of the metal.

Coordinates Location of an atom in the unit cell, found by moving the specified distances from the origin in the $x, y,$ and z directions.

Miller indices Descriptions of a plane found by taking reciprocals of the intercepts of the plane on the $x, y,$ and z axes and clearing fractions. Miller indices are enclosed in parentheses.

Indices of direction Descriptions of direction in a unit cell found by translating the direction passing through the origin, finding the coordinates of the point at which the arrow leaves the unit cell, and clearing fractions. These indices are enclosed in brackets.

Number of atoms per unit cell A number found by forming an imaginary box by connecting the corners of the cell. The fraction of each atom contained within the box is then determined and the fractions added. As an example, an atom at a corner position of a cube is counted as one-eighth because it is shared equally by eight unit cells meeting at the center of the atom.

Density Mass divided by volume, usually expressed as grams per cubic centimeter or kilograms per cubic meter.

Planar density The number of atoms that have their centers located within a given area of the plane. The planar area selected should be representative of the repeating groups of atoms.

Linear density The number of atoms that have their centers located on a given line of direction for a given length.

Atomic radius A measure calculated from the unit cell dimensions by finding a direction in which atoms are in contact, as for example the diagonal of the cube face in an FCC. Since the diagonal length can be calculated from a_0, the edge of the unit cell, the atomic radius can be found.

Phase A homogeneous aggregation of matter. In solid metals grains composed of atoms of the same unit cells are the same phase. Alloying atoms may also be present in a one-phase structure.

Solid solution A solution formed when the addition of one or more new elements still results in a single-phase structure.

Substitutional solid solution A solution formed when the dissolved element substitutes, i.e. replaces, an atom or atoms of the solvent element in its unit cell.

Interstitial solid solution A solution formed when the dissolved element fills holes (interstices) in the solvent element lattice.

Allotropy The occurrence of two or more crystal structures in the same chemical composition.

PROBLEMS

2.1 Write out all four quantum numbers for the following common elements.

K	(19)	$1s^2$, $2s^2$, $2p^6$, $3s^2$, $3p^6$, $4s^1$
Fe	(26)	$1s^2$, $2s^2$, $2p^6$, $3s^2$, $3p^6$, $3d^6$, $4s^2$
Cu	(29)	$1s^2$, $2s^2$, $2p^6$, $3s^2$, $3p^6$, $3d^{10}$, $4s^1$
Mo	(42)	$1s^2$, $2s^2$, $2p^6$, $3s^2$, $3p^6$, $3d^{10}$, $4s^2$, $4p^6$, $4d^5$, $5s^1$

2.2 A foundry was using selenium to control a certain property (the inclusion shape) in its castings. During a national emergency the selenium became scarce and tellurium was tried and found satisfactory. Give a basic reason for the similarity. (A humorous side issue is that exposure to selenium gave the workers garlicky breath. The wives did not appreciate the fascinating scientific fact that the substitution of tellurium was reproductive of the effects of selenium even to the breath.)

2.3 Another national emergency involved the shortage of tungsten used in producing high-speed steel. A typical analysis of this steel was 18% W, 4% Cr, 1% V, balance Fe. It was found that molybdenum could be substituted for more than half the tungsten. What basic relationships would lead you to believe that these elements can be substituted for one another? The new steel is still widely used today.

2.4 Identify the following elements as very active metals, less active metals, semiconductors, or nonmetals by examining their electronic structures. Then check the Periodic Table to find the actual elements and their positions.

Element number	Structure
3	$1s^2$, $2s^1$
6	$1s^2$, $2s^2$, $2p^2$
9	$1s^2$, $2s^2$, $2p^5$
28	$1s^2$, $2s^2$, $2p^6$, $3s^2$, $3p^6$, $3d^8$, $4s^2$
29	$1s^2$, $2s^2$, $2p^6$, $3s^2$, $3p^6$, $3d^{10}$, $4s^1$
32	$1s^2$, $2s^2$, $2p^6$, $3s^2$, $3p^6$, $3d^{10}$, $4s^2$, $4p^2$

2.5 Indicate the un-ionized and ionized short-hand electron notation for the following elements.

$$Cu \longrightarrow Cu^+ \text{ and } Cu^{2+} \qquad \text{atomic number} = 29$$
$$Fe \longrightarrow Fe^{2+} \text{ and } Fe^{3+} \qquad \text{atomic number} = 26$$
$$K \longrightarrow K^+ \qquad \qquad \qquad \text{atomic number} = 19$$
$$V \longrightarrow V^{3+} \text{ and } V^{5+} \qquad \text{atomic number} = 23$$
$$Ga \longrightarrow Ga^{3+} \qquad \qquad \quad \text{atomic number} = 31$$

2.6 Sketch the following directions and planes in a BCC unit cell and give the coordinates of the atoms they intersect. (We always deal only with intersections through atom centers.)

[100]	(100)
[110]	(110)
[111]	(111)

2.7 Repeat Prob. 2.6 for an FCC structure.

2.8 Calculate the linear density (atoms/Å) of atoms in the [100], [110], and [111] directions in BCC iron ($a_0 = 2.86$ Å).

2.9 Calculate the linear density of atoms in the [100], [110], and [111] directions in FCC copper ($a_0 = 3.62$ Å).

2.10 Calculate the atomic radius of zinc (HCP, $a_0 = 2.66$ Å, $c_0 = 4.95$ Å). Do the same for copper (FCC) and iron (BCC), using the data of Probs. 2.8 and 2.9.

2.11 Calculate the planar density of atoms in the (0001) plane of zinc. Why is the answer the same as for the (111) plane of copper? Will this be true for any HCP and FCC structures?

2.12 Calculate the planar density of atoms (atoms/Å²) in BCC iron ($a_0 = 2.86$ Å) in the (100), (110), and (111) planes.

2.13 As pointed out in the text, pure iron undergoes an allotropic transformation at 910°C (1670°F). The BCC form is stable at temperatures below 910°C, while the FCC form is stable above 910°C. Calculate the volume change for the transformation BCC → FCC, if at 910°C, $a = 3.63$ Å for FCC and $a = 2.93$ Å for BCC.

2.14 Calculate the density of BCC iron at room temperature from the atomic radius of 1.24 Å. Compare your results with the experimental value of 7.87 g/cm³ (7.87 × 10³ kg/m³).

2.15 A specimen of silver (FCC, $r_0 = 1.44$ Å) is placed in an x-ray camera and irradiated with molybdenum-characteristic radiation (0.709 Å). It is observed that θ for 111 planes decreases by 0.11° as the silver is heated from room temperature to 800°C (1470°F). Given that the crystal structure remains the same upon heating, find the change in a due to the heating. (*Hint*: $d_{111} = a/\sqrt{3}$.)

2.16 An investigator is trying to determine whether a structure in an 18% Cr, 18% Ni stainless steel is FCC or BCC. In this composition both structures are encountered at room temperature. From x-ray diffraction he finds that the distance between 111 planes is 2.1 Å. For BCC iron $a_0 = 2.86$ Å, and for FCC iron $a_0 = 3.63$ Å. Which structure is present?

2.17 In one area of the recent accelerated development of titanium alloys for aircraft and underwater research, it was decided to concentrate on alloys that would be single phase. From the following data decide which of the metals listed would be expected to form extensive solid solutions with titanium.

$$\begin{array}{lll}
\text{Ti is HCP} & a_0 = 2.95 \text{ Å} \\
\text{Be is HCP} & a_0 = 2.28 \text{ Å} \\
\text{Al is FCC} & a_0 = 4.04 \text{ Å} \\
\text{V is BCC} & a_0 = 3.04 \text{ Å} \\
\text{Cr is BCC} & a_0 = 2.88 \text{ Å}
\end{array}$$

Defining extensive solid solutions as 10 atomic percent, calculate the value in weight percent that would correspond to 10 atomic percent in titanium for the elements that form extensive solid solutions.

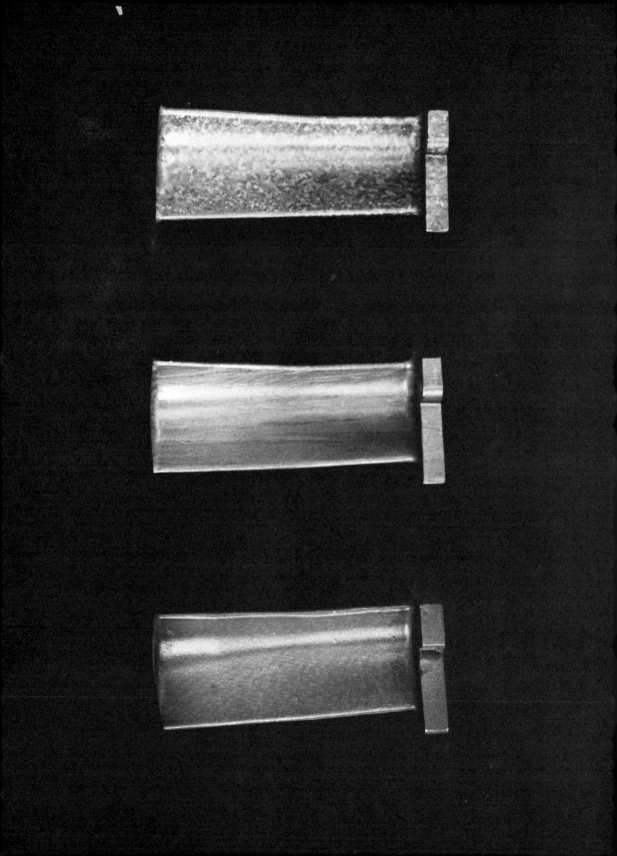

3

EFFECTS OF STRESS AND TEMPERATURE ON SIMPLE METAL STRUCTURES

THE blades shown in the illustration are typical of those used in the most severe application in modern jet aircraft, the combustion section of the turbine. These castings illustrate the rewards of careful development of the proper structure to resist fracture. The lowest-strength blade shows a randomly oriented grain structure. The first step in improvement was to grow the grains by controlled solidification so that a columnar structure formed and no grain boundaries were formed at right angles to the axis of the blade, since rupture in this case initiates at grain boundaries. Finally it was found possible to eliminate the grain boundaries by growing a single grain which has the direction of greatest strength oriented to resist the condition of maximum stress.

In this chapter we will study the effect of stress on the structures described in Chap. 2. We will see that both elastic and plastic strains are produced. Finally we will consider the combined effects of stress and temperature.

3.1 General

Now that we have described the nature of metal structures we can advance to observe the effects of stress and temperature on them. These effects are important for two reasons.

1. To form parts into the required shape and to obtain the desired properties, we use combinations of stress and heat, as in rolling, forming, annealing, heat treating.
2. In service all parts are subjected to stress and in many cases to elevated temperature as well, as in an automobile engine.

Let us consider first the effects of stress, then temperature, and finally the combined effects in the phenomena called "creep" and "stress relief." We will deal with only one-phase structures in this chapter and will take up more complex effects in two-phase structures in Chap. 4.

3.2 Effects of stress on metal structures

If we examine the objects about us in the street or in our homes we see that the most useful property of metals is their ability to resist or transmit stress. There are dozens of illustrations of this in an automobile. The frame and fenders resist the many imposed stresses, while the power train transmits the stress from the pistons to the wheels. In this case all the parts deform a little and then spring back when the stress is removed. This deformation is *elastic* strain.

Another important effect of stress on metals is permanent deformation or *plastic* strain. If we exceed the proper load in an automobile, the springs may sag permanently, or if we hit a tree, the fender will deform. This ability to deform and not shatter is an important feature of metals, not merely in case of automobile accidents, but so that we can form metal sheets and bars into desired shapes such as automobile bodies, structural columns, beams, and aircraft wings.

We will now take up these vital characteristics of metal deformation under stress: elastic strain and plastic strain.

Elastic Strain. Let us first review the engineering aspects of elastic strain and then discuss the relation of the structure to the observations.

A common experiment in elementary physics is to suspend a long steel wire from the ceiling and then hang increasingly heavy weights from the end. The student notes that the length of the wire increases proportionately with the load and that upon removal of the load the wire returns to its original length. In general terms, we define

$$\text{``Engineering'' stress} = \sigma = \frac{\text{load}}{\text{original area of wire cross-section}}$$

$$= \frac{P}{A_0} \quad \text{units:} \begin{cases} \text{pounds per square inch (psi) or} \\ \text{newtons per square meter (N/m}^2\text{)} \end{cases}$$

$$\text{``Engineering'' strain} = \varepsilon = \frac{\text{change in length}}{\text{original length}} = \begin{cases} \dfrac{\text{in.}}{\text{in.}} \quad \text{(units dimensionless)} \\[2mm] \dfrac{\text{m}}{\text{m}} \end{cases}$$

We use the terms "engineering" stress and "engineering" strain because later we shall define *true* stress and strain. In this illustration of elastic deformation the difference is very small.

If we calculate the stress in the wire and the strain produced, we find the relation

$$E = \frac{\sigma}{\varepsilon} \quad \text{(units: psi or N/m}^2\text{)}$$

where E is called the "modulus of elasticity." As long as we are careful not to reach too high a stress, the deformation is principally elastic and E is a constant. For each group of materials E has a characteristic value. For example, $E = 30 \times 10^6$ psi $(2.07 \times 10^5 \text{ MN/m}^2)$† for all steels and 10×10^6 psi $(0.69 \times 10^5 \text{ MN/m}^2)$ for aluminum alloys. The modulus is basically related to the bonding between atoms.

This is the macroscopic picture of elastic strain and it can be deceivingly simple. Let us go to the other extreme and test a single crystal of iron in place of the wire we have just discussed, which contains thousands of grains or crystals.

If we stress the single crystal along different crystal directions, we obtain values quite different from 30 million psi $(2.07 \times 10^5 \text{ MN/m}^2)$, as follows:

Crystal direction:	[111]	[100]
$E \times 10^6$ psi:	41	18
$E \times 10^5$ MN/m²:	2.83	1.24

Although this seems astonishing at first, we recall that in BCC structures (iron at room temperature) the atomic packing is densest in [111], the direction of highest E. We would expect that the interatomic forces would be greatest along this direction and therefore that the stress required to produce a given strain would be highest.

How do we explain the practically constant value of 30×10^6 psi $(2.07 \times 10^5 \text{ MN/m}^2)$ for steel? This value is obtained when there are many

---

† 1 psi $= 6.9 \times 10^3$ N/m² $= 6.9 \times 10^{-3}$ MN/m² $= 7.03 \times 10^{-4}$ kg/mm² (MN = meganewtons).

crystals of different orientations and our measurement gives the average value. There are occasional important exceptions which make it essential for us to remember the properties of the single crystal. Let us consider some examples.

A number of identical dentures for teeth were cast of Vitallium, a cobalt alloy, and tested for deflection under constant stress. Different deflections were obtained although the cross-sections were the same. It was found that because of the thin cross-section only one or two grains were present at the highly stressed region. Since the modulus E varies with direction in a grain, the orientation of these grains determined the amount of deflection.

In some cases it is desirable to produce a part with the crystals oriented in one direction. This is called development of "preferred orientation." An outstanding example is the production of transformer steel sheet with what is termed a cubic texture. By special processing the {100} directions of BCC iron are aligned in the plane of the sheet. In this case the ease of magnetization, like the modulus, varies with direction in the crystal. Thus when transformer laminations are stamped from sheet with this preferred orientation, the hysteresis losses (magnetic energy losses manifested as heat) are lower in the finished part.

In a more recent example a process has been developed to cast blades for aircraft gas turbines with controlled directional properties, so that the best value of strength is in the direction in which the operating stress is highest (see the frontispiece for Chap. 3).

3.3 Plastic strain, permanent deformation

We have just described elastic strain using the case of the physics experiment with a wire suspended from the ceiling and loaded with weights at the bottom. Occasionally an inventive student adds a really massive weight from some other equipment and finds (1) that the extension in length is more than expected from the equation $\varepsilon = \sigma/E$, and (2) that when the load is removed the wire does not return to its original length. The portion of the total deformation under load that does not disappear on removal of the load is called "plastic" or "permanent" deformation. Let us now discuss what is occurring within the structure during plastic deformation.

3.4 Critical resolved shear stress for plastic deformation

To illustrate how plastic deformation occurs, let us first grow a number of single crystals of zinc in the form of rods. This can be done simply by melting the zinc in a test tube in a vertical furnace and then lowering the tube very slowly out of the bottom of the furnace. Freezing will start and if we are careful, a single crystal will form and the rest of the metal will freeze with this as a nucleus. In this way we form a single crystal in the form of a rod. After breaking away the glass, let us grip the rods at the ends in a tensile machine and pull until appreciable permanent deformation takes place. We will find that there is a great difference among the specimens in the axial stress required to cause this permanent deformation or strain. In all cases, however, the flow occurs by shearlike movement (called "slip") on the (0001) planes. The specimens that require the least stress for slip (plastic flow) are those with both the normal to the (0001) planes and the [110] directions at 45° to the axis of the specimen.

To explain this we need an important but simple derivation.

Consider the rod in Fig. 3.1a, which is a single crystal of zinc. We locate the orientation of the (0001) planes, such as A_2, by x-rays. The (0001) planes are the basal planes of the HCP structure in the rod crystal. It is observed that slip takes place on these planes and in the [110] direction.

We first find the component of the *force* in the slip plane. This is $F \cos \lambda$. The shear stress in the slip plane will be this force divided by the area A_2. Next A_2 may be related to the known area A_1 by trigonometry: $A_2 = A_1/\cos \varphi$. Therefore, the shear stress τ is

$$\tau = \frac{F}{A_1} \cos \lambda \cos \varphi$$

If we take the different axial stresses required to cause slip and the angular measurements and calculate the resolved shear stress in each case, we obtain the same value (within experimental error). This is the *critical resolved shear stress, τ_c.* When $\varphi = \lambda = 45°$, the maximum shear stress is obtained from a given axial stress.

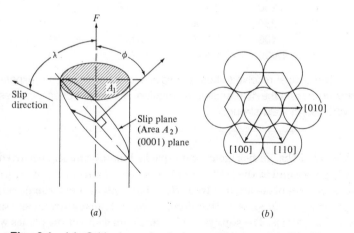

Fig. 3.1 (a) *Critical resolved shear stress model.* (b) *Slip directions in HCP.*

We finally have the answer to the unexpected behavior of the grains of copper in the bar shown in Fig. 2.2. Although slip occurred in some regions away from the region of maximum tensile stress, these were regions of maximum resolved shear stress. That is, because of the favorable *orientation* of these grains, τ_c was reached sooner than in the other grains.

EXAMPLE 3.1 An investigator prepares four single crystals of magnesium in cylinders of the same cross-sectional area. He finds that different axial stresses are required to produce permanent deformation (0.2 percent plastic strain). Furthermore, by x-rays he determines the orientation of the slip plane and of the slip direction relative to the axis.

Are the differences in yield strength due to imperfections in the crystals?

What are the critical resolved shear stresses?

The data are as follows.

Crystal	Yield stress, g/mm^2	φ	λ
1	200	45°	54°
2	230	30°	66°
3	400	60°	66°
4	1,000	70°	76°

ANSWER

Crystal	F/A	×	$(\cos \varphi)$	×	$(\cos \lambda)$	=	τ_c, g/mm^2
1	200		0.707		0.587		83
2	230		0.865		0.407		81
3	400		0.500		0.407		81
4	1,000		0.342		0.242		83

The critical resolved shear stresses are the same (within experimental error) and the difference in orientation, not imperfections, causes the variation in yield strength.

We noted that in hexagonal close-packed metals the slip occurred in the (0001) planes and in the [110] direction. As illustrated in Fig. 3.1*b*, three directions and one plane are involved. We define a plane and a slip direction in the plane as a *slip system;* therefore, at room temperature magnesium has three slip systems. The slip system is usually made up of the planes with highest atomic density (closest packing) and the directions with highest linear density. In the FCC metals the {111} planes and the ⟨110⟩ directions attain maximum density, and there are no exceptions to the rule. In BCC the important feature is that the $\langle 111 \rangle$ directions are closest packed and slip always occurs in these directions. There is no plane of maximum packing density, and therefore a number of planes such as (110) are involved. In the HCP metals the c/a ratio (see Fig. 2.12) is never the ideal of 1.633, and when this value is lower than 1.633, planes other than the (0001) planes may slip.

3.5 Twinning

Another type of plastic deformation is called "twinning." This is particularly important in hexagonal crystals because normally slip can occur on only one plane, (0001). If this plane is normal to the specimen axis there is no shearing stress, and brittle fracture would occur if the phenomenon of twinning could not take place. The essential difference between slip and twinning is that in slip each atom on one side of the slip plane moves a constant distance, whereas in twinning the movement is proportional to the distance (Fig. 3.2). Twinning is most common in BCC and HCP metals and can take place much more rapidly than slip. Therefore, mechanically formed twins instead of slip bands are often encountered when shock loading takes place. In FCC structures twins are usually formed only on heating (annealing) of cold-worked structures (Fig. 3.17).

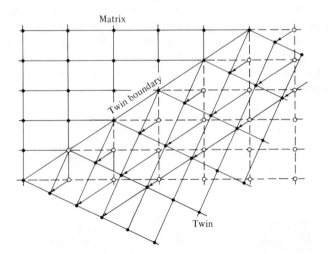

Fig. 3.2 *Formation of a twin in a tetragonal lattice by a uniform shearing of atoms parallel to the twin boundary. The dashed lines represent the lattice before twinning, the solid lines after.*

(H. W. Hayden, W. G. Moffatt, and John Wulff, "The Structure and Properties of Materials," vol. 3: Mechanical Properties, John Wiley & Sons, Inc., New York, 1965, Fig. 5.10, p. 111. By permission of John Wiley & Sons, Inc.)

3.6   Engineering stress-strain curves

Now that we have discussed the basic aspects of deformation, let us take up the manner in which the strength of commercial materials is tested and specified. If we look through the specifications for materials of the principal engineering groups such as the American Society for Testing Materials, The American Iron and Steel Institute, or the Society of Automotive Engineers, we find that the most common basis for testing is the 0.505-in.- (1.28-cm-) diameter tensile specimen (Fig. 3.3). In other words, whether we are buying material for a bridge or a crankshaft, a common tensile test is used to evaluate the mechanical properties. Usually modulus of elasticity, tensile strength, yield strength, percent elongation, and reduction of area are specified. Let us discuss how these are determined.

The "0.505 bar," as it is called from its diameter, is machined from a portion of the stock being bought or from separately cast specimens from the same ladle of metal. The bar is screwed into a pair of grips, which in turn are part of a tensile testing machine (Fig. 3.4).

Next the grips are pulled apart by mechanical means, and the load on the specimen is recorded continuously. The load can be converted to engineering stress by multiplying by 5, since the area of the 0.505-in.-diameter specimen is 0.2 in.² and engineering stress is load divided by the original area.

It is important to measure the extension or strain of the bar at the same time. A variety of devices, ranging from mechanically operated extensometers to electric strain gages, are used. In the usual mechanical extensometer the gage is anchored to a 2-in. gage length at the start of the test. We must note carefully that the raw data are in terms of *load* and *extension*.

These are converted as follows:

$$\text{Engineering stress} = \sigma = \frac{\text{load}}{\text{original area}}$$

$$\text{Engineering strain} = \varepsilon = \frac{\text{change in length}}{\text{original length}}$$

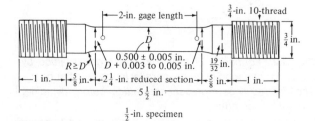

Fig. 3.3   *Design of a tensile test specimen. (To convert to centimeters, multiply by 2.54.)*

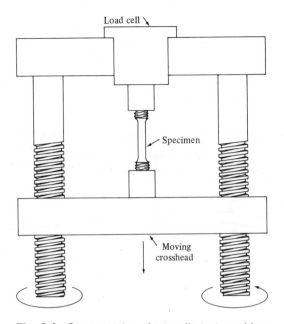

Fig. 3.4 *Cross-section of a tensile test machine.*

(H. W. Hayden, W. G. Moffatt, and John Wulff, "The Structure and Properties of Alloys," vol. 3: Mechanical Properties, John Wiley & Sons, Inc., New York, 1965, Fig. 1.1, p. 2. By permission of John Wiley & Sons, Inc.)

We then plot the stress vs. strain and obtain curves, as shown in Fig. 3.5*a* and *b*.

The following data are used in specifications and are obtained from a tensile test:

Modulus of elasticity (psi or MN/m^2)
= (stress/strain) in elastic range (slope of stress-strain curve)
Tensile strength (psi or MN/m^2)
= maximum stress on the stress-strain curve
Yield strength at 0.2 percent offset (psi or MN/m^2)
= stress at which 0.2 percent permanent or plastic strain is present
Percent elongation at fracture
= $(l_f - l_o)/l_o \times 100$, where l_f = final length
l_o = original length
Percent reduction of area
= $(A_o - A_f)/A_o \times 100$, where A_o = original area
A_f = final area

Breaking stress, the engineering stress at fracture, is also noted but not included in specifications for ductile materials. For brittle materials it is hard to distinguish from the tensile strength.

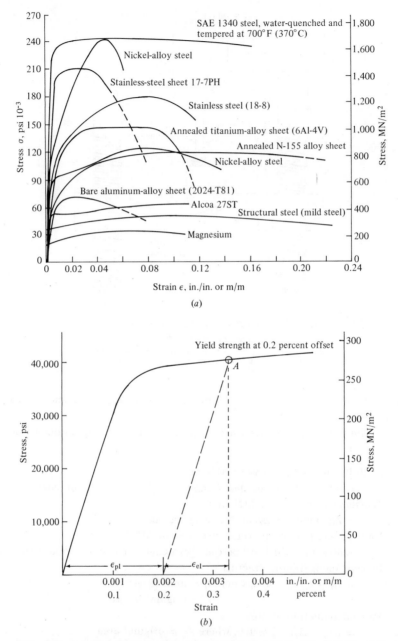

(a)

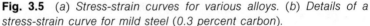

(b)

Fig. 3.5 (a) *Stress-strain curves for various alloys.* (b) *Details of a stress-strain curve for mild steel (0.3 percent carbon).*

[Part (a) from Joseph Marin, "Mechanical Behavior of Engineering Materials," © 1962, Fig. 1.10, p. 24. Reprinted by permission of Prentice-Hall, Inc., Englewood Cliffs, N.J.]

The *modulus of elasticity* is used to calculate the deflection of a given part under load. For example, a crankshaft will bend a certain amount between bearings, and clearance must be allowed for this. Also, a material of the same strength may not be substituted for another without consideration of the modulus. For instance, an aluminum crankshaft will deflect three times as much as a steel one because the modulus is 10×10^6 psi in place of 30×10^6 psi. It is important to note as well that the modulus does not change with strength. Characteristic moduli are given in Table 3.1.

The *tensile strength* is an index of the quality of the material. It is not used much in design for ductile materials because considerable plastic strain has occurred by the time this load-carrying capacity has been reached. However, it is a good indication of defects because if flaws or harmful inclusions are present, the bar will not reach the same maximum stress.

The *yield strength* is the most important value for design. The significance of the phrase "0.2 percent offset" requires explanation. As the tensile specimen is loaded, two types of strain are encountered: elastic and plastic.

As we mentioned earlier, the elastic strain will disappear upon unloading, while the plastic strain resulting from permanent deformation by slip will remain. Years ago it was thought that the bar behaved elastically up to a certain point called the "yield point," and from there on plastic strain began. However, we know now from precise strain measurements as well as from microscopic observation of slip that plastic strain begins at low stresses. The question then becomes: How much set (plastic elongation) can the designer tolerate? The percentage of set that can be tolerated in the spring of a delicate balance is different from that in the boom of a steam shovel. However, for most engineering uses a plastic strain of 0.2 percent can be tolerated, and the stress at which this occurs is called the yield strength.

Table 3.1 MODULI OF ELASTICITY FOR SEVERAL METALLIC MATERIALS

Material	Modulus of Elasticity,* psi $\times 10^6$
All steels, alloyed and unalloyed	30
Nickel alloys	26 to 30
Copper alloys	15 to 18
Aluminum alloys	10 to 11
Magnesium alloys	6.5
Cast iron, depending on amount and type of graphite	15 to 22
Ductile iron, depending on amount of graphite	22 to 25
Malleable iron, depending on amount of graphite	26 to 27
Molybdenum	47

* Multiply by 6.9×10^{-3} to obtain MN/m^2 or by 7.03×10^{-4} to obtain kg/mm^2.

To calculate this value we use the following construction (Fig. 3.5b). First 0.002 strain (0.2 percent strain) is laid off from the origin on the x axis. Next a line is drawn through this point parallel to the straight line portion of the stress-strain curve until it intersects this curve. This intersection gives the value of the yield strength, 40,000 psi (276 MN/m^2). The logic behind this construction is that if we follow normal testing procedures with a tensile specimen to this point and then remove the load, we will have 0.2 percent set, which is tolerable in most cases. In general, to find the amount of permanent and elastic strain at any point on a stress-strain curve, we draw a line parallel to the straight line portion to the strain axis, as shown in Fig. 3.5b. The total strain at A is approximately 0.0033 in./in. (m/m), the elastic component ε_{el} is 0.0013 in./in. (m/m), and the plastic component ε_{pl} is 0.002 in./in. (m/m). In cases where less plastic strain can be tolerated the yield strength is specified at 0.1 percent offset or lower.

In other alloys such as copper, aluminum, and magnesium the stress-strain graph begins to curve at low stresses. In this case the yield strength is usually specified as the stress at 0.5 percent *total* strain, which is read directly from the graph as the sum of the elastic and plastic components.

The *percent elongation* at fracture serves several purposes. It is possibly a better index of quality than the tensile strength because if inclusions or porosity are present, the elongation is drastically lowered. Second, the elongation multiplied by the tensile strength is an index of toughness at low rates of strain. The toughest steel available, Hadfield's 12 percent manganese steel, is used for railroad crossings, safe parts, and in ore crushers, as discussed in Chap. 6. It has elongation of over 40 percent and tensile strength of over 100,000 psi (690 MN/m^2). The gage length over which the percent elongation is calculated must be noted.

EXAMPLE 3.2 The following data were obtained for a high-strength aluminum alloy (7075-T5). Plot the engineering stress-strain curve. A 0.505-in.-diameter tensile specimen with a 2-in. gage length was used.

Load, lb	Stress, psi	Gage length, in.	Strain
0	0	2.0000	
4,000	20,000	2.0041	0.002
8,000	40,000	2.0079	0.004
10,000	50,000	2.0103	0.005
12,000	60,000	2.0114	0.006
13,000	65,000	2.0142	0.007
14,000	70,000	2.0202	0.010
16,000	80,000	2.0503	0.025
16,000 (maximum)	80,000	2.0990	0.050
15,600 (fracture)	78,000	2.1340	0.067

Calculate the modulus of elasticity, yield strength at 0.2 percent offset, percent elongation, and percent reduction of area. (No appreciable necking down—local yielding at one section of the bar—was encountered.)

ANSWER The data for load are converted to stress by multiplying by 5, since the diameter of a 0.505-in.-diameter tensile test bar is 0.2 in.². The data are then plotted, as shown in Fig. 3.6. Note that to obtain the modulus of elasticity and yield strength a magnified scale is used on the x axis.

The percent reduction of area is obtained by calculating first the area of the test section after testing. The volume after testing will equal the volume

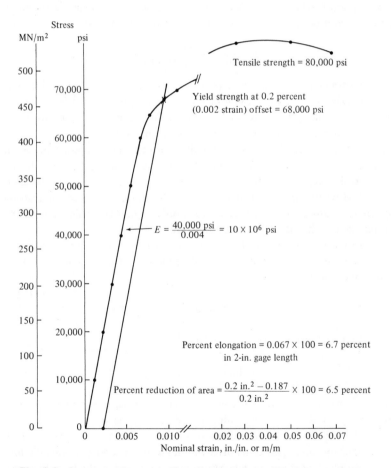

Fig. 3.6 *Stress-strain curve. (See Example 3.2.) Consider an original volume as* $V_1 = 2$ *in. by* 0.2 *in.² = 0.4 in.³. The final volume will be the same:* $V_2 = 2.134$ *in. by X in.² = 0.4 in.³. Final area = X = 0.187 in.².*

before testing, and since we know the length of the cylinder before and after testing and the area before testing, the only unknown is the area after testing. If necking down takes place, it is necessary to measure the diameter of the failed section to obtain the area after testing.

3.7 True stress–true strain relations

True stress–true strain curves are used in development and research more than in routine testing because the data are more difficult to obtain and plot than engineering stress and strain. We will discuss the basic features here and add further material in the problem section. True stress is simple to understand because it follows the real definition of stress as

$$\sigma_t = \frac{\text{load}}{\text{true area at the time}}$$

In calculating engineering stress we used the original area throughout. The area decreases only a small amount during testing in the elastic range but decreases significantly in the plastic range. To obtain true stress we must measure the diameter at several intervals during testing.

True strain ε_t is a slightly more difficult concept. Suppose we stretch the gage length from 2 to 3 in. Now let us further stretch the bar 0.1 in.

In the engineering strain definition the added strain would be $\Delta l/l_o$ or 0.1/2. However, if we consider strain as the change in length divided by the length *at the time,* the added strain would be 0.1/3.05, using 3.05 as the average length during the additional straining.

To obtain true strain, therefore, we need to sum up a succession of Δl's divided by the length at the time of producing the Δl. The solution is found by calculus to be

$$\varepsilon_t = \ln \frac{l}{l_o}$$

This expression is equivalent to

$$\varepsilon_t = \ln \frac{A_o}{A} = 2.3 \log_{10} \frac{A_o}{A}$$

Note that we use the symbols σ = engineering stress, σ_t = true stress, ε = engineering strain, ε_t = true strain. A typical curve is shown in Fig. 3.7.

The chief value of the true stress–true strain data is in our understanding of forming operations involving high plastic deformation. We find that if we plot log stress vs. log strain, we obtain a straight line (Fig. 3.8). From the graph we can then read the amount of stress needed to produce a given plastic deformation.

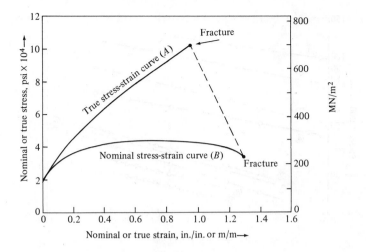

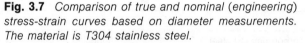

Fig. 3.7 *Comparison of true and nominal (engineering) stress-strain curves based on diameter measurements. The material is T304 stainless steel.*

(From Joseph Marin, "Mechanical Behavior of Engineering Materials," © 1962, Fig. 1.12(a), p. 34. Reprinted by permission of Prentice-Hall, Inc., Englewood Cliffs, N.J.)

As a practical illustration, the public often clamors for automobile bodies of stainless steel to prevent corrosion. The added cost is not merely a question of alloy expense. The stainless steel is more difficult to work (higher stress for a given strain), as shown by the true stress curve (Fig. 3.8) and gives greater die wear.

3.8 Dislocations

There used to be an afternoon colloquium in Cambridge, Massachusetts, in which metallurgists and physicists of leading institutions would meet after tea to present new data. Often a budding metallurgist, beaming with success, would present data for a new steel showing greatly improved tensile strength, to 300,000 psi (2.07×10^3 MN/m^2), for example. Invariably, at the end of the presentation a pipe-puffing physicist would puncture his younger colleague with a statement such as: "Really interesting my dear chap, but when are you going to reach at least the order of a million psi predicted by simple theoretical calculations of strength?"

This scene was re-enacted many times, but finally one day some scientists at the Bell Telephone Laboratories received samples of tiny tin crystals called "whiskers" which were shorting out capacitors. Out of curiosity they built a microtesting device and found that these whiskers did indeed have the theoretical strength level of over 1 million psi (6.9×10^3 MN/m^2).

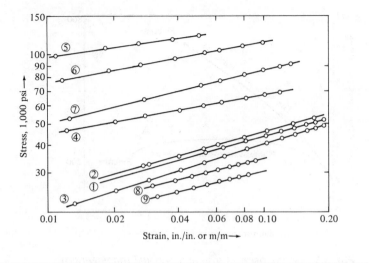

	Material	Treatment
1.	0.05 percent carbon rimmed steel	Annealed
2.	0.05 percent carbon killed steel	Annealed and temper-rolled
3.	Same as 2, completely decarburized	Annealed in wet hydrogen
4.	0.05 to 0.07 percent phosphorus low-carbon steel	Annealed
5.	SAE 4130 steel	Annealed
6.	SAE 4130 steel	Normalized and temper-rolled
7.	Type 430 stainless steel (17 percent chromium)	Annealed
8.	Alcoa 24-S aluminum	Annealed
9.	Reynolds R-301 aluminum	Annealed

Fig. 3.8 *Logarithmic true stress–true strain tensile relations for various materials. Tests are by J. R. Low and F. Garafalo. (Multiply psi by 6.9 × 10⁻³ to obtain MN/m² or by 7.03 × 10⁻⁴ to obtain kg/mm².)*

For a long time it was suspected that the reason for the difference between actual and theoretical values was the presence of microdefects. Indeed Griffith had shown that by testing glass fibers of fine diameters (where the probability of a defect was low) he could obtain strengths of 500,000 psi (3.45×10^3 MN/m²). It was postulated that these defects could be:

1. Point defects or missing atoms (vacancies)
2. Line defects or rows of missing atoms (dislocations)
3. Area defects (grain boundaries)
4. Volume defects as in actual cavities

The first type is not of great importance where strength is concerned, but is significant in diffusion, as discussed later.

We are already familiar with the third type at the face of the grain boundary.

The fourth type may also be dismissed at present because it results only from improper processing of the material, leading to voids, or from uneven diffusion conditions.

The second type, the line defect or dislocation, is of primary importance in understanding the reason for the gap between commercial and theoretical strengths. To illustrate this, suppose we have a block of metal which has an extra plane of atoms extending halfway through it (Fig. 3.9).

We see that this will result in a core of unstable material going back into the block, as shown by the five circled atoms in the face section. This core is a line defect called an "edge dislocation." An important characteristic is that the atoms in the upper side are in compression, those in the lower side in tension. It is apparent that the bonding forces are not as strong as in a perfect lattice.

Now let us see what happens if we apply a shearing stress (Fig. 3.10). In the perfect lattice if we are to displace the A atoms, we need to attain a uniformly high stress level; this is the theoretical yield strength obtained in whiskers. However, where we have a dislocation the atoms nearby are not so firmly held and the dislocation moves easily to the right. In other words, we obtain a lower value of yield strength.

In the actual case there are millions of dislocations which result from casting, rolling, etc. Although they are on an atomic scale, evidence of their presence and movement can be seen with an electron microscope. There is a great amount of literature on the subject, but we shall use the concept to understand rather simple phenomena.

By now it should be clear that the high strength of the whiskers is due to the absence of dislocations in the direction in which they were tested. It

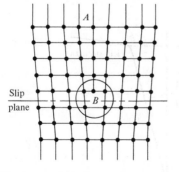

Fig. 3.9 *Atomic packing near edge dislocation. The dislocation B is circled. The atoms at A are in a normal configuration.*

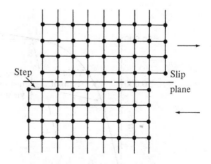

Fig. 3.10 *Correlation of slip with a line dislocation.*

is questionable whether material can be prepared that is completely free of dislocations. However, whiskers of many materials have been tested and found to be exceptionally strong. As a matter of fact, sapphire whiskers are now being used as strengtheners in composite materials.

We will see that the concept of dislocation and dislocation movement will be valuable as a "thinking tool" in explaining many phenomena.

3.9   Cold work or work hardening

We have already been introduced to this phenomenon in the stress-strain curve. To illustrate, let us take a bar of metal and apply stress until we reach the point A on the curve far above the yield strength of 50,000 psi (345 MN/m²), but below the tensile strength of 100,000 psi (690 MN/m²) (Fig. 3.11a).

Now let us unload the sample, take it from the machine, and give it to another technician for testing, who will obtain a yield strength of over 80,000 psi (483 MN/m²) (point B in Fig. 3.11b) in place of 50,000 psi (345 MN/m²). We can check this effect by recalling that on unloading (Fig. 3.11a) the points would follow the line AA' and on restressing would follow the same line.†

Now let us visualize basically what is happening. As we stress into the region of plastic strain, slip takes place on the favorably oriented planes, with dislocation production and movement. However, as more and more slip occurs, dislocations interact, pile up, and dislocation tangles form. This makes it more and more difficult for further slip to take place. This is shown by the rising stress-strain curve; to produce more strain, more stress is needed.

---

†There are small "aftereffects" of a lesser order of magnitude which we need not consider at this time.

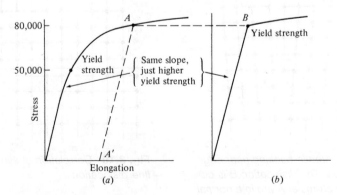

Fig. 3.11 *Correlation of work hardening with the stress-strain curve. The units of stress are psi.*

We come now to the key point of the argument. When we reached point *A* in Fig. 3.11*a*, all the planes and dislocation sites for easy slip were used up. When the load was removed, there was no change in this situation. Therefore, when the load was reapplied, no plastic strain was possible until the stress level of point *A* was reached. Consequently, we encountered only elastic strain to a much higher level than in Fig. 3.11*a* and the yield strength was correspondingly higher.

This phenomenon is called "work hardening," "strain hardening," or "cold work." We shall see soon that the term "cold" is relative. It means working at a temperature that does not alter the structural changes produced by the work. In other words, cold work causes atoms to move and dislocation tangles to form, and as we shall see in Sec. 3.13, this effect can be removed by working at higher temperatures.

3.10   Methods of work hardening

Naturally if we wish to raise the yield strength by work hardening, we do not need to put the part in a tensile machine. From our basic reasoning we see that it is necessary only to produce slip. Recalling our experiences with the copper bar, we know that slip is produced in the same way (by shearing) under compressive stress. So a variety of methods are available, as shown in Fig. 3.12. Note also that at the same time we can be producing the final shape we desire.

Other processes include *spinning,* in which a rotating sheet is forced into a rotating backing form with a tool; *swaging,* in which small die-shaped

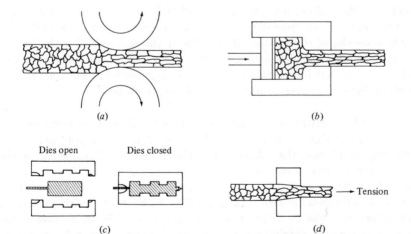

Fig. 3.12 *Methods for work hardening during processing. (a) Cold rolling bar or sheet. (b) Cold extrusion. (c) Cold forming, stamping, forging. (d) Cold drawing.*

hammers beat on a rod to reduce it in diameter; and *shot peening,* in which metal shot is thrown at the surface.

In all processes the percent cold work is defined as $(A_o - A_f)/A_o \times 100$, which is the same as percent reduction of area in the tensile test.

In all these cases the part is strengthened by the cold working. Another important effect is that the part is hardened, making it more resistant to wear and galling. Since the actual operation of hardness testing involves cold work, this is a good opportunity to digress briefly to analyze hardness measurements. Hardness tests are as important as tensile tests in many specifications.

3.11 Hardness testing

Hardness is usually defined as resistance to penetration. Let us review a few of the most common tests and see how closely they fit this definition.

Brinell Hardness Number (*BHN*). This is one of the oldest tests and is still the most common standard (Fig. 3.13).

The specimen with a flat upper surface is placed on the anvil, and a steel or tungsten carbide ball is pressed into the sample with a load of either 500 or 3,000 kg. The lighter load is used for the softer nonferrous metals such as copper and aluminum, and the heavier load is used for iron, steel, and hard alloys. The load is left in place for 30 sec and then removed. The diameter of the impression is then read in millimeters with a low-power microscope with a filar (measuring) eyepiece. Next the observer reads the Brinell hardness number (BHN) opposite the impression diameter from a table of values for the load used. We will not analyze the method of derivation of the numbers but merely note that the more difficult the penetration, the higher the BHN. The table is developed so that the BHN is about the same whether the 500- or 3,000-kg load is used, although obviously the impression diameter is different. The reason for the lighter load is that in very soft materials the 3,000-kg load will continue to penetrate until the ball is deeply sunken.

Vickers Hardness Number (*VHN*). This is an improvement on the Brinell test. Here a diamond pyramid is pressed into the sample under loads that are much lighter than those used in the Brinell test. The diagonal of the square impression is read, and the Vickers hardness number (VHN) is located on a chart. As shown in Fig. 3.14, the VHN is close to the BHN from 250 to 600. The figure does not show that the VHN climbs steadily with strength at higher values, while the BHN is not used above 750. The advantages of the Vickers test are in obtaining hardness measurements at high levels and in measuring the hardness of a small region. On the other hand, the BHN gives a better averaging effect because of the larger impression.

Test	Indenter	Shape of Indentation		Load	Formula for Hardness Number
		Side View	Top View		
Brinell	10-mm sphere of steel or tungsten carbide			P	$BHN = \dfrac{2P}{\pi D(D - \sqrt{D^2 - d^2})}$
Vickers	Diamond pyramid	$136°$		P	$VHN = 1.72\, P/d_1^2$
Knoop microhardness	Diamond pyramid	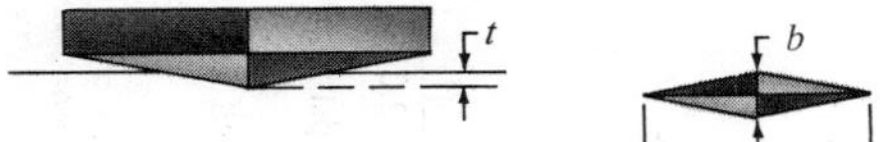 $l/b = 7.11$ $b/t = 4.00$		P	$KHN = 14.2\, P/l^2$
Rockwell A C D	Diamond cone	$120°$		60 kg 150 kg 100 kg	$R_A =$ $R_C =$ 100 − 500t $R_D =$
B F G	$\frac{1}{16}$-in.-diameter steel sphere			100 kg 60 kg 150 kg	$R_B =$ $R_F =$ 130 − 500t $R_G =$
E	$\frac{1}{8}$-in.-diameter steel sphere			100 kg	$R_E =$

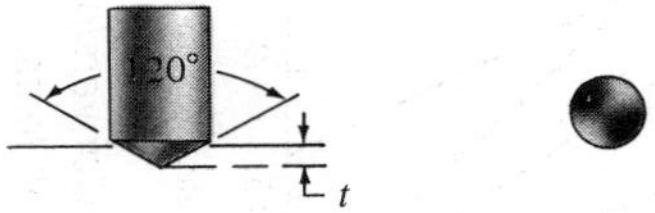

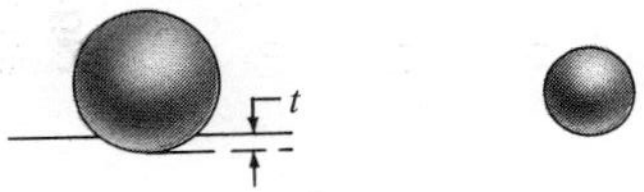

Fig. 3.13 *Hardness testing methods.*

(H. W. Hayden, W. G. Moffatt, and John Wulff, "The Structure and Properties of Alloys," vol. 3: Mechanical Properties, John Wiley & Sons, Inc., New York, 1965. By permission of John Wiley & Sons, Inc.)

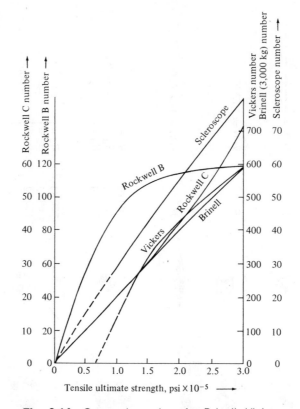

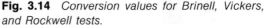

Tensile ultimate strength, psi × 10^{-5} ⟶

Fig. 3.14 *Conversion values for Brinell, Vickers, and Rockwell tests.*

(From Joseph Marin, "Mechanical Behavior of Engineering Materials," © 1962, Table 10.2, p. 450. Reprinted by permission of Prentice-Hall, Inc., Englewood Cliffs, N.J.)

Rockwell Hardness Testing (R_A, R_B, etc.). The chief advantage of the Rockwell test is that the hardness is read directly from a dial. The indenter for the R_C test is a diamond cone or "brale" suitably supported. The observer first turns a handle which presses the diamond cone a slight standard amount into the sample. This is called the "preload." Next he releases the standard R_C load of 150 kg (Fig. 3.13), which forces the diamond farther into the sample. Then with the same lever he removes the load. At this point he reads the R_C hardness from the dial and then unloads the specimen. The principle of this test is that the dial, through a lever system, records the depth of penetration between the preload and the 150-kg load and reads directly in R_C. The R_C is approximately $\frac{1}{10}$ BHN (Fig. 3.14). The R_B scale is used for softer materials, employs a $\frac{1}{16}$-in.-diameter ball and a 100-kg load, and is also direct reading. There are also Brinell test machines that are direct reading.

The scleroscope hardness test is used chiefly for checking large rolls on which it is difficult to use the other tests. The value is obtained by measuring the height of rebound of a small weight under standard conditions.

3.12 Correlation of hardness, tensile strength, and cold work

All the hardness tests depend on resistance to plastic deformation. It is no surprise, therefore, to find that the hardness of a bar is higher after cold working, because all the sites for easy slip have been used up. Thus there is good correlation between hardness and strength. For steel a simple relation to remember is that the tensile strength in pounds per square inch is 500 times the BHN (derived from Fig. 3.14).

3.13 Effects of temperature on work-hardened structures

We do not always want maximum strength and hardness in a part, because as the hardness increases the ductility decreases, as shown by the percent elongation (Fig. 3.15). Furthermore, let us suppose we are fabricating a simple metal cup 4 in. (10.16 cm) high. When we try to press the cup to this depth, we find that the sheet cracks at the corners when formed only to a 2-in. (5.08-cm) depth. We have "used up" the plastic elongation. We encounter similar situations in the other cold-forming operations. What can we do to permit deeper drawing?

Basically we wish to restore the original structure by eliminating the extensive slip and dislocation tangles. Here we need to recall that atoms are not rigidly fixed but can diffuse from their positions. Diffusion increases rapidly with rising temperature. Therefore, if we heat the part, the atoms in severely strained regions can regroup and move to unstrained positions. This is

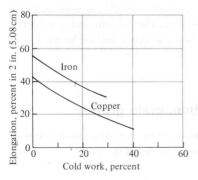

Fig. 3.15 *Correlation of cold work and elongation in tensile testing.*

(L. H. Van Vlack, "Elements of Materials Science," 2d ed., Addison-Wesley Publishing Company, Inc., Reading, Mass., 1964, Fig. 6.24, p. 150.)

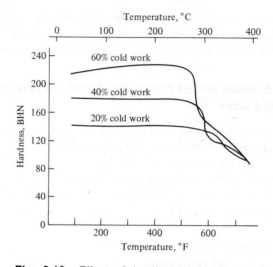

Fig. 3.16 *Effect of heating on hardness of cold-worked 65% Cu, 35% Zn brass, 1 hr.*

(L. H. Van Vlack, "Elements of Materials Science," 2d ed., Addison-Wesley Publishing Company, Inc., Reading, Mass., 1964, Fig. 6.28, p. 153.)

adjustment to strain on a microscopic scale, and the partly formed shape does not change dimensions.

Therefore, we take the cold-worked part and heat treat it in a furnace in the process called "annealing." We find that the metal softens as a function of heating temperature (Fig. 3.16) and time at temperature. If we observe the microstructure, we see that profound changes take place (Fig. 3.17). At elevated temperatures new small grains of equal dimensions in all directions (*equiaxed* grains) grow inside the old distorted grains and at the old grain boundaries. At higher temperatures the grain size is larger. Hardness and strength decrease with grain size, but elongation increases. This can be explained from a microscopic view. Slip on a given plane stops when we reach a grain boundary, where a grain boundary is an area of high dislocation density and entanglements. Thus the more grain boundaries, the greater the limitation of slip and the higher the strength. Ductility (elongation) is lower because slip is limited by the larger grain boundary area.

3.14 Recovery, recrystallization, grain growth

It is useful to divide the effects of temperature on cold-worked material into three regions, in order of increasing temperature (Fig. 3.18).

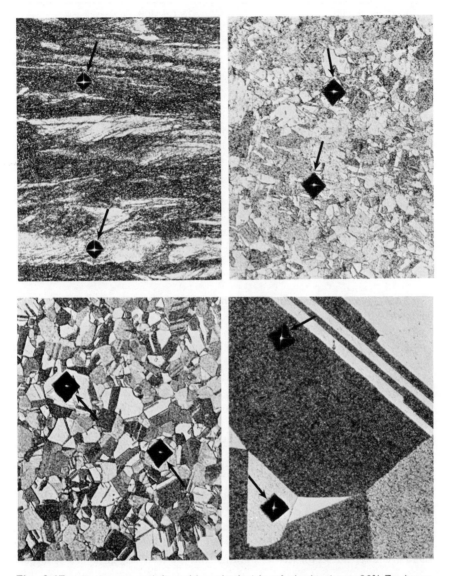

Fig. 3.17 *At the upper left: cold-worked strip of single-phase 30% Zn brass, 50 percent reduction, microhardness 140 to 170 VHN, 25-g load. At the upper right: partly recrystallized after 1 hr at 600°F (316°C), microhardness (see arrows) 93 to 109 VHN. At the lower left: recrystallized at 880°F (471°C) for 1 hr, small grains, microhardness 90 to 100 VHN. At the lower right: recrystallized at 1400°F (760°C) for 1 hr, large grains, microhardness 65 to 87 VHN. Magnification 500X in all cases, peroxide etch.*

	Copper, OFHC			70% Cu, 30% Zn Brass*				
	Prior Cold Work			Prior Cold Work		Tensile Strength, psi × 10^3	Percent Elongation in 2 in.	Grain Size, mm
	30%	50%	80%	50% F.G.	50% C.G.			
Initial 30 min	$86R_H$	$91R_H$	$95R_H$	$99R_X$	$97R_X$	80	8	
150°C	85	90	94	101	98	81	8	
200°C	80	88	93	102	100	82	8	
250°C	74	75	65	103	101	82	8	
300°C	61	54	42	82	98	76	12	
350°C	46	40	34	66	80	60	28	0.02
450°C	24	22	27	50	58	46	51	0.03
600°C	15	17	22	38	34	44	66	0.06
750°C				20	14	42	70	0.12
Final grain size	0.15	0.12	0.10	0.08	0.12			

*F.G. = originally fine-grained; C.G. = originally coarse-grained; R_H = Rockwell scale, $\frac{1}{8}$-in. ball, 60-kg load; $R_X = \frac{1}{16}$-in. ball, 75-kg load.

Fig. 3.18 *Recovery, recrystallization, and grain growth as produced on heating cold-worked material.*

(After Brick and Phillips.)

Recovery. This is the temperature range just below recrystallization. With the electron microscope it can be shown that stresses are relieved in the most severely slipped regions. Dislocations move to lower energy positions, giving rise to subgrain boundaries in the old grains; this process is called "polygonization." Hardness and strength do not change greatly in this period, but corrosion resistance is improved.

Recrystallization. In this temperature range the formation of new stress-free and equiaxed crystals leads to lower strength and higher ductility. There is an interesting relation between the amount of previous cold work and the grain size of the recrystallized material. With a lower amount of cold work there are fewer nuclei for the new grains, and the resulting grain size is larger.

Grain Growth. As the temperature is raised further, the grains continue to grow. This is because the large grains have less surface area per unit of volume. There are fewer atoms leaving a unit grain boundary area than with a small grain, and the larger grains grow at the expense of the smaller.

It should be added that all three effects are time dependent as well (see Prob. 3.14). A summary of these effects is shown in schematic form in Fig. 3.19.

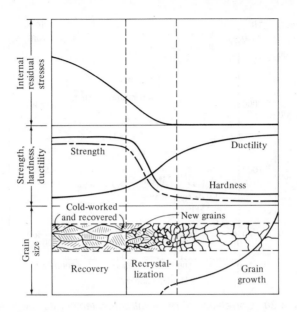

Fig. 3.19 *Summary of annealing effects (schematic representation).*

(After Sachs.)

3.15 Selection of annealing temperature

The temperature required for recrystallization varies with the metal and is approximately one-third to one-half the melting temperature of a pure metal, expressed on an absolute scale (Fig. 3.20). Therefore, we see steel being annealed at red heat (1600 to 1800°F) (870 to 980°C), while aluminum is still liquid at this temperature. On the other hand, lead can hardly be cold-worked at room temperature because the recrystallization temperature is so low. Furthermore, we should point out that the recrystallization temperature is really a range, not a sharp point. Also, the greater the amount of cold work, the lower the recrystallization temperature because of the stored energy. The sudden drop in hardness for the 65% Cu, 35% Zn brass in Fig. 3.16 shows the order of magnitude of the drop in recrystallization temperature to be expected with increased cold work.

3.16 Effect of grain size on properties

From the preceding discussion we see that we can control the grain size of the final part by the combination of cold work and annealing. With large amounts of cold work and rapid annealing the grain size will be small, because many nuclei for growth of new grains are present. We realize, furthermore, that

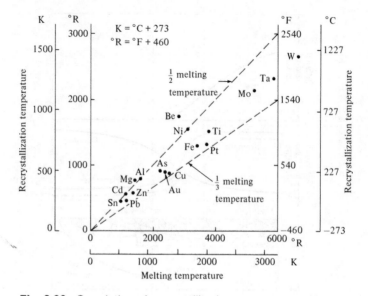

Fig. 3.20 *Correlation of recrystallization temperature with melting point.*

(L. H. Van Vlack, "Elements of Materials Science," 2d ed., Addison-Wesley Publishing Company, Inc., Reading, Mass., 1964, Fig. 6.29, p.154.)

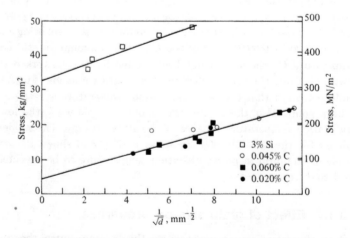

Fig. 3.21 *Relation between yield stress and grain diameter of iron alloys.*

(W. J. M. Tegart, "Elements of Mechanical Metallurgy," Macmillan Publishing Co., Inc., New York, 1966, Fig. 6.14, p. 175.)

grain boundaries are an impediment to slip. We would expect then that a fine-grained material would have higher strength, and this is given in the Petch relation: $\sigma = \sigma_0 + kd^{1/2}$, where σ is the yield strength, d is the grain size, and σ_0 and k are constants for a particular material. An added advantage to fine-grained material is that the plastic flow under stress is more even and a smoother surface is obtained than with coarse-grained material in producing stamped or formed parts. An example of the effect of grain size on various alloys is given in Fig. 3.21.

3.17 Engineering application of cold work and annealing

The cold work–annealing sequence may be repeated as desired. We can start with a simple sheet of metal, roll or form it extensively, and each time the material becomes difficult to work or in danger of cracking we can anneal it. Furthermore, we can ensure the right combination of strength and hardness on the one hand and ductility on the other by the final sequence of operations. There are two possibilities:

1. Finish the part with higher hardness and lower ductility than desired and then anneal it to obtain the desired combination.
2. Use the proper amount of cold work for the final operation to reach the required hardness level.

3.18 Hot working; hot rolling, forging, extrusion

At this time we may well ask: Instead of cold working plus annealing, what would happen if we worked the part *at* the annealing temperature? Would not recrystallization take place, so that we could continue to work the part without interruption? The answer is yes, and therefore we can conduct practically all the operation of working at the annealing temperature or above. Steel, for example, is passed through a set of rolls at high temperature and in some cases is actually increased in temperature from the heat produced by the mechanical work! The disadvantages of hot working are oxidation of the metal surface and shorter die life caused by heating from the part. A better surface can be produced by cold working because the part can be annealed in a furnace with a protective atmosphere after working. Also, if the part is finished by hot working, the strength and hardness will be lower.

In many cases a combination of hot and cold working is used. First a large shape such as an ingot is cast, hot-rolled to bar stock, and then cleaned. The surface is then preserved by cold rolling and annealing in protective atmospheres.

EXAMPLE 3.3 In the manufacture of practically all stamped and cold-formed parts it is necessary to balance the properties desired in the final part with the economic and engineering problems of forming the part. Let us recall the following aspects of the problem.

Effect of:	Strength	Hardness	Percent elongation	Percent reduction of area
Increased work before testing	Increased	Increased	Decreased	Decreased
Increased annealing temperature	Decreased	Decreased	Increased	Increased

Therefore, for maximum workability we want a well-annealed part, but for highest strength and hardness we want maximum cold work. Suppose we have some annealed 70% Cu, 30% Zn brass bar stock (cylindrical rod) of 0.35-in. (0.889-cm) diameter.

PROBLEM Produce bar stock of 0.21-in. (0.533-cm) diameter with a tensile strength of over 60,000 psi (414 MN/m²) and an elongation of over 20 percent.

ANSWER One way to solve the problem is first to determine what the final step in cold working should be. We read from Fig. 3.22 that to obtain 60,000 psi (414 MN/m²) tensile strength we need more than 15 percent cold work. Also, to obtain the required elongation we should cold-work less than 23 percent. Let us choose to cold-work 20 percent as the final working step, which will meet the tensile strength and elongation specifications. Then

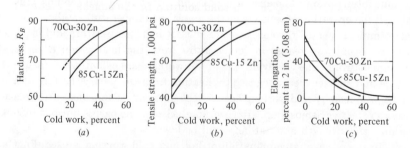

Fig. 3.22 *Effect of amount of cold work on mechanical properties. (For the tensile strength multiply psi by 6.9 X 10⁻³ to obtain MN/m².)*

(L. H. Van Vlack, "Elements of Materials Science," 2d ed., Addison-Wesley Publishing Company, Inc., Reading, Mass., 1964, Fig. 6.26, p. 151.)

$$\text{Percent cold work} = \frac{\Delta A}{A} = \frac{\frac{1}{4}\pi d^2 - \frac{1}{4}\pi(0.21)^2}{\frac{1}{4}\pi d^2} \times 100 = 20$$

where d is the diameter at which we wish to start the final deformation (starting of course with an annealed structure). Solving, we have

$$d = 0.235 \text{ in.} \quad (0.596 \text{ cm})$$

Therefore, we will obtain 20 percent cold work going from a 0.235- to 0.21-in. (0.596- to 0.533-cm) diameter.

Now we have the problem of how to reduce from a 0.35- to 0.235-in. (0.889- to 0.596-cm) diameter. With a ductile material this can be done cold in several steps. In this case it would be necessary to anneal before the final step; otherwise the elongation would be too low. (The effect of cold working is cumulative.) An alternative would be to hot-work (at or above the recrystallization temperature) to a 0.235-in. (0.596-cm) diameter. After hot working, a recrystallized structure would be present which could then be cold-worked. The most economical solution would depend on the surface quality desired and the equipment available.

3.19 Effects of cold work and annealing in solid solution alloys

We find that we can strengthen and harden or anneal a solid solution alloy such as cupronickel or brass by cold work, just as we harden pure metals. Therefore, we can use together the two strengthening mechanisms of work hardening and solid solution hardening. Upon heating the material we encounter recovery, recrystallization, and grain growth, as we did in the pure metals but at a somewhat higher temperature range. Typical data for a 65% Cu, 35% Zn brass are given in Fig. 3.16. Table 3.2 shows the effect of alloying on properties of pure copper and 30% Ni, 70% Cu solid solution in the annealed condition.

Table 3.2 COMPARISON OF MECHANICAL PROPERTIES OF ANNEALED COPPER AND 70% Cu, 30% Ni CUPRONICKEL

Property	Copper	Cupronickel
Tensile strength,* psi $\times 10^3$	30	55
Yield strength,* psi $\times 10^3$	10	20
Modulus of elasticity,* psi $\times 10^6$	17	22
Percent elongation in 2 in.	50	45
Hardness, R_F†	40	84

*Multiply by 6.9×10^{-3} to obtain MN/m².
†Rockwell F scale.

3.20 Creep and stress relief

To conclude this chapter we should consider the effect of temperature on stress-strain relations. These data are important in designing parts for operation at higher temperatures and in the operation called "stress relief."

If we apply a stress of 10,000 psi (69 MN/m^2) to a typical bar of structural steel, measure the elongation, and then leave the specimen under load overnight, we will not find any appreciable additional length change in the morning. However, if we surround the specimen with a resistance furnace and conduct the same experiment with the specimens at 1000°F (540°C), we find that plastic elongation occurs as a function of time. This phenomenon is called "creep," and its measurement is very important for parts subjected to elevated temperatures. For example, it would be useless to develop a turbine blade material of exceptionally high tensile strength if at operating temperature it would elongate and contact the housing after a few hours of operation. We will discuss this phenomenon further in the general review of mechanical properties, Chap. 11.

Stress relief is an important concept related to creep. As a result of uneven temperature distribution during processing, it is possible to develop high internal or residual stresses in components. The residual stress is actually present as an elastic strain, just as if the section were held in a tensile test machine under load. If we heat the entire part to a temperature at which creep takes place, the elastic strain is converted to plastic strain. If the part has been cold-worked, recrystallization may also occur. The part is then slowly cooled to avoid setting up new elastic strains resulting from temperature gradients in the material during cooling.

SUMMARY

When stress is applied to a metallic material, the strain at low stresses is principally elastic and varies with direction in a crystal. At higher stresses both elastic and plastic (permanent) strain are encountered. For design purposes the yield strength is defined as the stress at which the plastic strain reaches 0.2 percent.

The modulus of elasticity is an important property in design because it gives the elastic strain ε produced by a given stress σ: $\varepsilon = \sigma/E$. Other design values are tensile strength, percent elongation, and percent reduction of area.

A basic understanding of plastic deformation is obtained from the critical resolved shear stress, which is a constant for a given material. While different values of tensile stress may be required to cause yielding, there is a constant critical resolved shear stress to be reached. A further basic consideration is the role of dislocations in promoting slip. In material that is free from dislocations the yield strength is over 10^6 psi (6.9×10^3 MN/m^2).

The hardness of a material is determined by the size of the cavity produced by a hardened indenter and the prominent scales of hardness are Brinell (BHN), Rockwell (R_A, R_B, etc.) and Vickers (VHN).

After a material has undergone plastic deformation, as by cold rolling for example, the increase in hardness and decrease in ductility may be removed by an annealing treatment. As the specimen is heated, recovery, recrystallization, and grain growth take place at increasingly higher temperatures. By a combination of cold working and annealing the desired balance between strength and hardness on the one hand and ductility, as measured by percent plastic elongation, may be attained.

Creep occurs when a metal is heated to a temperature at which it is relatively hot and plastic. Care must be exercised to avoid placing metals in service where substantial deformation will occur with time. On the other hand, if elastic strain is to be removed, as in residual stresses, the part is heated to allow creep to occur, and the elastic strain is replaced by plastic.

DEFINITIONS

Elastic strain The elastic displacement of atoms from their normal positions, as for example by applying a tensile or compressive stress. When the stress is removed, the atoms return to the normal spacing.

Plastic strain The permanent displacement of atoms from a given starting position, as by slip or twinning.

Work hardening An increase in hardness and strength due to plastic deformation.

Engineering strain, ε Change in length divided by original gage length, $\Delta l/l_o$.

Engineering stress, σ Load divided by original area, P/A_o.

True stress, $σ_t$ Load divided by area at the given load, P/A.

True strain, $ε_t = \ln (l/l_o) = 2.3 \log_{10} (l/l_o)$.

Modulus of elasticity, E Stress divided by strain, $σ/ε$.

Critical resolved shear stress, $τ_c$ The applied stress resolved in the slip direction in the slip plane, $(F/A) \cos λ \cos φ$.

Slip system A combination of a slip direction and a slip plane containing the direction.

Twinning The shifting of the atoms on one side of a twinning plane by an amount proportional to the distance from the plane. The twin produces a mirror image of one part of the grain with the other.

Engineering stress-strain curve The results, usually of a tensile test, plotted with σ as the y axis and ε as the x axis.

Tensile strength The maximum engineering stress encountered during a tensile test.

Yield strength The stress at which a specified amount of plastic strain is produced, usually 0.2 percent.

Percent elongation at fracture Plastic (engineering) strain times 100,

$$\frac{l_f - l_o}{l_o} \times 100$$

Percent reduction of area $\dfrac{\text{Change in area}}{\text{Original area}} \times 100 = \dfrac{A_o - A_f}{A_o} \times 100$

Dislocation A line of missing atoms, leading to a region of easy slip.

Percent cold work $\dfrac{\text{Change in cross-sectional area}}{\text{Original area}} \times 100$

Brinell hardness number, BHN The value obtained from a Brinell indentation hardness test.

Vickers hardness number, VHN The value obtained from a Vickers test.

Rockwell hardness number, R_A, R_B, R_C, etc. The number obtained from a Rockwell hardness test.

Annealing In general, a heat treatment in which a part is heated to soften the material. In this chapter the treatment leads to the recrystallization of cold-worked material.

Recovery The relief of elastic strain taking place in early stages of annealing.

Recrystallization The growth of new stress-free equiaxed crystals in cold-worked material.

Grain growth The development of larger grain size by preferential growth of larger crystals.

Cold working The deformation of material below its recrystallization temperature.

Hot working The deformation of material at or above its recrystallization temperature.

Creep Permanent strain that increases as a function of time under load.

Stress relief The disappearance of elastic strain, especially during annealing.

Equiaxed grains Grains that are of equivalent dimensions in all directions, i.e. not longer in one direction than in another, as is common in cold working.

PROBLEMS

3.1 Re-examine the bent microspecimen of copper (Fig. 2.2b), and note the most common angle that the slip planes make with the surface. Note, however, that some appear to be nearly at 90°. Explain both observations.

3.2 Old specifications frequently refer to a *yield point,* which is defined as the stress required to cause the onset of permanent deformation in a specimen. Discuss the difficulty of measuring this quantity in a commercial metal.

3.3 Slip planes and directions for the three common metal structures are found to be:

Structure	Directions	Planes
FCC	⟨110⟩	{111}
BCC	⟨111⟩	{110}
HCP	⟨110⟩	{0001}

Are these the planes and directions of densest packing (greatest atomic density)?

3.4 A slip system is defined as a combination of a plane and a direction. Show that the number of slip systems in FCC is 12 and in HCP is 3, using only the most densely packed planes and directions.

3.5 Given that the critical resolved shear stress is 82 psi ($0.566 \ \mathrm{MN/m^2}$), plot axial stress for slip as a function of $\cos \varphi \cos \lambda$.

3.6 From the data of Example 3.2 draw the true stress–true strain curve. Above what engineering stress do the true stress–true strain data differ appreciably (by more than 5 percent) from the engineering stress-strain data? A simple way to calculate true strain ε_t from engineering strain ε is $\varepsilon_t = \ln (1 + \varepsilon)$, since

$$\varepsilon = \frac{l - l_o}{l_o} = \frac{l}{l_o} - 1$$

and therefore

$$\frac{l}{l_o} = \varepsilon + 1$$

The area for true stress calculations may be found from the relation

$$A = \frac{l_o}{l} A_o$$

where A = area at time i

A_o = original area

l = length at time i

l_o = original length

3.7 From the following data plot the engineering stress-strain curve and calculate the modulus of elasticity, yield strength at 0.2 percent offset, percent elongation, and percent reduction of area.

Load, lb	Gage length, in.
0	2.0000
5,000	2.00167
10,000	2.00340
12,000	2.00425
13,000	2.0047
14,000	2.0052
13,900	2.0065
15,000	2.0079
18,000	2.0118
20,000	2.0156
26,000	2.056
39,000 (maximum)	2.472
32,000 (fracture)	2.820 (after fracture)

Original diameter: 0.505 in.
Diameter at maximum load: 0.454 in.
Diameter at fracture region: 0.284 in.

(To convert stress to MN/m^2, multiply psi by 6.9×10^{-3}.)

3.8 Derive the relation between original area A_o and the area at σ_{true} and true strain. (*Hint:* Take a small volume at the region of fracture. With no load the volume would be $l_o A_o$. After fracture the volume would be $l_f A_f$ and the volumes would be equal. Now substitute for the ratio l_f/l_o in the formulas for true strain.)

3.9 Calculate the true stress–true strain curve using the data of Prob. 3.7 and the relation

$$\text{True strain} = \ln \frac{A_o}{A_f} = 2.3 \log \frac{A_o}{A_f}$$

for the strain at fracture.

3.10 A 70% Cu, 30% Zn brass wire with a minimum tensile strength of 60,000 psi (414 MN/m^2), a hardness of 75 R_B minimum, and an elongation of over 10 percent are needed in some 0.1-in.- (0.254-cm-) diameter wire. You have some 0.25-in.- (0.635-cm-) diameter stock available which has been cold-rolled 40 percent. Specify your procedure. Assume that 70% Cu, 30% Zn brass fails at cold work in excess of 60 percent.

3.11 A bar of 85% Cu, 15% Zn alloy (red brass) that has a 0.5-in. (1.28-cm) diameter is to be cold-rolled to a bar of 0.125-in. (0.317-cm) diameter. Specify the procedure needed to obtain a final tensile strength of 60,000 psi (414 MN/m^2) minimum and an elongation of 10 percent minimum.

3.12 A rolled 70% Cu, 30% Zn brass plate 0.500 in. (1.28 cm) thick has 2 percent elongation as received from the supplier. The final thickness desired is

0.125 in. (0.317 cm), with a tensile strength of 70,000 psi (483 MN/m²) minimum and an elongation of 7 percent minimum. Assume the rolling is conducted so that the width of the sheet is unchanged. (This means that in calculating the area, $A = wt$, where the width w is constant.) Specify all steps in the procedure, including heat treatments.

3.13 Sketch the microstructures (longitudinal and transverse to the rolling direction) you would expect to see in the material of Prob. 3.12 as received, after intermediate annealing, and after final working. Show the differences in grain shape, slip bands, and annealing twins.

3.14 The times for 50 percent recrystallization of pure copper are as follows:

Temperature:	203°C	162°C	97°C	72°C
Time:	6 sec	1 min	100 min	1,000 min

Consider the value

$$\frac{1}{\text{Time for 50 percent recrystallization}}$$

as representing the rate. Plot ln [1/time (50 percent)] against $1/T$, where T is degrees absolute (°C + 273). From the plot estimate the time for 50 percent recrystallization for the case where the temperature is 20°C (68°F). Under these conditions would you worry about the loss of strength in a cold-worked copper cable over a period of 10 yr?

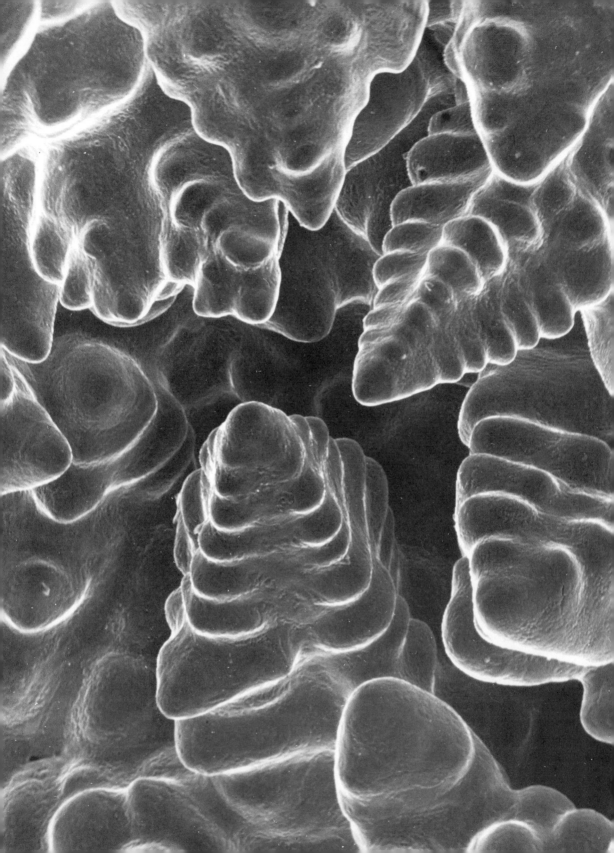

4

CONTROL OF POLYPHASE STRUCTURES IN METALS: UNDER EQUILIBRIUM CONDITIONS (PHASE DIAGRAMS); UNDER NONEQUILIBRIUM CONDITIONS (PRECIPITATION REACTIONS)

THE somewhat fernlike growths in the illustration are called "dendrites" and are actually crystals of a copper alloy growing into a cavity in a casting, photographed with a scanning electron microscope (100X). This is the most common type of growth in metal parts, but of course it is not necessarily accompanied by cavity formation.

This chapter is devoted entirely to the formation of metal structures from the liquid, as in the case of dendrites, and from changes taking place in the solid state.

4.1 General

So far we have discussed the properties of only *single-phase* materials, such as pure metals, solid solutions, and intermetallic compounds. These are useful materials, but the metallic structures of highest strength, hardness, and wear resistance are composed of two or more phases in a well-controlled dispersion. This microstructure is usually not obtained in the original casting or ingot, but is formed by carefully controlled processing involving hot working and heat treatment. As an example, the properties of a common aluminum alloy 2014 used for structural members in aircraft are:

	Yield strength at 0.2 percent offset, psi × 10³†	Tensile strength, psi × 10³	Percent elongation
Alloy 2014,‡ annealed	14	27	18
Alloy strengthened by heat treatment	60	70	13

We see, therefore, that if an engineer had to design on the basis of yield strength of the annealed material, an aircraft would have to be 4 times heavier and perhaps it would never get off the ground!

At this time it is well to introduce a general statement of the greatest importance for understanding the relations between structure and properties, not only for metals but for all materials. The properties of a material depend on the nature, amount, size, shape, distribution, and orientation of the phases. This point is illustrated in large-scale fashion in Fig. 4.1a, which shows how the properties of a reinforced concrete slab may vary. Let us review this figure and at the same time consider the equivalent effects in different microstructures.

---

†To obtain MN/m², multiply by 6.9 × 10⁻³.
‡4% Cu, 0.8% Si, 0.8% Mn, 0.6% Mg, balance Al.

Fig. 4.1 (a) *Schematic representation of how the nature, size, shape, amount, distribution, and orientation of phases control the physical properties (as applied to steel and concrete). The photographs show the microstructure of the aluminum-silicon alloy used in an aluminum automobile engine block. The composition is 17% Si, 4% Cu, 0.5% Mg, balance Al. The large gray crystals are β (almost pure silicon) and are precipitated from the liquid. The balance of the structure is a mixture of fine α, β, and a trace of a copper-rich phase θ (light gray). (b) Silicon crystals 1070 VHN; matrix 130 VHN. 500X, etched. (c) Silicon gray, θ light gray, matrix white. 1,500X, unetched.*

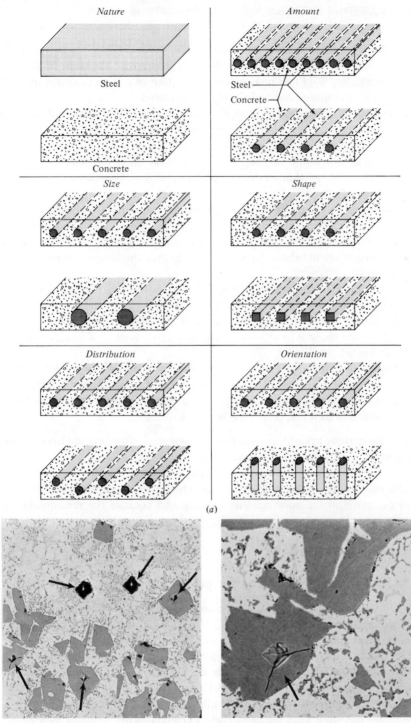

Nature

Steel

Concrete

Amount

Steel

Concrete

Size

Shape

Distribution

Orientation

(a)

(b)

(c)

1. *Nature* of the phases. In the concrete slab the individual properties of the concrete and the steel affect the overall strength of the slab. In the aluminum alloy 2014 two phases, a solid solution of copper in aluminum and an intermetallic compound $CuAl_2$, are involved.

2. *Amount* of the phases. Just as the relative amounts of steel and concrete are important, the amounts of the two phases in the alloy are significant.

3. *Size* of the phases. If all the steel reinforcing rods were in a few large bars, the structure would not be as strong as it is when the same amount of steel is used in thinner bars. Similarly, a fine dispersion of $CuAl_2$ is preferred.

4. *Shape* of the phases. If the steel is in square bars, stress concentration will occur at the corners, causing cracking of the concrete; hence round bars are preferred. In the case of a precipitate in the microstructure, different properties are obtained depending on whether the shape is spheroidal or platelike.

5. *Distribution* of the phases. If the slab is to encounter bending stresses, it is better to place the steel closer to the surface rather than at the center. Similarly, in a microstructure there will be a difference in properties if a precipitate phase is localized at grain boundaries instead of being uniformly distributed.

6. *Orientation* of the phases. If the slab is to be used to resist typical traffic, a horizontal orientation of the steel is preferred. Similarly, we will see later that for optimum magnetic properties a certain alignment of particles or precipitates in a magnetic tape is very important.

From these examples we see that it is essential to understand the methods of control of polyphase structures.

We will begin with a discussion of the structural changes that take place under equilibrium (slow cooling), because these treatments are useful in softening a material for machining or forming and because the nonequilibrium treatments discussed next are usually based on the control or suppression of the equilibrium reactions.

4.2 Phase diagrams for polyphase alloys

Up to this point we have discussed only the simple but important case of the single-phase alloys. In these all the grains in the metal have the same crystal structure, the same analysis, and the same properties.

As an example of a two-phase alloy, let us look at the microstructure of an aluminum-silicon alloy which is used in many engine blocks (Fig. 4.1b and c). It is apparent that two different major phases are present. (Recall that a phase is a homogeneous and physically distinct region of matter.)

We can show that the properties of the phases are different by a *micro-*

hardness test. In this procedure we locate a diamond indenter (which is carefully ground to a pyramid-shaped point) above the phase we wish to test. A load of 25 g, for example, is applied to the indenter to press it into the sample, as in the Vickers test discussed earlier, but with a much lower load. The diagonal of the impression is measured and related to the hardness; the smaller the value, the harder the phase.

In the sample shown, the original melt or liquid contained 17 percent silicon with the balance essentially aluminum. On cooling, it separated into an aluminum-rich phase, α, which has some silicon in solid solution, plus a silicon-rich phase, β, with very little aluminum. (Greek letters are used to distinguish the different solid phases in a given alloy. It must be realized that the solid phases are seldom pure, but rather are solid solutions or compounds. The symbol L is used for the liquid phase; L_1 and L_2 are used if there are two different liquid phases.) It is apparent from the hardness tests that the alloy contains a hard structure in the softer background or continuous phase, usually called the "matrix." We notice further that some of the silicon phase is in fine needles, whereas other particles are large. The wear resistance of the cylinder walls of an engine block depends on the combination of these hard particles and the soft matrix.

These observations are a good introduction to the case in which two or more phases are present—the polyphase alloys. Most metallic materials are of this type.

4.3 Phase diagrams

The most useful tool in understanding and controlling polyphase structures is a knowledge of *phase diagrams*. The phase diagram is simply a map showing which structures or phases are present as the temperature and overall composition of the alloy are varied. We shall therefore discuss these in some detail at this point.†

The Aluminum-Silicon Diagram

Let us construct the very important aluminum-silicon phase diagram by direct observations from several experiments. If we heat pure aluminum in a crucible, we find that it melts sharply at 660°C. Similarly, on cooling the crystallization from the liquid takes place at the same constant temperature.

Now let us remelt the aluminum and add 5 percent silicon to the bath. The silicon dissolves just like sugar in water. It is important to avoid saying

---

†In the sections concerning nonequilibrium conditions (precipitation reactions) we will discuss the driving force leading to solidification.

that the silicon "melts," for the melting point of silicon is far above the temperature of the liquid aluminum. The silicon *diffuses* from the surface of the solid into the liquid. After the silicon is dissolved, we cool the melt. We find two important differences compared with the freezing of the pure aluminum.

1. The alloy starts to freeze or crystallize at a lower temperature, 624°C compared with 660°C for pure aluminum.
2. The alloy is in a mushy condition (liquid plus solid) over a range of temperatures, rather than freezing at a constant temperature. More and more solid precipitates from the liquid until the alloy is finally solid at 577°C. We find the same type of behavior at 10 percent silicon, but the start of freezing is delayed until 588°C.

Now we have enough data to begin constructing the aluminum-silicon phase diagram (Fig. 4.2). As we said earlier, this may be considered simply a map showing which phases are present at different compositions as the temperature is changed. We lay off percentage silicon by weight on the x axis and temperature (°C) on the y axis and plot the points we have obtained.

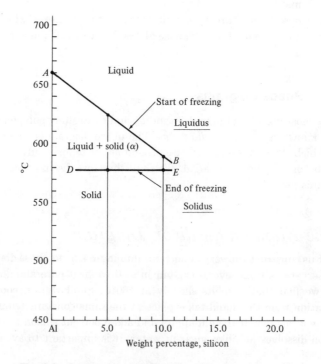

Fig. 4.2 *Construction of a portion of the aluminum-silicon phase diagram.*

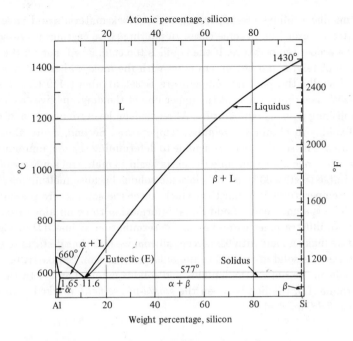

Fig. 4.3 *Complete aluminum-silicon phase diagram.*

(American Society for Metals Handbook, 8th ed., vol. 8: "Metallography, Structures, and Phase Diagrams," Metals Park, Ohio, 1973.)

Above line *AB* the melt is entirely liquid, and *AB* is called the "liquidus." Below *DE* the melt is entirely solid, and *DE* is called the "solidus." Between *AB* and *DE* we label the region liquid plus solid (L + α).

If we continue our experiments, we can complete the diagram as shown in Fig. 4.3. This diagram contains many new lines which we will now discuss. First let us take up the fact that as the percentage of silicon increases, the liquidus temperature falls until 11.6 percent silicon is reached; then the temperature rises. This pattern is quite common in alloys, and the composition at the minimum freezing point has a special name, *eutectic* (*E*), meaning "of low melting point." (In usage, the temperature at which a liquid of eutectic composition freezes is called the eutectic.) It is important to note that although silicon has a melting point twice that of aluminum, its addition *lowers* the melting range of aluminum until 11.6 percent silicon is reached. Note also that the phases present to the right of the eutectic and above the solidus are β + L instead of α + L. In other words, if we cool an alloy with 16 percent silicon, the first phase to crystallize is β (crystals of almost pure silicon). Only after crossing the solidus at 577°C do α crystals begin to form.

Let us interrupt our discussion of the aluminum-silicon phase diagram to point out the practical significance of what we have learned. First, the aluminum-silicon alloys are important engine-block materials and have been substituted for cast iron in some cases. In pouring these castings it is essential to use the proper temperature. If the liquid is too cold, it will not fill the mold cavity completely; if too hot, it can react with the mold wall and give a poor surface. Usually the pouring temperature is set at about 110°C above the liquidus. We see, therefore, that the upper line of the diagram gives us a guide for establishing pouring temperature. At any silicon level we merely add 110°C to the liquidus to obtain the required temperature. Second, many alloys are heat-treated. The *solidus* gives us a guide to determining the maximum heating temperature, since above this line the parts begin to melt. (As a matter of fact, it is advisable to stay 30 to 60°C below the solidus because melting may start at lower temperatures if impurities, which lower the solidus, are present.)

There is a good deal of additional information to be obtained from this diagram. To illustrate, let us discuss the determination of line *AD* in Fig. 4.4.

If we make a melt with 0.2 percent silicon, we find that it starts to freeze at point 1 and is solid at point 2 in the enlarged diagram. Similarly, using 0.4 percent silicon, we obtain the liquidus and solidus points 3 and 4. In this way we determine *AD*, the line between the *single-phase field,* marked α, and the *two-phase field,* marked α + L.

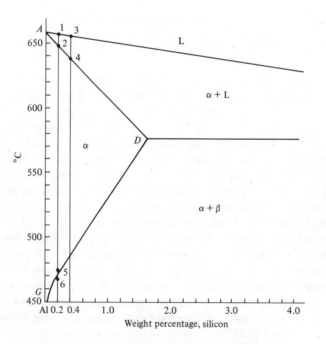

Fig. 4.4 *Enlarged region of the aluminum-silicon phase diagram (the high aluminum end).*

Now the question of how DG is determined develops. If we follow the cooling of our sample containing 0.2 percent silicon, we see that it is in a single-phase α field from 648 to 470°C. However, on crossing DG at 470°C we see that the specimen enters a two-phase field, meaning that β will begin to precipitate from solid α. This is the key to our problem.

To answer the question we must digress to explain an important tool of the metallurgist, the heat treatment cycle involving heating and quenching. If we observe a sample of our 0.4 percent silicon alloy which was slowly cooled to 20°C, we find two phases. Now let us heat the sample to 550°C, where only one phase is stable according to the diagram. We let the sample soak at temperature for about 1 hr to let the $\alpha + \beta$ change to α only. Now we quench the sample in cold water. We find only α under the microscope. The reason for this is that for β to precipitate, a certain time is required for silicon atoms to diffuse from their dispersed solid solution positions in the α phase to form the new 99 percent silicon β phase. In this case diffusion is so slow at 20°C that the alloy contains the same structure (α) it showed at 550°C. In other words, by quenching a sample from a given temperature we tend to fix the structure that was present at that temperature. There are important exceptions we will discuss later, but the technique can be used successfully in many cases.

Now we are ready to consider the location of line DG. This time let us take five samples of the alloy, heat all to 550°C, and soak them for 1 hr. We then slow-cool a sample to 525°C and quench it. We slow-cool the other samples to 500, 475, 450, and 425°C, respectively, and quench them. We find the structures shown in Fig. 4.5.

Since no β was present under slow-cooling (equilibrium) conditions at 500°C but appeared in the 450°C sample, the line DG is between 450 and 500°C. By further testing we could locate the line with greater accuracy.

4.4 Phase compositions (phase analyses)

If we have two phases present, we really have two different materials making up the structure, each with a different chemical analysis. For example, if we dissolve 15 percent silicon in liquid aluminum, we have a single liquid phase but find two solid phases after freezing, similar to the illustration in Fig. 4.1*b* and *c*. If we separate these phases at room temperature and analyze them, we find that α contains over 98 percent aluminum and β over 99 percent silicon. (With a new device called the "electron microprobe" we can analyze the individual phases in the sample. The procedure is related to the use of x-rays discussed in Chap. 2. A tiny beam of electrons is focused on the particular phase, and these cause x-rays to be emitted by the atoms of the elements that are present. Since each element gives off x-rays of its own characteristic wavelength, it is necessary only to record this x-ray spectrum, just as the visible spectrum

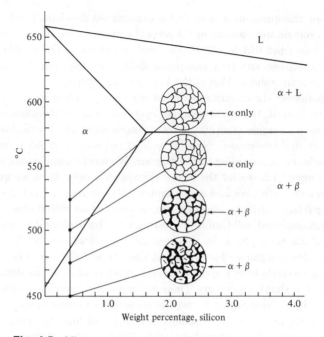

Fig. 4.5 *Microstructure of heat-treated 0.4 percent silicon-aluminum alloys (slowly cooled from 550°C to the indicated temperature and then quenched to room temperature).*

is recorded in a spectrograph. The percentage of a given element in the phase is related to the intensity of radiation of its characteristic wavelength.)

Let us suppose we have an alloy of 97 percent aluminum and 3 percent silicon at 600°C. This is definitely in the two-phase field $\alpha + L$. To find the phase compositions we simply draw a horizontal line across the two-phase field until it touches the single-phase fields (Fig. 4.6). Then we drop a perpendicular from each contact point to the x axis and read off the particular phase composition. Thus we see that the phase composition of α in the 3 percent silicon alloy at 600°C is 1.2 percent silicon. Using the same construction, we find that even if the overall composition of the alloy changes from 3 to 8 percent silicon, the α phase at this temperature still contains 1.2 percent silicon. Similarly, using the same horizontal, we see that the composition of the liquid in all these alloys is 8.3 percent silicon. We should realize that there is nothing artificial about this procedure; it is merely a way of recovering the information that we found experimentally. This method of determining the phase compositions in a two-phase field is so important that we will state it as a simple rule.

RULE 1. FOR DETERMINING PHASE COMPOSITIONS (CHEMICAL ANALYSIS OF PHASES). In a two-phase field draw a horizontal line (called a "tie line") at the temperature desired and touching the single-phase fields. Drop perpendiculars from the points where the tie line meets these fields, and read the phase compositions on the x axis.

Note that in a phase diagram single-phase fields are always at the ends of the tie line in a two-phase field. This is a consequence of the phase rule discussed later.

4.5 Amounts of phases

It is important to know the *amounts* of phases because the properties of a two-phase mixture will depend on these percentages. These amounts can be found by calculation or by a simple graphic treatment of the phase diagram. Although the graphic method is faster, the calculation will be presented first because it deepens our understanding of what is meant by *overall* composition compared to *phase* composition.

Let us consider again the two-phase region α + L of the aluminum-silicon diagram (Fig. 4.6).

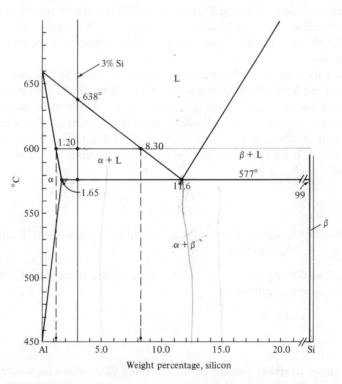

Fig. 4.6 *Determination of phase compositions from a phase diagram.*

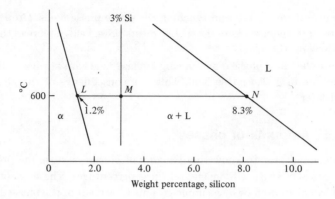

Fig. 4.7 *Calculation of phase amounts from a phase diagram.*

Let us ask the question: What are the relative amounts of solid and liquid in the alloy containing 97 percent aluminum and 3 percent silicon overall composition when it is at equilibrium at 600°C? In other words, it is evident from the diagram that solid and liquid are present. If we pour off the liquid and weigh the liquid and solid separately, what percentage of the total weight will each be?

Let us start by making a homogeneous melt at 700°C using 97 g of aluminum and 3 g of silicon. Now let us cool the all-liquid alloy to 600°C, where two phases are present. Even though there are two phases, there must still be 3 g of silicon present in these phases, so that the amount of silicon in α plus the amount of silicon in the liquid equals 3 g. We add to this the information obtained from a horizontal at 600°C (Fig. 4.7), that from Rule 1 the compositions of the phases are

$$\alpha: 98.8\% \text{ Al, } 1.2\% \text{ Si}$$
$$\text{L: } 91.7\% \text{ Al, } 8.3\% \text{ Si}$$

Now let the grams of α phase equal x. Then the grams of the liquid phase equal $100 - x$. The total amount of silicon is

(grams of α)(fraction of silicon in α) + (grams of L)(fraction of silicon in L) = total grams of silicon

or
$$(x)(0.012) + (100 - x)(0.0830) = 3$$
$$-0.071x = -5.3$$
$$x = 74.6 \text{ (grams of } \alpha)$$

Therefore, we have 74.6 percent by weight α and 25.4 percent by weight liquid.

Now let us use a simpler graphic method called the "inverse lever rule."

RULE 2. LEVER RULE TO DETERMINE AMOUNTS (PERCENTAGES) OF PHASES PRESENT. We again draw a tie line across the two-phase field at the desired temperature 600°C and touching the two-phase fields (Fig. 4.7). Next we draw a vertical line at the overall composition of the melt (3 percent silicon). The inverse lever rule (which can be proved by geometry) states that the percentage of α is $(MN/LN) \times 100$ and the percentage of L is $(LM/LN) \times 100$. Note that the amount of α is found by taking the length of the tie line on the *far* side of the overall composition, hence the expression *inverse* lever rule.

Making the calculation, we have

$$\text{Percentage of } \alpha = \frac{8.3 - 3}{8.3 - 1.2} \times 100 = 74.6 \text{ g of } \alpha$$

$$\text{Percentage of L} = \frac{3 - 1.2}{8.3 - 1.2} \times 100 = 25.4 \text{ g of L}$$

These are the weight percentages of the phases, and are the same as those obtained by the algebraic method.

4.6 Phase fraction chart of amounts of phases present as a function of temperature

To summarize our interpretation of a phase diagram, a phase fraction chart (Fig. 4.8) is quite useful. This chart shows how the percentages of phases present in an alloy of fixed overall composition change as the temperature changes. Let us use the 3 percent silicon alloy we have just discussed (Fig. 4.6) as an example. In this case the vertical axis represents the percentage of phases present and the horizontal axis the temperature. A separate line graph is used for each phase. Taking the data for α as an example, if we follow the vertical overall composition line, we can calculate and show the following:

1. α appears at 638°C (liquidus).
2. At 600°C we have 74.6 percent α and 25.4 percent liquid, as calculated before.
3. At just above the eutectic, say at 578°C, we have, by the inverse lever rule,

$$\text{Percentage of } \alpha = \frac{11.6 - 3}{11.6 - 1.65} \times 100 = 86.4$$

$$\text{Percentage of L} = 13.6$$

4. At 576°C (just below the eutectic) we have $\alpha + \beta$ present and no liquid.

$$\text{Percentage of } \alpha = \frac{99 - 3}{99 - 1.65} = 98.6$$

$$\text{Percentage of } \beta = \frac{3 - 1.65}{99 - 1.65} = 1.4$$

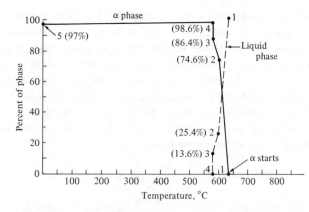

Fig. 4.8 *Phase fraction chart calculated from the phase diagram for 3 percent silicon in aluminum. Only α and liquid are shown. The β phase is 1.4 percent at 576°C and increases to 3 percent at 20°C.*

The change in α from 86.4 percent to 98.6 percent is abrupt because the 13.6 percent of liquid of eutectic composition that was present at 578°C formed α + β at the eutectic.

5. At 450°C there is more β.

$$\text{Percentage of } \alpha = \frac{99 - 3}{99 - 0} = 97$$

$$\text{Percentage of } \beta = \frac{3 - 0}{99 - 0} = 3$$

It is important to our later discussions dealing with the control of the shape of phases to note that the α precipitated under two different conditions. In the temperature interval 638 to 578°C the α precipitated alone from a liquid. This is called the "primary" α, and it forms large crystals or dendrites. The second period of importance was at the eutectic, and here α and β precipitated together, giving a mixture of fine α and β crystals. Therefore, 86.4 percent of the α will be in the form of large primary grains, and 12.2 percent will be small grains in the eutectic mixture. [Total α (98.6 percent) = primary α (86.4 percent) + eutectic α (12.2 percent).]

This concludes our simplified discussion of phase diagrams.

EXAMPLE 4.1 Calculate the volume percentages of phases present in an alloy of 16 percent by weight silicon and 84 percent by weight aluminum. [Density of silicon = 2.35 g/cm³ (2.35 × 10³ kg/m³); density of aluminum = 2.70 g/cm³ (2.70 × 10³ kg/m³).]

ANSWER From the phase diagram we find that we will have approximately 16 g of β and 84 g of α because at room temperature these phases are essentially pure silicon and aluminum.

$$\text{Volume of } \alpha = \frac{84 \text{ g}}{2.70 \text{ g/cm}^3} = 31.2 \text{ cm}^3$$

$$\text{Volume of } \beta = \frac{16 \text{ g}}{2.35 \text{ g/cm}^3} = 6.8 \text{ cm}^3$$

$$\text{Total volume} = 31.2 + 6.8 = 38.0 \text{ cm}^3$$

$$\text{Volume percentage of } \alpha = \frac{31.2 \text{ cm}^3}{38 \text{ cm}^3} \times 100 = 82$$

$$\text{Volume percentage of } \beta = \frac{6.8 \text{ cm}^3}{38 \text{ cm}^3} \times 100 = 18$$

Note: The physical properties are more easily related to the volume percentages of phases than to the weight percentages.

4.7 The phase rule

Now that we have acquired a "feel" for the use of phase diagrams in calculating the composition and quantity of phases as functions of temperature and overall or total composition, we should take one further step to complete our understanding of this important tool of the materials engineer.

The essence of this powerful concept, the phase rule, is that nature imposes certain restrictions on the variations that are possible in a phase diagram. These restrictions are very helpful to the engineer in planning experiments for determining phase diagrams and in exposing false data. For example, we would now, from our limited experience, be ready to criticize an observer who reported the existence of liquid plus solid over a temperature range for a pure metal. But suppose the observer showed an area (or field) indicating three phases in equilibrium for a two-metal diagram? Furthermore, what type of diagram should we expect when we have more than two elements in an alloy?

The phase rule, developed by J. Willard Gibbs, states:

$$F = C - P + 2$$

where F = degree of freedom

   C = number of components

   P = number of phases in equilibrium

(The degree of freedom is the number of variables not controlled by nature in a given situation, or, in other words, variables that we can change, such as temperature, phase composition, etc. The components are the materials at the endpoints of the phase diagram, such as aluminum and silicon in the diagrams just discussed. However, one or both can be a compound, such as Fe and Fe_3C.)

The number "2" in the equation refers to the normal phase variables, temperature and pressure, which we would expect to change the nature of our phase diagrams.

In the usual case we are investigating a phase diagram at a fixed pressure of 1 atm. This means that we have reduced the variables F that we can change, and we must rewrite the equation as

$$F = C - P + 1 \qquad (\text{pressure} = 1\,\text{atm})$$

In effect, the degree of freedom F tells us how many variables must be fixed for us to describe the system completely. As an analogy, in mathematics if we have four independent equations and five independent unknowns, we need one more relation to enable us to fix the values of the variables. We say, therefore, that we have one degree of freedom.

Now let us apply the phase rule to the aluminum-silicon diagram we have discussed. First, consider the case of a one-phase field (liquid as an example). From the phase rule we have

$$\begin{aligned} F &= C - P + 1 \\ &= 2 - 1 + 1 \\ &= 2 \end{aligned}$$

This means that to describe fully to a colleague the inherent characteristics of a liquid in the one-phase liquid field, we would need to fix two things. For example, if we specified the composition of the liquid and the temperature, our colleague could obtain a liquid of the same density, conductivity, or any other property he might measure. If we only specified the composition, this would not be true, because without fixing the temperature of the liquid the other properties would vary.

Now let us look at the two-phase field $\alpha + L$. The phase rule states:

$$\begin{aligned} F &= C - P + 1 \\ &= 2 - 2 + 1 \\ &= 1 \end{aligned}$$

Therefore, in a two-phase field we have only one degree of freedom. Let us test this by using up our one degree of freedom.

1. If we fix temperature, we also fix the *phase* compositions. These are given by the intersections of the horizontal tie line with the single-phase fields at this temperature. (We are interested in specifying the *nature* of the phases *not* their amounts. The overall composition can be any value in the two-phase field.)

2. Instead of temperature, let us fix the phase composition of α in equilibrium with L. This fixes temperature and the composition of the L phase, again by the tie line relationship.

Suppose an investigator reports a *field* where α, β, and L are present.

$$F = C - P + 1$$
$$= 2 - 3 + 1$$
$$= 0$$

The phase rule states that if two components and three phases are present, there is no degree of freedom. This is illustrated by the eutectic. At 577°C, the temperature fixed by nature, not by us, we have three phases of fixed phase composition in equilibrium (α: 1.65 percent silicon, L: 11.6 percent silicon, β: 99 percent silicon). Thus if we say $P = 3$, then $F = 0$, and all conditions are already fixed. Although we may change the *overall* composition horizontally along the line, the composition of the individual phases (the phase rule variables) does not change. Therefore, $\alpha + \beta + L$ does not occur in a field but rather along a unique three-phase tie line—the eutectic in our example.

4.8   Complex phase diagrams

The most complex binary diagram is made up of just two types of lines, horizontal and nonhorizontal. Let us take up the nonhorizontal lines first. All the nonhorizontal lines are very simple, since they merely give the boundary between a one-phase field and a two-phase field. Let us check, for example, the copper-zinc diagram (Fig. 4.9) to convince ourselves of this. We never find two two-phase fields touching along a vertical line, say $\alpha + \beta$ and $\beta + \gamma$, because this would mean that three phases, α, β, and γ, could exist over a range of temperature at the composition of the dividing line. Varying the temperature would leave one degree of freedom to be specified, which is contrary to our previous phase rule analysis for two components and three phases in equilibrium at 1 atm ($F = 0$). Note therefore that as we go across the phase diagram at a given temperature, we go from one- to two-phase fields alternately.

Now let us consider the horizontal lines in the copper-zinc diagram. At 903°C we have the situation shown in Fig. 4.10. Let us consider an alloy with 36.8 percent zinc at 902°C containing β. If this is heated to 904°C we find $\alpha + L$. Thus one solid phase has transformed to a new solid phase plus liquid. On cooling the same material we have a solid α (S_1) reacting with liquid to form one new solid S_2:

$$S_1 + L \xrightarrow{\text{cooling}} S_2$$
$$(\alpha) \qquad\qquad\qquad (\beta)$$

This is different from the *eutectic* reaction in the 11.6 percent silicon alloy in

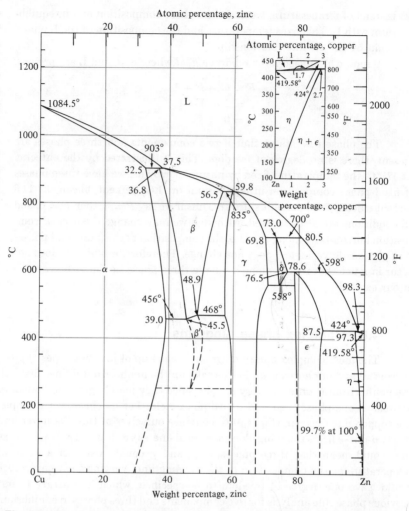

Fig. 4.9 *Copper-zinc phase diagram.*

(American Society for Metals Handbook, 8th ed., vol. 8: "Metallography, Structures and Phase Diagrams," Metals Park, Ohio, 1973, p. 301.)

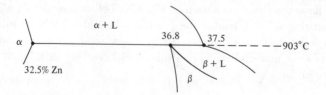

Fig. 4.10 *Peritectic transformation in copper-zinc diagram.*

the aluminum-silicon system which reacts on cooling:

$$L \xrightarrow{\text{cooling}} \underset{(\alpha)}{S_1} + \underset{(\beta)}{S_2}$$

We therefore give the $S_1 + L \xrightarrow{\text{cooling}} S_2$ reaction a different name, *peritectic*.

At 835°C we encounter the same type of reaction, $\beta + L \rightarrow \gamma$, at 59.8 percent zinc, and at 598°C we have $\delta + L \rightarrow \varepsilon$ at 78.6 percent zinc.

Another type of reaction, $\delta \rightarrow \gamma + \varepsilon$, is encountered on cooling an alloy of 74 percent zinc at 558°C. This is analogous to the eutectic reaction but since it involves an all-solid transformation, it has the special name *eutectoid*.

There are a few other types of horizontal lines involving three phases which we will cover in the problems.

4.9 Ternary diagrams

So far we have discussed only systems with two components, leading to binary diagrams. In many alloys we have three principal elements present, as in 18/8 stainless steel which contains 18 percent chromium, 8 percent nickel, and about 74 percent iron. Let us compare the degrees of freedom of this system with the binary system:

Binary: $F = C - P + 2 = 4 - P$

Ternary: $F = C = P + 2 = 5 - P$

If we fix pressure at 1 atm in the binary system, $F = 3 - P$. Thus if we specify that three phases are in equilibrium, we have nothing else to specify; the temperature and phase compositions are fixed by nature.

However, in a ternary system under the same conditions (fixed pressure, $F = 4 - P$), we can encounter three phases over a range of temperatures and four phases at a given temperature.

Obviously we need some sort of three-dimensional map to represent this added variable. This is done with a triangular graph (Fig. 4.11).

First let us consider the representation of composition at constant temperature, then the added dimension for varying temperature. Perhaps the quickest way to visualize the net shown is as a three-sided football field. The point at the iron corner is 100 percent iron. Any point on the 90 percent iron line contains that amount of iron. Next let us locate an 18 percent chromium, 8 percent nickel, and 74 percent iron stainless steel called 18/8. The 74 percent iron line is shown, and the composition must lie somewhere on this line.

Next we recognize that the chromium corner represents 100 percent chromium. We move away to find the 18 percent chromium line. The point

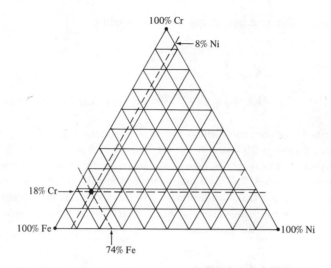

Fig. 4.11 *Location of a point in the iron-nickel-chromium ternary diagram.*

where this intersects the 74 percent iron line is 18 percent chromium and 74 percent iron. We do not need to draw the 8 percent nickel line except as a check, because this is determined by the difference from 100 percent (that is, % Ni = 100 − 74% Fe − 18% Cr).

RULE. To find a composition, locate the proper isocomposition lines for each element starting from that element's corner of the triangle.

4.10 Representation of temperature in ternaries

A complete three-component or ternary diagram at constant pressure but varying temperature is shown in Fig. 4.12.

In some cases the determination of the liquidus can be a million-dollar research problem. For example, in the production of pig iron in a blast furnace it is necessary to keep the slag liquid, or a very expensive shutdown will occur. It is known that the principal components of the ternary diagram for the slag are three oxides, CaO, SiO_2, and Al_2O_3, instead of three metals. The million-dollar liquidus is shown in Fig. 4.13. The significance is that a low melting combination occurs in the range 49 percent CaO, 39 percent SiO_2, 12 percent Al_2O_3 at 1315°C. Knowing this, the metallurgist adds the right amounts of oxides to the ore to avoid a freeze-up and, equally important, to give a fluid slag which is active in removing impurities such as sulfur.

The ternary diagram has equal importance in dealing with metals. A third element can be added to lower the melting point of an alloy. For example, an alloy of 51 percent bismuth, 40 percent lead, and 9 percent

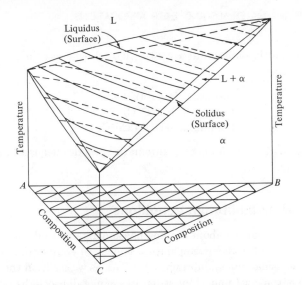

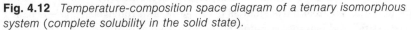

Fig. 4.12 *Temperature-composition space diagram of a ternary isomorphous system (complete solubility in the solid state).*

(F. Rhines, "Phase Diagrams in Metallurgy," McGraw-Hill Book Company, New York, 1956. Used with permission of McGraw-Hill Book Company.)

Fig. 4.13 *Liquidus surface of the ternary system CaO-SiO_2-Al_2O_3. The constituent with the highest melting point is CaO in the right rear of the photograph. The horizontal lines or planes would be constant temperature contours.*

cadmium will melt in boiling water (212°F, 100°C), whereas the individual melting points are:

> Bismuth: 520°F (271°C)
> Lead: 621°F (327°C)
> Cadmium: 609°F (320°C)

This type of alloy is used in plugs in automatic sprinkler heads which melt when fire occurs.

4.11 Nonequilibrium conditions

All we have learned about reading phase diagrams to determine which phases are present at a given temperature applies to equilibrium conditions. We must also consider nonequilibrium conditions, as we see from the following example. One of the most important structures in metallurgy—the microstructure in hardened steel parts such as drills, saws, and roller bearings—is called "martensite" and cannot be found on any phase diagram. On the other hand, it cannot be produced under controlled conditions without a knowledge of the iron-carbon phase diagram, because it is necessary first to produce a stable phase called "austenite" (FCC iron with carbon in interstitial solid solution) at high temperature.

There are three fundamentals to be presented which bear directly on nonequilibrium conditions:

1. Diffusion
2. Nucleation and growth
3. Segregation effects

4.12 Diffusion phenomena

Diffusion is the migration of atoms; it depends on time and temperature. Let us look at examples in the gaseous, liquid, and solid states.

If we open a partition between a glass chamber with nitrogen gas and another with bromine, we can see the brown color of the bromine become fainter as the gas molecules migrate and intermix. If we carefully pour a layer of water and a layer of alcohol in a vertical cylinder, the two liquids will become one after a time. Similarly, in the solid state, if we take a U.S. quarter (made up of layers of copper and a copper-nickel alloy) and heat it at an elevated temperature, the atoms will interdiffuse.

There are two principal types of diffusion: interstitial and vacancy or substitutional (Fig. 4.14). In interstitial diffusion small atoms such as hydrogen, carbon, and nitrogen are involved, and these jump from interstitial position to position. This is the most rapid type of diffusion, because most interstitial

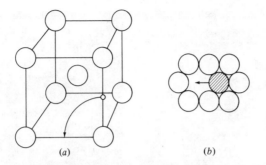

(a) (b)

Fig. 4.14 *Mechanisms of (a) interstitial and (b) substitutional diffusion. In substitution the atom moves into a vacancy normally considered to be a substitutional atom position and not an interstitial position.*

sites or positions are empty. Diffusion of substitutional atoms is slower, since the atoms have to wait for a vacancy to jump into. Even in the most carefully prepared single crystal there is an equilibrium number of vacancies which increases with temperature. In addition, grain boundaries and dislocations provide diffusion paths.

Let us review some of the many practical cases of diffusion before treating it quantitatively. First, the case of diffusion in bimetals is quite important. It is possible to roll a metal sandwich such as a layer of nickel over a layer of copper-nickel alloy to obtain the familiar U.S. quarter. Because of diffusion, the bond between the layers is not mechanical but a true metallurgical bond in which the composition changes gradually across the interface (Fig. 4.15). Later we shall discuss the "aluminum-clad" materials used in aircraft; in these a layer of pure aluminum protects the less corrosion-resistant alloy beneath.

Another somewhat different example is a carburized gear (Fig. 4.16). In this case there is a gradual change from high carbon at the surface to low-carbon

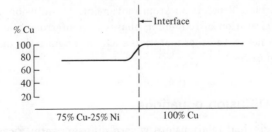

Fig. 4.15 *Analysis of the interface of a U.S. quarter after diffusion.*

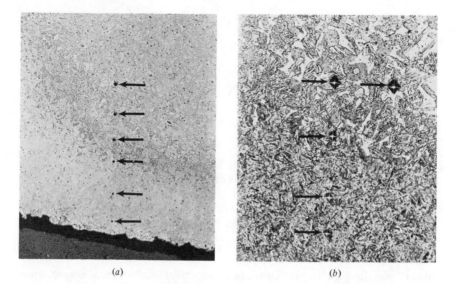

(a)   *(b)*

Fig. 4.16 *Microstructure and microhardness survey of a steel gear that was carburized and then quenched. The surface layers are martensite, but the soft regions contain ferrite. (a) VHN (reading upward from bottom surface): 740, 710, 390, 220, 193, 171. 100X; 2 percent nital etch. (b) Transition region: martensite at bottom, martensite and ferrite at top. 500X; 2 percent nital etch.*

steel in the interior. This gives high hardness and good wear resistance at the surface, supported by the tough low-carbon steel beneath the surface. The carbon gradient is produced by heating the part in a carbon-rich atmosphere. The carbon atoms dissolve and build up in the surface layers, and a portion diffuses to the layers beneath.

Diffusion processes are also important in many cases in which we are changing the internal structure of a part, not merely the surface layers. The most important effect of the heat treatment of steel is the change from a one-phase structure to a two-phase structure. The precipitation of the second phase, iron carbide, depends on atom movements by diffusion. Similarly, the controlled precipitation of hardening phases in nonferrous alloys depends on diffusion movements. In other materials such as ceramics the diffusion of ions is of great importance.

4.13 Diffusion equations

Three principal relationships govern diffusion rate: Ficks' first and second laws, and the variation of the diffusion constant with temperature.

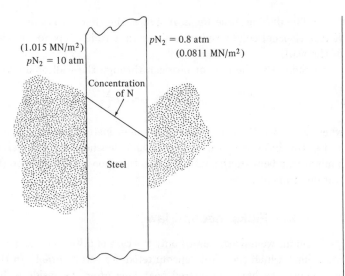

Fig. 4.17 *Diffusion of nitrogen through a steel wall.*

Ficks' First Law. Ficks' first law describes the diffusion of an element under steady-state conditions.

A simple case is the rate of loss of nitrogen under constant pressure through the side walls of a container (Fig. 4.17). This is analogous to the more familiar case of the wall of a heated house in winter (Fig. 4.18). The rate of heat loss or the thermal flux through the wall is

$$\text{Heat flux} = J = K\frac{\Delta T}{\Delta X}$$

Heat flux is defined as energy per square foot (or square meter) of area per second.

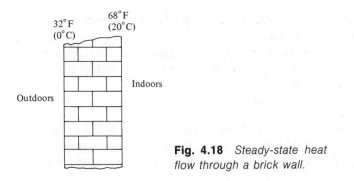

Fig. 4.18 *Steady-state heat flow through a brick wall.*

The driving force for heat flow is the thermal gradient $\Delta T/\Delta X$, and K (the thermal conductivity) is a constant which depends on the material of the wall.

Similarly, the flow of nitrogen through the wall of thickness X is

$$N_2 \text{ flux} = -D\frac{\Delta C}{\Delta X}$$

where $\Delta C/\Delta X$ is the concentration gradient instead of temperature gradient and D, the diffusivity, is a constant which depends on the material. (We use a minus sign because the diffusion is in the opposite direction to the concentration gradient.)

4.14 Ficks' second law

So far we have discussed only the case of diffusion under steady-state conditions, which are rarely encountered. To return briefly to the analogy of a house, we have considered heat flow *after* the inside wall has been brought up to constant temperature (20°C, 68°F). Suppose now we take the condition in which we light the fireplace on our return from a long vacation. The wall surface of the room will reach 20°C (68°F) rather quickly, but the interior of the wall will take some time to heat up (Fig. 4.19a).

In the same way, if we place a steel gear in a carbon-rich furnace atmosphere, the surface will reach a high level quickly, but the carbon will diffuse inward as a function of time (Fig. 4.19b).

If we desire a carbon content C_x at a depth x to provide adequate wear resistance and strength, we need to wait a time t at the furnace temperature used.

The relation governing these conditions is a solution of Ficks' second law:

$$\frac{C_s - C_x}{C_s - C_0} = \text{erf}\left(\frac{x}{2\sqrt{Dt}}\right)$$

where C_s = surface concentration of carbon produced immediately by the atmosphere (a constant)

C_0 = initial uniform concentration of carbon throughout the steel

C_x = concentration of carbon at distance x from the surface at time t

x = distance from surface

D = diffusivity (as described earlier)

t = time

The quantity erf is called the "error function," and it can be found from standard tables (Table 4.1) or a graph. In other words, we substitute numerical values for x and Dt and then find the error function of the result.

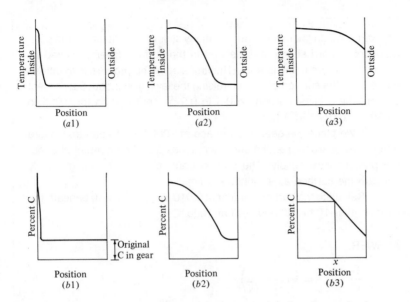

Fig. 4.19 (a) Unsteady-state heat transfer in heating the wall of a house. (a1) After fire lit, t_0. (a2) After time t_1. (a3) After time t_2. (b) Unsteady-state diffusion in carburizing a gear. (b1) Time = t_0. (b2) Time = t_1. (b3) Time = t_2.

A key point is that if $C_0 = 0$, then

$$\frac{C_x}{C_s} = 1 - \text{erf}\left(\frac{x}{2\sqrt{Dt}}\right)$$

Then as D or t increases, the erf expression becomes smaller and C_x/C_s approaches 1. In other words, C_x approaches the surface concentration.

Table 4.1 TABLE OF THE ERROR FUNCTION

Z	erf Z	Z	erf Z	Z	erf Z
0	0	0.50	0.521	1.30	0.934
0.025	0.028	0.60	0.604	1.40	0.952
0.05	0.056	0.70	0.678	1.50	0.966
0.10	0.113	0.80	0.742	1.60	0.976
0.15	0.168	0.90	0.797	1.80	0.989
0.20	0.223	1.00	0.843	2.00	0.995
0.30	0.329	1.10	0.880	2.40	0.999
0.40	0.428	1.20	0.910	∞	1.0

EXAMPLE 4.2 In the manufacture of many sliding and rotating parts such as gears, a hard structure is desired in the surface layers, backed by a tough structure in the interior. The first step in the process is to diffuse carbon into the surface of a steel, raising the level from about 0.2 percent carbon, the original carbon, to 0.5 to 0.9 percent carbon for 0.005 to 0.050 in. (0.0127 to 0.127 cm).

If we place the gear in a furnace at 1000°C with an atmosphere rich in hydrocarbon gas, the surface reaches a carbon content of about 0.9 percent very rapidly. The carbon content will then rise gradually beneath the surface as a function of time.

Calculate the carbon content at 0.010 in. (0.0254 cm) beneath the surface after 10 hr (36,000 sec) at 1000°C.

ANSWER

$$\frac{C_s - C_x}{C_s - C_0} = \text{erf}\left(\frac{x}{2\sqrt{Dt}}\right)$$

$$C_s = 0.9$$

$$C_x = \text{desired value}$$

$$C_0 = 0.2 \quad \text{(assume an SAE 1020 steel is used)}$$

$$D = 0.298 \times 10^{-6} \text{ cm}^2/\text{sec}$$

Let $x = 0.01$ in. $= 0.0254$ cm, since D is in cm^2/sec. Then

$$\frac{0.9 - C_x}{0.9 - 0.2} = \text{erf}\left(\frac{0.0254 \text{ cm}}{2\sqrt{(0.298 \times 10^{-6} \text{ cm}^2/\text{sec})(3.6 \times 10^4 \text{ sec})}}\right)$$

$$= \text{erf } 0.123$$

$$\frac{0.9 - C_x}{0.7} = 0.135 \quad \text{from Table 4.1}$$

$$C_x = 0.805 \text{ percent carbon}$$

4.15 Effects of temperature

In Ficks' first and second laws we see that the movement of the diffusing material is proportional to the diffusivity D. We would expect D to increase with temperature since the atomic motion and number of vacancies both increase. The relationship is found to obey the equation

$$D = Ae^{-Q/(RT)}$$

where A = constant

Q = constant for the diffusing substance and solvent involved

R = gas constant (1.987 cal/mole-K)

T = absolute temperature (K)

Therefore, taking $\log_e$ of both sides, we have

$$\ln D = \ln A - \frac{Q}{RT}$$

which is the equation of a straight line if we plot $\ln D$ vs. $1/T$.

The importance of this approach is that if we determine D for only two temperatures, we can solve for A and Q and obtain a general relation for any temperature. In a practical case, if we wish to estimate the effect of changing a furnace temperature for carburizing a gear, we can calculate the time needed at the new temperature to obtain equivalent results.

EXAMPLE 4.3 Calculate D for 1000°C as used in the previous example, given $A = 0.25$ cm^2/sec and $Q = 34{,}500$ cal/mole for the diffusion of carbon in γ iron.

ANSWER

$$D = (0.25 \text{ cm}^2/\text{sec})e^{-\frac{34{,}500 \text{ cal/mole}}{(1.987 \text{ cal/mole-K})(1273 \text{ K})}}$$

$$= \frac{0.25}{e^{13.64}}$$

since

$$\ln e^{13.64} = 13.64$$

$$\log_{10} e^{13.64} = \frac{13.64}{2.3} = 5.92$$

$$\text{antilog } 5.92 = 8.38 \times 10^5$$

$$e^{13.64} = 8.38 \times 10^5$$

then

$$D = \frac{0.25}{8.38 \times 10^5}$$

$$= 0.298 \times 10^{-6} \text{ cm}^2/\text{sec}$$

4.16   Other diffusion phenomena

It is interesting to analyze the value of D for different combinations of materials (Fig. 4.20). Note in particular that the elements of small diameters which form interstitial solid solutions diffuse very rapidly. The faster diffusion rate of carbon in BCC iron compared to FCC iron seems paradoxical until we recall that the BCC cell is a less dense structure. Therefore,

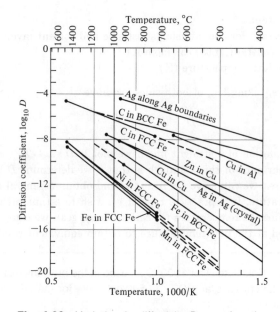

Fig. 4.20 *Variation in diffusivity D as a function of temperature for various materials.*

(From L. H. Van Vlack, ''Elements of Materials Science,'' 2d ed., Addison-Wesley Publishing Company, Inc., Reading, Mass., 1964.)

although there is more room in the FCC cell centers for the carbon atoms, the passageways through the unit cell are tighter for carbon movement.

The role of dislocations and grain boundaries is also of interest. The looser packing leads to a more rapid movement of the diffusing substance, and this can lead to a factor as high as 10:1 in penetration along grain boundaries.

4.17 Nucleation and growth

It is important to re-emphasize that we frequently find phases that are not indicated by the phase diagram, i.e. phases that are out of equilibrium. In discussing the aluminum-silicon diagram we mentioned that we could quench an all-α alloy from a high temperature and retain the same structure at room temperature even though the equilibrium diagram called for $\alpha + \beta$ at 0.4 percent silicon at 20°C. The reason is that silicon atoms need to diffuse from their random solid solution in the FCC lattice to form the silicon crystals with a new structure. In addition to requiring time for diffusion, the material must form a nucleus.

The problem of nucleation has wide application. For example, everyone is familiar with experiments in cloud seeding in which silver iodide crystals are supplied as nuclei. As another example, pure water can be cooled to $-30°C$ in a special container without freezing. Also, common metals such as nickel can be cooled to 200°C below the equilibrium freezing point without solidifying. The usual impression is that this supercooling or undercooling is done by fast cooling, but it is important to understand that these liquids can be maintained for a period of time at these low temperatures.

The reason for these effects lies in the important phenomenon of nucleation. Once a liquid is cooled below its equilibrium freezing temperature, there is a driving force for solid to precipitate. This force is measured as the difference in bulk free energy between the liquid and the solid. For instance, for a *spherical* volume the difference is [(free energy of liquid/cm³) − (free energy of solid/cm³)] × (volume of sphere, cm³) or $\Delta F_V \frac{4}{3}\pi r^3$.

Bulk free energy is a quantity precisely defined in chemistry, and for those unfamiliar with the concept a mechanical analogy may be useful. Suppose we have an old-fashioned screen door which slams shut when released because of an overhead spring. As we open the door, tension builds up in the spring. The further the door is opened, the greater the energy in the spring for shutting. In the case of cooling a liquid below the equilibrium freezing temperature, the greater the undercooling, the greater the driving force for solidification (or the greater the bulk free energy).

However, there is a force to be overcome in growing a sphere of solid of radius r in place of the liquid. A new surface has to be formed, so we need something like the energy required to blow a soap bubble against surface-tension effects. The energy required to create a new surface for the sphere is equal to the surface area times the surface tension or $4\pi r^2 \sigma$, where σ is the surface tension in ergs per square centimeter of surface area.

Now if we plot these two energies and their sum, we obtain the curve shown in Fig. 4.21. Since the difference in free energy or bulk free energy is a cubic function, whereas the surface energy is a quadratic equation, their sum results in an inflection point which we call r^*. We recall that in nature the tendency is for a system to minimize its energy. Therefore, for r values less than r^*, the easiest way to go to lower energy is to have r values become still smaller, or the nucleus dissolves. On the other hand, for values of r^* or greater, minimum energy is obtained by an increase in r so that the nucleus will grow. Here we encounter two important cases.

CASE 1. If no nuclei are present in the melt, it will have to be supercooled greatly for nucleation to occur. The lower the temperature, the greater the difference in bulk free energy and the greater the driving force to transform. (The surface tension does not change appreciably with temperature.) With great undercooling, r^* becomes very small and *homogeneous* nucleation takes place.

CASE 2. If we introduce a solid nucleus of radius greater than r^*, it will grow. This nucleus may be another material that is "wet" by the liquid

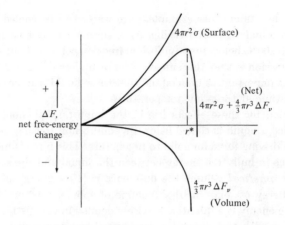

Fig. 4.21 *Change in free energy as a function of the radius of the nucleus.*

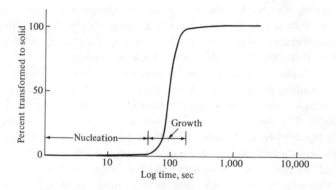

Fig. 4.22 *Representation of a liquid transforming to a solid as a function of time.*

and hence acts like a nucleus of the metal. This is called *heterogeneous* nucleation. Because this nucleus is usually relatively large, it produces freezing close to the equilibrium temperature.

Once a nucleus or nuclei are present, growth occurs. The typical growth curve is shown in Fig. 4.22. At first the rate of growth is slow because there is limited solid surface. Then the growth curve rises rapidly as the surface increases. Finally the rate of growth slows down because solid surfaces come in contact, reducing the area of the solid-liquid interface.

These nucleation and growth phenomena are also present in solid-solid reactions. There is one additional nucleation effect of importance which affects the structure of the solid precipitate. We notice often that the solid comes out in the form of needles or plates rather than spheres or cubes. The reason is that an additional effect operates against the growth of the nucleus, and this effect is related to the volume change in going from one solid to another. If the second solid is more voluminous than the first, the nucleus builds up compressive stress. This effect is diminished if the shape of the nucleus is needlelike instead of spherical. Therefore, in transformations that occur at low temperatures, where the structure is high strength, the precipitate is usually in the form of needles or plates.

A practical application is found in the control of the size and shape of the silicon crystals in the aluminum-silicon engine block. For many years the hypereutectic (16 percent silicon) alloys could not be used because large hard bladelike silicon crystals formed from the liquid. These were difficult to machine and also gave a rough surface. Finally it was found that if phosphorus was introduced, many small aluminum phosphide crystals formed. These in turn nucleated many small silicon crystals, and the desired refinement was obtained.

4.18 Segregation

The key to understanding the phenomenon of segregation is first to realize that the crystals that freeze initially are different in composition from the liquid. Second, to satisfy equilibrium conditions, the composition of the solid that freezes originally must change by diffusion as freezing continues. These points are illustrated in Fig. 4.23.

If we start freezing overall composition (1) at T_1 (Fig. 4.23b), the first crystals will have composition A. Under equilibrium at a lower temperature T_2 *all* the solid will have composition B. However, if there is not time for diffusion to occur throughout the solid, the average composition of the solid will be C at T_2. The solid is made up of a cored structure, as shown in Fig. 4.23a. One consequence is that freezing will not be complete when the

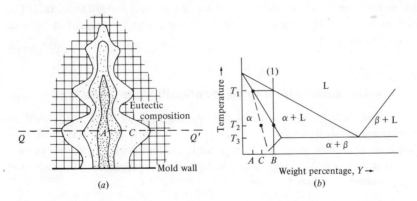

Fig. 4.23 (a) Segregation in dendrite growth. (b) Phase diagram illustrating segregation. The dashed line shows the new solidus temperature.

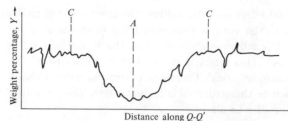

Fig. 4.24 *Variation in composition in Fig. 4.23 due to nonequilibrium cooling.*

temperature T_2 is reached, and we can reach the eutectic temperature T_3 with eutectic liquid present. This phenomenon is called "segregation." A survey of chemical analysis would show the features indicated in Fig. 4.24.

At times the last liquid to freeze will be forced between the crystals from the interior to the surface of the ingot or casting, causing "inverse segregation." Often droplets of eutectic are exuded from the surface.

In cast ingots that are subsequently rolled, there is an opportunity to improve on the as-cast structure. It is important first to avoid shrinkage cavities in the ingot. This is done in one of two ways. First a "hot top," which is similar to a riser on a casting, may be used. (These are liquid metal reservoirs which freeze last and supply liquid metal to fill voids caused by liquid-solid shrinkage.) Second, the ingot is allowed to freeze with a deliberate evolution of gas (called "rimming"). The ingot then contains a number of sealed-in gas voids in place of shrinkage cavities. These voids are later closed by the mechanical action of the rolls. Segregation is still present in the rolled material, and for specialty steels several processes have been developed to minimize it. In one case an electrode of the desired analysis is prepared and then an ingot is cast by striking an arc between the electrode and the base of the ingot mold. The metal is then cast almost drop by drop as the electrode melts and the metal falls into the water-cooled mold.

4.19 Strengthening by nonequilibrium reactions

From the preceding discussion we see that we need not accept the combinations of structures that are obtained under equilibrium conditions following the phase diagram. It is possible to obtain improved structures by proper processing at a number of points in the manufacture of a component. These methods include:

1. Control of the liquid-to-solid reaction (nucleation and chilling effects)
2. Control of distribution and fineness of products in reactions involving

precipitation of one or more phases from a *solid* matrix (age hardening, dispersion hardening, martensite-type reactions)

Let us consider these in order.

4.20   Control of liquid-to-solid reactions

We have already referred to the nucleation of silicon in aluminum. Another change of great engineering importance is the change in the shape of graphite particles in iron alloys. In normal gray cast iron, the graphite precipitates in flakelike particles (Fig. 4.25*a*). The matrix of the material is similar to steel with good ductility, but the flakes act as notches and the ductility is below 1 percent. Adding a small amount of magnesium (0.05 percent magnesium) causes the graphite to crystallize in the form of spheres (Fig. 4.25*b*), and the ductility is raised to as high as 20 percent and the strength is increased several times. This material is known as "ductile" or "nodular" iron and is discussed in detail in Chap. 6.

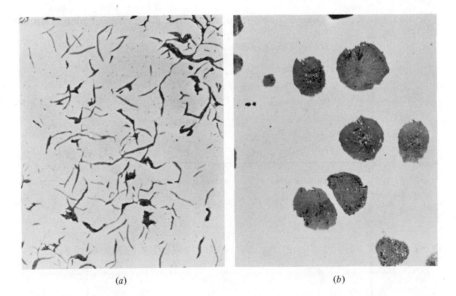

(*a*)   (*b*)

Fig. 4.25 (a) Flake graphite in gray cast iron. 100X, unetched. (b) Spheroidal graphite in ductile cast iron. 100X, unetched. Both irons have approximately the same analysis (3.5 percent carbon, 2.5 percent silicon), but the graphite shape is spheroidal in (b) because of the presence of 0.05 percent magnesium. As a result, the strength and ductility of this iron are more than twice as great as those of (a).

4.21   Control of solid-state precipitation reactions

As examples of this effect, let us discuss the very important control of precipitation of iron carbide in steel and the age hardening of aluminum alloys by the $CuAl_2$ phase.

There are two heat treatment cycles for the hardening of steel. Both result in refinement of the dispersion of iron carbide, but the mechanisms are quite different. The first is a time-dependent nucleation and growth reaction; the second is a rapid shearlike change in structure. Let us consider both these reactions in a typical steel containing 0.8 percent carbon.

Pearlite Formation. The pertinent section of the iron–iron carbide phase diagram is shown in Fig. 4.26. It is important to note that at 0.8 percent carbon a single phase, γ, is present above 1333°F (723°C). This consists of FCC iron with all the carbon in interstitial solid solution. Under equilibrium conditions the diagram shows that on cooling below 1333°F (723°C), γ changes to two phases—α, which is essentially pure, soft BCC iron, and iron carbide, which is hard. This is called a "eutectoid reaction." The two-phase mixture is called "pearlite." Under slow-cooling conditions the dispersion of plates of Fe_3C in

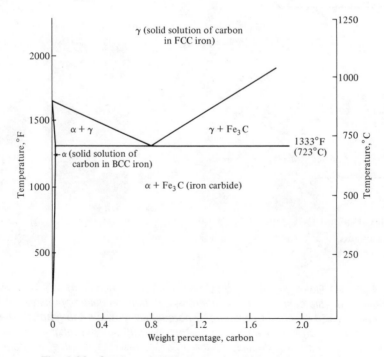

Fig. 4.26 *Section of the iron–iron carbide phase diagram.*

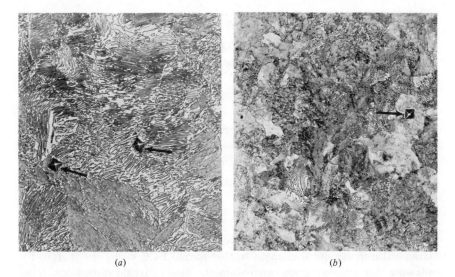

(a)   (b)

Fig. 4.27 (a) *Coarse iron–iron carbide distribution (coarse pearlite), VHN 196 to 266. 500X, 2 percent nital etch.* (b) *Fine iron–iron carbide distribution (fine pearlite), VHN 270 to 320. 500X, 2 percent nital etch. The finer carbide distribution results in higher hardness.*

α is coarse (Fig. 4.27a), and the hardness low, VHN 230. If we heat a piece of the same steel into the γ range and then quench it in a bath of liquid lead or salt at 1000°F (538°C), the reaction will take place at this temperature over a period of less than 1 min, and a much finer dispersion (Fig. 4.27b) will be obtained with a higher hardness, VHN 300. Still finer dispersion and further increases in hardness and strength are obtained by transforming the γ to α plus carbide at still lower temperatures. The effect of these changes on properties is shown in Fig. 4.28. The strength increases as the spacing† between carbide particles decreases.

Martensite Reaction. The second heat treatment cycle is more complex. We find that if the 0.8 percent carbon steel which is originally all γ is quenched quickly below 400°F (205°C), a very rapid shearlike change in structure occurs. Instead of α plus carbide, a single new phase, called "martensite," is formed. The structure is BCT (body-centered tetragonal), which can be considered an elongated BCC (Fig. 4.29). The carbon is still distributed in atomic form as in γ, but in a supersaturated solid solution. When the martensite is reheated at a relatively low temperature [400°F (205°C), for

---

†The average spacing is called the "mean free path," and a straight-line relationship is obtained by plotting the log of the mean free path against yield strength.

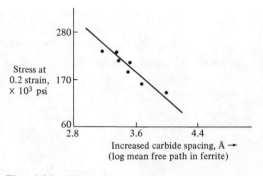

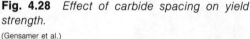

Fig. 4.28 *Effect of carbide spacing on yield strength.*

(Gensamer et al.)

example], the carbon precipitates as very fine iron carbide particles and the tetragonal iron structure changes to the normal BCC structure of iron. By following this processing we can obtain the finest carbide dispersions. This reaction therefore involves two steps: quenching to low temperature to form martensite and reheating (called "tempering") to form fine iron carbide.

Age Hardening. As an example of this process, consider the hardening of aluminum-copper alloys by a three-step process which involves a solution treatment at about 950°F (510°C) in which (1) the $CuAl_2$ phase is dissolved in the matrix, (2) the alloy is quenched to room temperature, retaining the solid solution formed at high temperature, and (3) following the quench, the alloy is either aged naturally, i.e. held at room temperature, or aged artificially by heating at a relatively low temperature (200 to 400°F, 95 to 205°C). During

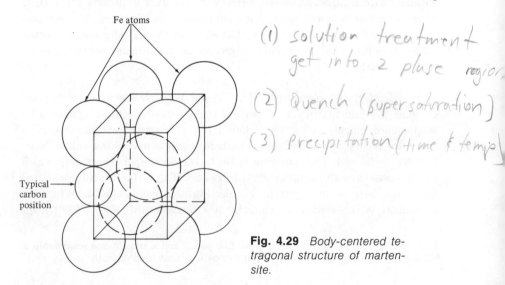

Fig. 4.29 *Body-centered tetragonal structure of martensite.*

the aging step the $CuAl_2$ precipitates and hardens the alloy. The final dispersion of the precipitate is much finer than the dispersion in the original ingot. The relation of these steps to the aluminum-copper phase diagram is shown in Fig. 4.30. The line between the α and $\alpha + \theta$ fields (called the "solvus") tells us that a great deal more copper will dissolve in the α phase at elevated temperatures

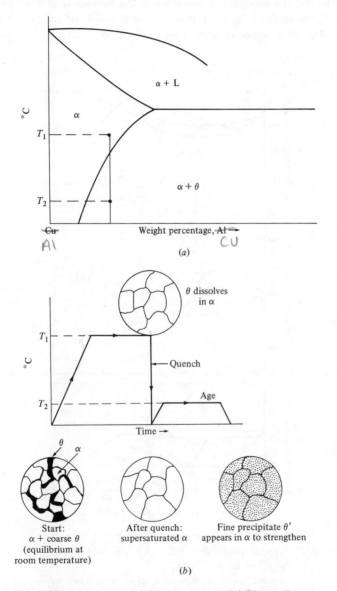

(a)

(b)

Fig. 4.30 *Age-hardening heat treatment. (a) Phase diagram. (b) Heat treatment chart and microstructures.*

than at room temperature. After quenching to room temperature, the degree of supersaturation can be read from the diagram. Since diffusion is slower at room temperature, the equilibrium amount of θ is slow to precipitate and the rate of hardening is slow. At an intermediate temperature we obtain more rapid hardening. However, if the alloy is heated for extended periods or at too high a temperature, the precipitate coarsens and the hardening is diminished. This is called "over-aging" (Fig. 4.31). It should be pointed out that not all precipitates produce this degree of hardening in the aged condition. It is only when

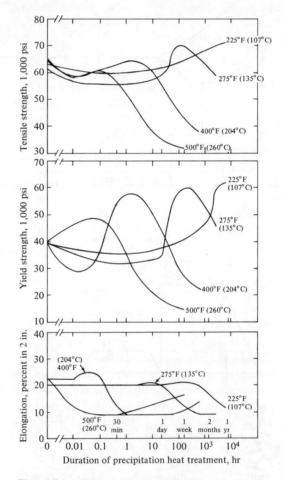

Fig. 4.31 *Effect of age hardening on the mechanical properties of a 4.5 percent copper-aluminum alloy (2014).*

(American Society for Metals Handbook, 8th ed., vol. 2: "Heat Treating, Cleaning and Finishing," Metals Park, Ohio, 1964.)

some atom spacing in the precipitate is close to a spacing in the matrix that a so-called "coherent" precipitate is obtained. A region of strained material then surrounds the precipitate particle, making it difficult for slip to occur in this region, and enhanced hardening results.

SUMMARY

The optimum mechanical properties are obtained from polyphase alloys by control of the nature, amount, size, shape, distribution, and orientation of the phases. The two principal tools of the metallurgist for accomplishing this are the equilibrium phase diagram and the relations governing the kinetics of nonequilibrium conditions.

The phase diagram may be used as a map showing the phases that are present as functions of temperature and composition. For any overall composition the order of separation of phases on cooling, as well as the *compositions* and *amounts* of the phases, may be read. Similarly, the phase compositions and amounts obtained on heating to a given temperature may be readily determined. This is a very powerful tool in itself, but, in addition, the step of heating to a known high temperature a structure such as γ iron is often the prelude to a step involving nonequilibrium conditions such as a rapid quench.

Before discussing the strengthening mechanisms involving nonequilibrium conditions, we described three important phenomena: diffusion, nucleation and growth, and segregation effects. Diffusion is important because it governs the time required for a strengthening precipitate to go into solution or to precipitate. Also, the hardening of a surface by diffusing carbon from the atmosphere, as in carburizing, is important.

Nucleation is important because in some cases the shape and size of a precipitate can be controlled by proper nucleation.

Segregation is significant because the development of composition differences or even nonequilibrium phases may result.

The principal strengthening mechanisms described were the control of liquid-solid reactions such as the production of spheroidal graphite in ductile iron, the pearlite reaction showing the effect of temperature on pearlite fineness, the martensite reaction, and age hardening. These are the reactions most often employed to strengthen the two-phase materials described in the chapters that follow.

DEFINITIONS

Single-phase material A material in which only one phase is present, such as BCC iron. Any number of elements may be present, but these must be in solid solution.

Two-phase material A material in which two different phases are present and grains or crystals of two different materials, with different unit cells for example, can be found.

Polyphase material A material in which two or more phases are present.

Microhardness test A test for measuring the hardness of a grain of a particular phase by using a very light load on the indenter.

Phase diagram A graph showing the phase or phases that are present for a given composition as a function of temperature (a collection of solubility lines).

Liquidus The temperature at which a liquid *begins* to freeze under equilibrium conditions (solid first forms).

Solidus The temperature at which the liquid phase disappears on cooling or at which melting begins on heating.

Eutectic temperature The temperature at which a liquid of eutectic composition freezes to form two solids simultaneously under equilibrium conditions. Also, the temperature at which a liquid and two solids are in equilibrium.

Eutectic composition The composition of the liquid that reacts to form two solids at the eutectic temperature. In the aluminum-silicon system this is 11.6 percent silicon, balance aluminum. Note, however, that some liquid of eutectic composition will be obtained during freezing of any alloy of overall composition between 1.65 and 99 percent silicon.

Phase rule $F = C - P + 2$, where F = degree of freedom, C = number of components and P = number of phases in equilibrium. This formula is usually written +1 instead of +2 because we have used up one degree of freedom in fixing the pressure at 1 atm.

Degree of freedom, F The number of variables that the experimenter can specify or control and still have the number of phases and components originally substituted in the phase-rule equation. For example, in the aluminum-silicon system, if two phases $(\alpha + L)$ are present, $F = 2 - 2 + 1 = 1$, and the temperature can be varied over a wide range without leaving the two-phase field.

Number of components, C The number of materials, elements or compounds, for which the phase diagram applies. For example, the iron-nickel and iron-iron carbide systems are both binary or two-component systems.

Number of phases, P The sum of all the solid, liquid, and gas phases. There can only be one gas phase since all gases are intersoluble in all proportions. (See Chap. 2 for a detailed discussion of phases.)

Peritectic reaction A reaction in which a solid goes to a new solid plus a liquid on heating, and the reverse occurs on cooling:

$$S_1 \xrightleftharpoons[\text{cooling}]{\text{heating}} S_2 + L$$

Ternary system A three-component system. The additional component gives an additional degree of freedom.

Isomorphous Of the same structure. Note in Fig. 4.12 that there is only one solid phase α, hence only one structure is present. However, the lattice parameter will change continuously with composition.

Diffusion Migration of atoms in a solid, liquid, or gas. Even in a pure material atoms will change position. This is called "self-diffusion."

Interstitial diffusion Migration of atoms through a space lattice using interstitial positions.

Substitutional diffusion Migration using substitutional positions.

Bimetal A two-layer structure made by rolling or plating a layer of one metal or an alloy on another.

Steady-state condition A system in which diffusion or heat transfer is taking place and there is no change in composition or temperature at different points in the system with the passage of time.

Unsteady-state condition A system in which temperature or composition is changing as a function of time.

Diffusivity, D The rate of diffusion of a substance at a given temperature. D varies exponentially with temperature. $D = Ae^{-Q/(RT)}$.

Carburizing A process for adding carbon to the surface layers of an iron part.

Nucleation The development of nuclei which act as centers of crystallization for a new phase.

Homogeneous nucleation The development of a new phase by the formation of nuclei of the new phase from the parent phase.

Heterogeneous nucleation The development of a new phase by the addition of foreign material (seeding).

Segregation The development of a concentration gradient or nonequilibrium structure as a result of freezing under nonequilibrium conditions.

Shrinkage cavity A void produced within a casting or ingot because the solid, which is denser, cannot fill the mold volume originally occupied by liquid.

Matrix The continuous phase in a two-phase material, usually forming the "background" of the microstructure.

Pearlite A two-phase mixture of α iron and iron carbide produced when γ iron transforms by a eutectoid reaction.

Martensite A metastable, body-centered tetragonal phase formed by quenching γ iron containing carbon.

Age hardening A three-step process consisting of heating an alloy to dissolve all or part of a second phase in the matrix phase, quenching to retain the solute in a supersaturated solid solution, and finally allowing the second phase to precipitate in fine particles that are coherent with the matrix.

PROBLEMS

4.1 Consider an alloy of 5% Si, 95% Al.

   a. What is the percentage of silicon in the α phase at 630, 600, 577, and 550°C?
   b. What is the percentage of silicon in the liquid phase at 630, 600, and 577°C?
   c. What is the percentage of silicon in the β phase at 550°C?

   To answer this last part draw the horizontal tie line until it touches the β field.

4.2 Consider the typical engine-block alloy which is 16% Si, 84% Al.

   a. At what temperature will the first crystals of solid appear on slow-cooling the melt?
   b. At what temperature will the alloy be completely solid?
   c. Just before the alloy is all solid, at say 578°C, what will be the analyses of the β and liquid, respectively?
   d. At this temperature will the analysis of the liquid be greatly different from that of the liquid in the alloy of Prob. 4.1?
   e. What will be the phase analyses of the α and β in this alloy at 550°C?

4.3 Suppose you are going to pour some castings of red brass (85% Cu, 15% Zn) and some of yellow brass (60% Cu, 40% Zn) (Fig. 4.9).

   a. What pouring temperature would you use in each case, assuming that a temperature 200°C above the liquidus is needed to give the metal sufficient fluidity to fill the molds?
   b. At what temperature would each alloy be completely solid?

4.4 Draw a graph showing the maximum temperature to which copper-zinc alloys can be heated during heat treatment. [*Hint:* About 20°C below the solidus (Fig. 4.9).]

4.5 A common brass, often called "Muntz metal," is 60% Cu, 40% Zn. Over what temperature ranges are two phases present? One phase? At what temperature is the analysis of one of the phases highest in zinc content (Fig. 4.9)?

4.6 The β brass phase is quite important and is given the formula CuZn. If the phase were exactly this formula, what would be the weight percent of copper and weight percent of zinc?

4.7 What are the percentages of α and liquid in a 5% Si, 95% Al alloy at 620, 600, and 578°C? What are the percentages of α and β in this alloy at 576 and 550°C?

4.8 What are the percentages of α and liquid in a 1% Si, 99% Al alloy at 630, 600, and 578°C?

4.9 In a *hyper*eutectic (more than eutectic) analysis, 16% Si, 84% Al, calculate the amounts of liquid and β at 578°C and of α and β at 576°C.

4.10 Prepare a fraction chart for the hypereutectic 16% Si, 84% Al alloy.

4.11 Prepare a fraction chart for a 60% Cu, 40% Zn alloy.

4.12 Prepare a fraction chart for a 65% Cu, 35% Zn alloy. What is the chief difference in the freezing of this alloy compared with the 60% Cu, 40% Zn alloy of Prob. 4.11?

4.13 Calculate for the cast-iron engine block the volume percent of graphite. Assume the weight percents of phases are 3 percent carbon as graphite and 97 percent α (essentially iron). Assume the density of graphite is 2.25 g/cm^3 (2.25 $\times$ 10^3 kg/m^3) and that of iron is 7.87 g/cm^3 (7.87 $\times$ 10^3 kg/m^3).

4.14 In the copper-zinc diagram (Fig. 4.9), locate the temperatures at which three phases can exist at equilibrium and name the reactions (eutectic, peritectic, etc.).

4.15 In the diagram in Fig. 4.32 the single-phase regions are labeled. Label the others. (*Hint:* A horizontal tie line drawn in a two-phase field gives the type and analyses of the phases at the points where it intersects the single-phase fields. This is the important iron–iron carbide diagram; carbide is a solid-phase Fe$_3$C and is also the component of the right-hand side.)

4.16 Given the single-phase fields in the horizontal section of the ternary diagram (Fig. 4.33), label the others. This is the important iron-nickel-chromium diagram, the basis of many stainless steels and superalloys. Note that the fields containing three phases are triangles (straight sides), that there is a two-phase field at each side, that the corners of the triangle touch single-phase fields, and that all the phases found in the fields touching the triangle are also found in the triangle.

4.17 A student attempts to apply the phase rule to a 5% Si, 95% Al alloy in the α + L field. He sees that he is in a two-phase field, and at a pressure of 1 atm the equation is $F = C - P + 1$, or $F = 2 - 2 + 1 = 1$. He then says that since this is an area where there are two variables, composition and temperature, the phase rule is incorrect in giving $F = 1$. Do you agree?

4.18 Another student applies the phase rule to the eutectic temperature for the aluminum-silicon system. She says quite correctly that there are

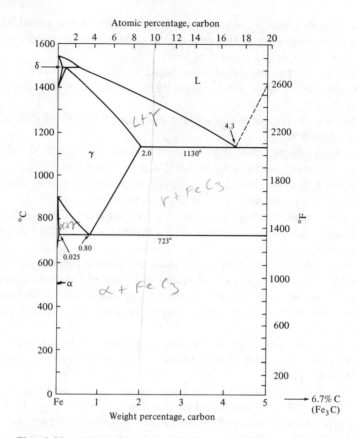

Fig. 4.32 *Unlabeled portion of the iron–iron carbide phase diagram (Prob. 4.15).*

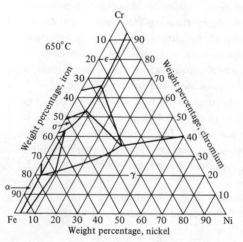

Fig. 4.33 *Unlabeled 650°C isothermal section of the iron-chromium-nickel ternary phase diagram (Prob. 4.16).*

(American Society for Metals Handbook, 8th ed., vol. 1: "Properties and Selection of Metals," Metals Park, Ohio, 1961, p. 214.)

three phases and two components, hence $F = 0$. She then complains that three phases will be in equilibrium from 1.6 to 99 percent silicon at 577°C, so there is still a variable to be fixed if she is to describe the system to another student. Can you help her?

4.19 The β phase is a random solid solution of copper and zinc, but β' is an ordered phase. The ordering temperature increases with the amount of zinc. Why is the line, which shows the temperature of β ordering, horizontal in both the $\alpha + \beta$ fields and the $\beta + \gamma$ fields but slanted in the pure β field?

4.20 In 5% Si, 95% Al alloys the β phase is in a fine intimate mixture with α in localized regions. However, in 15% Si alloys the β is in two distinct forms: large blocky crystals and a fine mixture. Explain from the phase diagram.

4.21 Assume that the heat-treatment foreman in charge of processing the gears described in Example 4.2 is on an economy program directed toward getting better life from his electric furnaces. He recommends reducing the carburizing temperature from 1000 to 900°C and says that the furnace life is much longer at the lower temperature. He also says that the carburizing time needed to get the same results will be only 1,000/900 times or about 10 percent longer. Show that he is sadly in error by calculating D for 900°C and calculating the time required to reach the same percentage of carbon as in Example 4.2.

4.22 In cases where segregation exists from solidification conditions we can use diffusion calculations to estimate the time required to homogenize the material. The simplest calculation is to obtain the time needed to reduce the difference in concentration to one-half. For example, suppose we have a carbon concentration C_s at one point and zero carbon at another. We calculate the time for C_x to equal 0.5 C_s.

a. Show that the homogenization time is approximately $t = x^2/D$.

b. Referring to Fig. 4.20, what would be the relative homogenization times for nickel in FCC iron compared to carbon in FCC iron at 1000°C?

5

ENGINEERING ALLOYS IN GENERAL; NONFERROUS ALLOYS—ALUMINUM, MAGNESIUM, COPPER, NICKEL, AND ZINC

THE hang glider is a good illustration of the use of high-strength lightweight alloys. For centuries people have attempted to fabricate wings with which they could glide like a bird. The success of the present equipment is related to the use of high-strength aluminum alloy 6061-T6 in the ribs, discussed in this chapter, and the high-strength polymer fabrics described in Chap. 9.

In this chapter we will put to use the concepts of structure and structural control that we discussed in the earlier chapters. We will be able to review the structures of the important nonferrous alloys and understand the interrelation of structure, mechanical properties, and processing.

5.1 General

We are now entering a new portion of the text devoted to the actual engineering alloys used in structures and components. Let us spend a few moments reflecting on what we have studied and how it applies to the new material. In the first chapter we agreed that if we were to reach a basic understanding of metals, we needed to know first the nature of the different atoms—how they were arranged in metallic structures and the effects of stress and temperature on these structures. As a result of this work we are now in the powerful position of being able to look over the entire field of thousands of metallic alloys and rapidly attain an understanding of their properties. The reason we can do this, despite the almost infinite number of combinations of analysis, is that the alloys are made up of just a few different metallic structures with which we are already familiar.

The simplest way to describe these commercial alloys is to arrange them into families based on the principal element present, such as the aluminum alloys, the copper alloys, and so forth. In passing we should mention the traditional division into nonferrous alloys and ferrous (iron-base) alloys. In some countries even the metallurgists are divided into these two specialized groups. This is really an unscientific point of view because, for example, the nickel alloys, while "nonferrous," have certainly more in common with iron-base alloys than with zinc alloys.

Therefore, let us take the more systematic approach of dividing the alloys into families. We find that over 95 percent of the tonnage is in the aluminum, magnesium, copper, nickel, and iron-base alloys (Table 5.1). In fact, over 90 percent is in the iron-base family alone, and although the percentages for magnesium and nickel alloys are small, these have special importance and can be discussed conveniently with the others.

There are several other features of the data that are important to our background. First we may ask: Since the iron-base alloys are low in price and highest in strength, why bother with the others? We must realize that strength is only one of the requirements of a component. Satisfactory service can depend on density, corrosion resistance, effects of temperature, as well as electrical and magnetic properties.

As examples, let us consider some parts for which the different alloys are especially suited.

Aluminum alloys: aircraft parts (high strength per pound)
Magnesium alloys: aircraft castings (competitive with aluminum)
Copper alloys: electrical wiring (high conductivity)
Nickel alloys: gas turbine parts (high strength at elevated temperatures)

Another question we may ask is: Why did we bother to survey so many other metals in Chap. 2 when only four or five are so important? The answer

Table 5.1 PROPERTIES OF IMPORTANT ELEMENTS AND THEIR ALLOYS

| Element | Properties of Element (Annealed) | | | Alloy No. | Properties of Common Alloy of Highest Strength | | | Use as Engineering Materials (U.S.), tons/yr |
	Yield Strength,* psi $\times 10^3$	Percent Elongation	BHN	Ingot,† dollars/lb	Yield Strength,* psi $\times 10^3$	Percent Elongation	BHN	
Aluminum	4	43	19	0.27	7178	78	10	160	2 million
Magnesium	Not used in pure form			AZ31B-H24	42	16	82	<100,000
Copper	10	50	25	0.90	172	140	7	380	1 million
Nickel	22	47	90	1.60	301	150	10	400	<100,000‡
Iron	20	48	70	0.05	4340	270	11	500	100 million

*To obtain MN/m², multiply psi by 6.9×10^{-3}. To obtain kg/mm², multiply psi by 7.03×10^{-4}.
†1974 prices.
‡As nickel-base alloy.

is that at least 20 others are of vital importance in controlling and modifying the structures of the principal metals. It is by this modification, combined with processing treatments, that we obtain the tremendously improved properties of the alloys. A glance at Table 5.1 will show that by experimenting with alloys and heat treatment, metallurgists have been able to raise the strength of pure aluminum 20 times, copper 14 times, and iron 13 times, while retaining satisfactory ductility in all cases.

5.2 Processing methods

As we study the specifications for commercial alloys, we will find that the processing method is given. For example, the properties will be accompanied by one of the following terms: cast, forged, wrought, rolled, extruded, and so on. There are two major divisions—cast and wrought—but recently powder metallurgy has also become important.

In the *casting* processes the liquid metal is poured into a mold of the desired shape and allowed to solidify. There are many different casting techniques.

In *sand casting* the mold is made of sand bonded with clay, and the cooling rate is relatively slow. This is used for all alloys and is quite versatile.

In *permanent mold casting* the mold is made of metal or graphite. The cooling rate is faster than for sand.

Die casting also uses a permanent mold, but here the metal is injected under pressure. This is particularly useful for metals with low melting points—such as zinc, aluminum, and magnesium—which do not attack the metal die rapidly.

Investment casting is a method for producing small precision castings. A wax replica of the desired part is made and a liquid, cementlike slurry is poured around it in a mold. After the cement (called an "investment") sets, the mold is heated to eliminate the wax and preheat the mold. Then liquid metal is poured into the cavity.

In the *wrought* processes an ingot is cast and then worked to the desired shape, usually above the recrystallization temperature at first. These processes were described in Chap. 3. In continuous casting the ingot step is eliminated and a continuous strand of metal issues from the bottom of a water-cooled die.

All alloys can be cast, but the wrought alloys are limited to the analyses that can be hot-worked without cracking. As examples of alloys that cannot be worked, we have in the aluminum alloys the low-ductility–high-silicon analyses and in the ferrous alloys the cast irons. These distinctions will be made as we consider the different alloys.

In *welded* structures combinations of cast and wrought parts can be used.

In *powder metallurgy* powders of the desired materials are mixed and then compressed in a die to the desired shape. Next the mass is treated at a high temperature to promote bonding of the grains of the powder by sintering, a type of diffusion bonding. This method is especially important for producing carbide tools for machining.

One final word. In covering the material ahead it is not necessary to memorize analyses. The important point is to recognize how the structure of a material is controlled by analysis and heat treatment and how the properties of the material respond to the changes. There are four important special cases of control of the structure which we discussed in the previous chapter. These are: (1) controlled crystallization from the melt, (2) controlled nucleation in solid, (3) martensite reaction, (4) age-hardened. These of course are all techniques for controlling the nature, size, shape, amount, distribution, and orientation of the phases, as pointed out in previous sections.

Aluminum Alloys

5.3 General

In discussing these materials which make up our first group of alloys for actual engineering applications, let us take special care to examine how even the complex specifications are based on the fundamental factors we have learned. In each alloy it is important to understand why certain alloying elements are used and why the processing results in the properties given.

Since we are going to build this family on an aluminum base, let us review briefly what we know about pure aluminum. The atomic structure is $1s^2$, $2s^2$, $2p^6$, $3s^2$, $3p^1$ or, more simply, a tightly bonded core surrounded by three valence electrons which are readily given up. This results in a light, reactive metal with good electrical and thermal conductivity. The unit cell is FCC, which leads to a number of plane and direction combinations for easy slip, $\{111\}$ and $\langle 110 \rangle$, so that we expect excellent ductility and formability. All these predictions are borne out by our everyday experiences with aluminum foil and wire.

Aluminum finds many uses in the unalloyed or relatively pure state. It is possible to raise the yield and tensile strengths by the cold-working method discussed in Chap. 3. Referring to Table 5.2, we see that two sets of properties are given for commercially pure aluminum (1060), for conditions marked 0 and H18. We will discuss these symbols in detail shortly, but for the present we may consider them as denoting annealed (recrystallized) and cold-worked conditions, respectively. The microstructures are similar to those of the cold-worked and annealed α brasses discussed in Chap. 3. As in the case of brass, we can obtain different combinations of strength, elongation, and hardness between the extremes listed in Table 5.2 by controlling the sequence and amount of annealing and cold working.

Now let us advance to the specifications involving the second method for improving the strength of a single-phase alloy, namely solid solution

Table 5.2 TYPICAL PROPERTIES OF ALUMINUM ALLOYS

Alloy Number	Chemical Analysis, percent*	Condition	Tensile Strength,† psi × 10³	Yield Strength,† psi × 10³	Percent Elongation	BHN	Typical Use
		Single-phase Wrought Alloys					
1060	99.6 minimum Al	0	10	4	42	19	Sheet, plate, tubing
		Hard H18	19	18	6	35	
3003	1.2 Mn	0	16	6	30	28	Truck panels, ductwork
		Hard H18	29	27	4	55	
5052	2.5 Mg, 0.2 Cr	0	28	13	25	47	Bus bodies, marine applications
		Hard H38	42	37	7	77	
5050	1.2 Mg	0	21	8	24	36	Sheet, trim, gas lines
		Hard H38	32	29	6	63	
		Two-phase Wrought Alloys					
2014	4.5 Cu 0.8 Si 0.8 Mn 0.5 Mg	Annealed	27	14	18	45	Airplane structures
		Heat-treated T6	70	60	13	135	
6061	1 Mg 0.6 Si 0.2 Cr 0.3 Cu	Annealed	18	8	30	30	Transportation equipment, pipe
		Heat-treated T6	45	40	12	95	
7178	7 Zn, 0.3 Mn, 3 Mg, 0.3 Cr, 2 Cu	Annealed	33	15	16	60	Structural parts in aircraft
		Heat-treated T6	88	78	10	160	

Two-phase Cast Alloys

Alloy	Composition*	Condition	[†]	[†]			Uses
B195	4.5 Cu	Solution heat-treated (T4)	37	19	9	75	Aircraft fittings, pump bodies
		Aged (T6)	40	26	5	90	
356	7 Si, 0.4 Mg	Aged T5	25	20	2	60	Auto transmission casings, wheels
		Aged T6	38	27	5	80	
40E	5.5 Zn 0.6 Mg 0.5 Cr 0.15 Ti	Aged T5	35	25	3	75	Machine parts
108	3 Si, 4 Cu	As cast (F)	21	14	2	55	General
380	8 Si, 3.5 Cu	As cast (F)	47	23	4	80	Die casting
390	17 Si, 1 Fe, 4.5 Cu, 0.5 Mg	As cast (F)	41	35	<3	120	Die casting

* Balance aluminum.

† Multiply psi by 6.9×10^{-3} to obtain MN/m^2 or by 7.03×10^{-4} to obtain kg/mm^2.

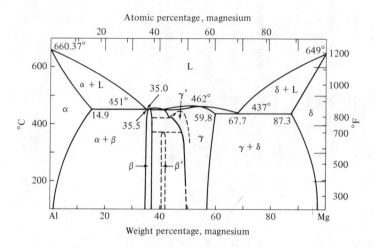

Fig. 5.1 *Aluminum-magnesium phase diagram.*

(American Society for Metals Handbook, 8th ed., vol. 8, "Metallography, Structures and Phase Diagrams," Metals Park, Ohio, 1973, p. 261.)

strengthening. This is the mechanism which operates in alloys 3003 and 5052 listed in Table 5.2. If we inspect the aluminum-magnesium phase diagram (Fig. 5.1), we see that we would expect to find the 2.5 percent magnesium of alloy 5052 in solid solution in the α (FCC) phase. Similarly, the manganese in 3003 is essentially in solid solution.

In both these alloys in the annealed (0) condition, the tensile and yield strengths are much higher than for 1060 aluminum with somewhat lower elongation. However, the microstructures still show the simple grains of a single phase. Basically, the solute atoms have raised the stress required for slip. It is important to note that when compared in the cold-worked conditions (marked H18 and H38), the alloys are still proportionately higher in strength than the cold-worked pure aluminum. This means that the effects of solid solution strengthening and work hardening can be used together; that is, the effects are *additive*. We will see this method of alloy design used in all the other systems as well. In other words, when we wish to retain the ductility and formability of the base metal, we add alloys that dissolve in solid solution to raise the strength in the annealed condition and then cold-work to the desired properties.

Now let us examine the *two-phase* wrought alloys. The most important element in 2014 is the 4.5 percent copper. Turning to the aluminum-copper phase diagram (Fig. 5.2), we see that we have a single-phase alloy at 550°C but a two-phase material at room temperature. As discussed in Chap. 4, this alloy is a candidate for age hardening if the precipitate is coherent, a condition which is found to exist. To develop a fine coherent precipitate, we give the alloy a two-step treatment. First the annealed material is heated to 500 to 550°C to dissolve the θ phase, and then quenched. This gives supersaturated κ. A fine precipitate is then produced by aging at 170°C for 10 hr; this is so fine it is not visible by conventional microscopy. A tensile strength of 70,000 psi (483 MN/m^2) develops (marked T6 condition) as compared to 27,000 psi (182 MN/m^2) (annealed condition).

Casting alloys are usually two-phase, but a notable exception is the casting of relatively pure aluminum cooling fins around iron laminations in electric rotors, where it is desired to capitalize on the high thermal conductivity of aluminum. In one group of heat-treatable casting alloys, such as 356, age hardening is produced as in the wrought alloys. In some cases the solution-treating step may be omitted if the cooling rate of the casting is fast enough to produce a supersaturated solid solution. In this case we need only to age the casting, and this is denoted by T5. (See the code given at end of this section.) In other cases the second phase acts simply as a hard dispersion to improve hardness and wear resistance. This is the case of the silicon-rich phase in the 390 alloy for engine blocks. The size and shape of the dispersion are controlled by the cooling rate and by nucleation with phosphorus (forming AlP). Fast cooling rates and abundant nuclei give a dispersion of fine equiaxed crystals of silicon in place of long bladelike crystals, thereby improving strength and ease of machining. It should be added that the alloy can be further hardened with a solution heat treatment and aging because of the presence of 4 percent copper as discussed for 2014.

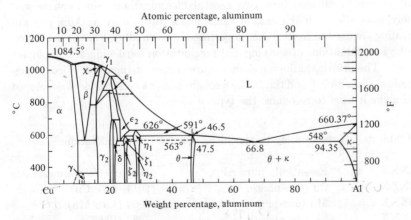

Fig. 5.2 *Aluminum-copper phase diagram.*

(American Society for Metals Handbook, 8th ed., vol. 8, "Metallography, Structures and Phase Diagrams," Metals Park, Ohio, 1973, p. 259.)

Now let us consider some details of the actual engineering specifications, which may seem complex after our simple discussions of working and heat treatment. For example, the specification for an ordinary aluminum frying pan would read:

Alloy 1100: 99.0% minimum Al, 1.0% maximum (Fe + Si), 0.20% maximum Cu, 0.05% maximum Mn, 0.10% maximum Zn, 0.05% maximum each of other elements, total of which shall be 0.15% maximum.

Mechanical properties: H16 temper, 20,000 psi (138 MN/m²) yield strength, 21,000 psi (145 MN/m²) tensile strength, 6 percent elongation.

To understand this we must learn to see through all the specification language and determine what structure we are talking about.

First, we recall that pure aluminum is an FCC structure and therefore has good ductility. If we consult the important phase diagrams for aluminum (Fig. 5.3), we see that the objective of specifying small maximum amounts of impurities is to avoid the formation of any quantities of hard second phases which would reduce ductility and increase corrosion, as discussed in Chap. 12. It would be possible to specify a purer material, but this would raise the cost.

Next we note the specification of "H16 temper." This denotes a certain amount of cold working, such as cold rolling, which raises the yield point and prevents denting of the frying pan. (The temperature encountered during normal heating of the pan would be too low for recrystallization.) In essence, then, the specification is written to give a cold-worked single-phase structure of commercially pure aluminum.

Let us now review the general specifications. We shall see that the alloys can be divided into two general classifications: single-phase and polyphase alloys. In the case of single-phase alloys only working and annealing are used to control the properties of a given alloy. For the polyphase alloys combinations of working and precipitation hardening are employed.

The principal alloying elements are copper, manganese, silicon, magnesium, zinc, nickel, and tin. In the wrought alloys a code has been developed to make it easy to recognize the type of alloy.

Code	Type of alloy (major element)	Example
1XXX	Essentially pure Al	1060 (99.6% minimum Al)
2XXX	Cu (two-phase)	2014 (4.5% Cu)
3XXX	Mn (one-phase)	3003 (1.3% Mn)
4XXX	Si (two-phase)	4032 (12.5% Si)
5XXX	Mg (one-phase)	5050 (1.2% Mg)
6XXX	Mg and Si (two-phase)	6063 (0.4% Si, 0.72% Mg)
7XXX	Zn (two-phase)	7075 (5.6% Zn)

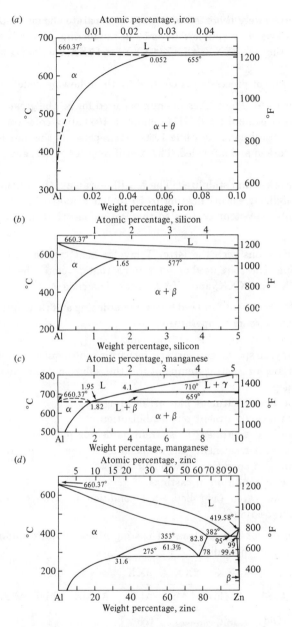

Fig. 5.3 *Portions of aluminum-rich phase dia-grams. (a) Aluminum-iron. (b) Aluminum-silicon. (c) Aluminum-manganese. (d) Aluminum-zinc.*

(American Society for Metals Handbook, 8th ed., vol. 8, "Metallography Struc-tures and Phase Diagrams," Metals Park, Ohio, 1973, pp. 260, 262, 263, 265.)

Unfortunately there is no attempt to separate the single-phase and two-phase alloys by a simple difference in numbering, and it is necessary to recognize that the 2XXX, 4XXX, 6XXX, and 7XXX series are two-phase alloys.

The type of processing is covered by the following code:

H1X, cold-worked. The higher the number used for X, the greater the cold working. For example, 1060-H14 denotes a 1060 alloy cold-worked about half of the total possible, while 1060-H-19 represents the maximum.

H2X, cold-worked and annealed. The X still represents the severity of cold work.

H3X, cold-worked and stabilized. The X still represents the amount of cold work. Stabilizing means heating to 50 to 100°F (30 to 55°C) above the maximum service temperature so that softening does not take place in service.

Specifications with the letter T involve heat treatment to produce age hardening. These are used only for alloys that develop coherent precipitates, the 2XXX, 6XXX, and 7XXX series. The code is:

T3 Solution treatment followed by strain hardening and then natural aging (i.e. held at room temperature).

T4 Solution treatment plus natural aging.

T5 Aged only. In special cases where the part is cooled quickly enough from the forging or casting temperature, the solution heat treatment is omitted.

T6 Solution heat treatment plus artificial aging.

T7 Solution heat treatment plus stabilization.

T8 Solution heat treatment plus strain hardening plus artificial aging.

T9 Solution heat treatment plus artificial aging plus strain hardening.

0 The annealed condition in wrought alloys.

T2 The as-cast condition in castings.

F The as-fabricated (as-rolled, etc.) condition in wrought alloys and as-cast in castings.

The code for the analyses of casting alloys is quite different and generally employs three digits:

Numeral	1 to 99	1XX	2XX	3XX	4XX	5XX	6XX	7XX
Family	Al-Si	Al-Cu	Al-Mg	Al-Si	Al-Mn	Al-Ni	Al-Zn	Al-Sn
	(Old symbol)			(New symbol)				

The same codes for specifying heat treatment are used where applicable. The T3, T8, and T9 designations, which call for strain hardening, apply only to wrought products, however.

EXAMPLE 5.1 The T5 designation means aged only. How can the alloy 40E attain properties close to 356-T6 if no solution treatment step is used after casting? (See Table 5.2.)

ANSWER A thin section in a casting cools rapidly enough to provide a solution treatment; i.e. supersaturated α is obtained. Next, at room temperature natural aging takes place and a coherent precipitate results in an age-hardened alloy. In a casting of grossly varying section sizes the physical properties in large sections may not attain those of lighter sections unless the material is solution-treated.

EXAMPLE 5.2 How much $CuAl_2$ (θ phase, Fig. 5.2) is present as a function of temperature in a 4.5 percent copper–aluminum alloy cooled under equilibrium conditions? It is experimentally found that there is more $CuAl_2$ present in an annealed alloy than in a naturally aged alloy but that the yield strength of the latter is greater. Why?

ANSWER

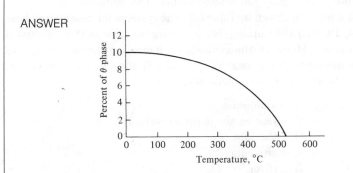

The approximate phase fraction chart is shown above. It is assumed from the phase diagram that no copper is soluble in aluminum at room temperature and 53 percent by weight copper is soluble in θ phase at all temperatures. This of course corresponds to 33.3 atomic percent copper to maintain the formula $CuAl_2$. The higher strength in the naturally aged alloy is due to a more desirable uniform distribution of the second phase rather than continuous $CuAl_2$ at the grain boundaries in the annealed alloy, which is not effective in preventing slip. Furthermore, the aged alloy has coherency which is not present in the annealed material.

Magnesium Alloys

5.4   General

The alloys of magnesium are competitive with those of aluminum because the density of magnesium is two-thirds that of aluminum. Therefore, many aircraft parts are made of magnesium alloys. Also, there has been an interesting competition between magnesium and aluminum alloys for the Volkswagen engine block, since both materials can be used successfully.

The basic atomic structure of magnesium ($1s^2$, $2s^2$, $2p^6$, $3s^2$) was discussed in Chap. 2. We see that beneath the two valence electrons we have a stable ring of eight, making magnesium a very active element. Despite the fact that fine magnesium powder will burn in air, a melt of liquid magnesium can be poured in air into castings or ingots with a few simple precautions. Parts made of magnesium will not ignite if heat-treated. However, magnesium will corrode more rapidly than aluminum in many environments, such as seawater, and this is why outboard motor parts are not made of magnesium.

Another major difference between magnesium and aluminum is that the unit cell of magnesium is HCP and therefore it has only three slip systems at room temperature compared to the 12 of aluminum. However, additional systems become operative at higher temperatures. To take advantage of this, magnesium alloys are usually hot-worked rather than cold-worked.

Typical alloys are shown in Table 5.3. The systems for designating heat treatment (T4, T6, etc.) and working (H24, etc.) are the same as those discussed for aluminum alloys. However, the methods for indicating the composition of the alloys are different. The letters AZ, etc., signify the two most important alloying elements according to the following code:

$$A = aluminum$$
$$E = rare\ earths\ (such\ as\ cerium)$$
$$H = thorium$$
$$K = zirconium$$
$$Q = silver$$
$$Z = zinc$$

The amounts of the elements are shown by the numbers following the letters. The first number gives the percentage of the element shown by the first letter, and the second number the percentage of the second. For example, AZ92 means a 9% Al, 2% Zn alloy.

The principal uses of the magnesium alloys are in aircraft and spacecraft, machinery, tools, and materials-handling equipment. The strength increases with aluminum level by solid solution strengthening and by the precipitation of a fine lamellar phase, $Mg_{17}Al_{12}$. In both the wrought and cast alloys rare earths are added to minimize flow (plastic deformation) at elevated

Table 5.3 TYPICAL PROPERTIES OF MAGNESIUM ALLOYS

Alloy Number	Chemical Analysis, percent	Condition	Tensile Strength,* psi × 10³	Yield Strength,* psi × 10³	Percent Elongation	BHN	Typical Use
Wrought Alloys							
AZ31B	3 Al, 1 Zn, 0.2 Mn	F	38	28	9	55	Sheet and plate
		H24	42	32	15	73	
AZ80A	8 Al, 0.2 Zn, 0.2 Mn	T5	50	34	6	72 }	Forgings and extrusions
ZK60A	6 Zn, 0.5 Zr	T5	44	30	16	82	Elevated temperatures
HK31A	3 Th, 1 Zr	H24	37	29	8	—	
Cast Alloys							
AZ63A	6 Al, 3 Zn, 0.13 Mn	As cast	29	14	6	55	General sand castings
		Solution heat-treated	40	13	12	50	
		T6	40	19	5	73	
AZ91C	9 Al, 1 Zn, 0.13 Mn	As cast	24	14	2	52	High-tensile castings
		Solution heat-treated	40	12	14	53	
		T6	40	19	5	66	
QE22A	2 Ag, 2 R.E.†	T6	40	30	4	77	Highest-strength uses
EZ33A	3 R.E., 3 Zn	T5	23	16	3	50 }	Elevated-temperature uses
EK31A	3 R.E., 1 Zr	T6	31	16	6	55	

* Multiply psi by 6.9×10^{-3} to obtain MN/m^2 or by 7.03×10^{-4} to obtain kg/mm^2.
† R.E. = rare earth elements.

temperatures (150 to 200°C). The rare earths produce a rigid grain boundary network in the microstructure, such as Mg_9Ce, which resists flow.

Careful control of the network is necessary since the second phases also limit room temperature ductility and if present in excessive amounts may even decrease the tensile strength.

It should also be pointed out that magnesium alloys should play an increasing role as materials because the amount of magnesium in seawater is tremendous compared to the dwindling supplies of other important metallic elements which are usually mined as complex oxides and sulfides.

Copper and Nickel Alloys

5.5 Copper alloys; general

The copper alloys have a unique combination of characteristics: high thermal and electrical conductivity, high corrosion resistance, generally high ductility and formability, and interesting color for architectural uses. While the hardness and strength of these alloys do not equal the properties of the hardest steels, tensile strengths of 150,000 psi ($1.035 \times 10^3 MN/m^2$) are obtained in some alloys.

The atomic structure of copper is $1s^2$, $2s^2$, $2p^6$, $3s^2$, $3p^6$, $3d^{10}$, $4s^1$. It is important to note that the outer electron $4s^1$ does not have a shell of eight beneath it, as in the case of the s electrons of aluminum and magnesium, which we have discussed. The energy of this electron is very close to the $3d$ electrons, and therefore the whole group is attracted to the positively charged nucleus. For this reason copper, instead of being an active metal similar to aluminum, is considered a noble, i.e. corrosion-resistant, metal, in the same vertical group in the periodic table as silver and gold. As discussed in Chap. 15, the unique red color of copper is due to selective absorption of the spectrum of white light by interaction with the $3d$ electrons.

At first glance it is easy to be confused by the variety of copper alloys. Over thousands of years the names bronze and brass have been used differently, and other names such as gun metal, admiralty metal, gilding bronze, manganese bronze, and ounce metal have added to the confusion.

We shall take a simple approach based on the microstructures of the alloys involved. As with the light metals, it is possible to divide all the copper alloys into two classes: single-phase and polyphase alloys. We would expect the single-phase alloys to exhibit good ductility because the unit cell is FCC, and this is confirmed by the data of Table 5.4. The mechanisms for strengthening the single-phase wrought alloys are the usual solid solution hardening and combinations of cold work and annealing. In the two-phase alloys, age hardening and other hardening by precipitates or second-phase dispersions are used.

In Table 5.4 we have chosen a few popular alloys to illustrate these points. We may begin with ETP (electrolytic tough pitch) copper. This is copper that has been refined electrolytically, and the term "tough pitch" refers to the oxygen level of about 0.04 percent (present as copper oxide) which gives a certain appearance to the ingot. This grade of copper is widely used, although it will embrittle if heated in an atmosphere containing hydrogen. The mechanism is that the hydrogen diffuses through the copper and encounters copper oxide at grain boundaries. Reaction takes place and water vapor is generated. These molecules are large compared to hydrogen. They do not diffuse appreciably and form voids at the grain boundaries, leading to embrittlement. To avoid this effect, two other grades of copper are available: phosphorus deoxidized copper (122), in which the oxygen is eliminated by a phosphorus addition to the melt, and OFHC (102), (oxygen-free high-conductivity) copper, which is melted under special reducing conditions to eliminate oxygen. The mechanical properties of all grades are comparable. The change in yield strength from 10,000 to 40,000 psi (69 to 276 MN/m^2) by cold working is especially important.

5.6   Solid solution copper alloys

An important solid solution effect is obtained with a silver content specified as 10 to 25 troy oz/ton. While this is only the order of 0.01 percent by weight silver, the softening temperature of the cold-worked copper is raised over 100°C. This enables the fabricator to soft solder (lead-tin alloy) cold-worked copper without lowering the strength by recovery and recrystallization. The electrical conductivity is not changed by the presence of silver.

The most widely used solid solution alloys are those with zinc, which are called "brass." The commonest range is between 65% Cu, 35% Zn and 70% Cu, 30% Zn. Since copper costs about $1/lb and zinc about $0.35/lb (1974 prices), the higher zinc brasses are cheaper. The phase boundary of the α field is quite important because the higher zinc alloys (above 35 percent) contain the β phase (Fig. 5.4). While this phase is stronger than α, it is more susceptible to a particular type of corrosion called "dezincification," which is discussed in Chap. 12. The cold working and recrystallization of brass were discussed in Chap. 3.

Considerable strengthening is also accomplished with aluminum and nickel, as shown in the 614 and 715 alloys. The 715 alloy is especially important in applications involving seawater, as in desalinization equipment.

5.7   Polyphase wrought copper alloys

The highest-strength copper alloy is produced by age-hardening a 2 percent beryllium alloy (172). By heating to 800°C a single-phase α solid solution is obtained (Fig. 5.5). After quenching to obtain supersaturated α, the alloy is aged for 3 hr at 315°C to precipitate the γ_2 phase CuBe, which gives a coherent

Table 5.4 TYPICAL PROPERTIES OF COPPER ALLOYS

Alloy Number	Chemical Analysis, percent	Condition	Tensile Strength,* psi × 10³	Yield Strength,* psi × 10³	Percent Elongation	Hardness	Typical Use
		Single-phase Wrought Alloys					
110	ETP, 99.9 Cu	Annealed	32	10	45	40 R_F	Architectural, electrical
		Cold-worked	50	40	6	85 R_F	
268	65 Cu, 35 Zn Yellow brass	Annealed	46	14	65	88 R_F	Plumbing, Grill work
		Cold-worked	74	60	8	80 R_B	
614	91 Cu, 7 Al, 2 Fe Aluminum bronze	Cold-worked	82	40	35	90 R_B	Condenser tubing
715	70 Cu, 30 Ni Cupronickel	Annealed	44	20	40	37 R_B	Desalinization tubing
		Cold-worked	75	68	12	85 R_B	
		Polyphase Wrought Alloys					
172	98 Cu, 2 Be Beryllium copper	Annealed	70	30	42	57 R_B	Springs, tools
		Precipitation-hardened	175	140	7	38 R_C	

Alloy Number	Chemical Analysis, percent	Condition	Tensile Strength,* psi $\times 10^3$	Yield Strength,* psi $\times 10^3$	Percent Elongation	Hardness	Typical Use
			Cast Alloys				
801	Cu	As cast	25	9	40	BHN 44	Electrical conductors
836	85 Cu; 5 Sn, 5 Zn, 5 Pb	As cast	37	17	30	BHN 60	Valves, bearings
937	80 Cu, 10 Sn, 10 Pb	As cast	35	18	20	BHN 60	Bearings, pumps
964	70 Cu, 30 Ni	As cast	68	37	28	BHN 140	Marine valves
824	98 Cu, 2 Be	Hardened	150	140	1	38 R_C	Dies, tools
916	88 Cu, 10 Sn, 2 Ni	As cast	44	22	6	BHN 85	Gears
953	89 Cu, 10 Al, 1 Fe	As cast	75	27	25	BHN 140	Gears, bearings
		Heat-treated	85	42	15	BHN 174	

*Multiply psi by 6.9×10^{-3} to obtain MN/m² or by 7.03×10^{-4} to obtain kg/mm².

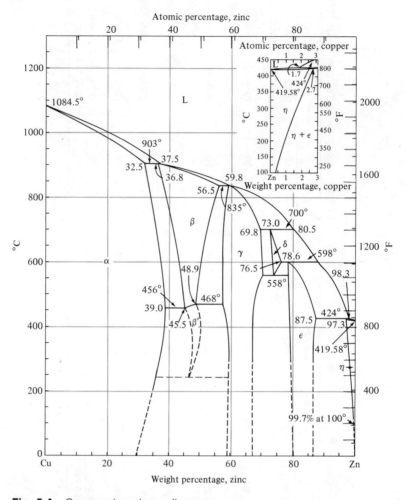

Fig. 5.4 *Copper-zinc phase diagram.*

(American Society for Metals Handbook, 8th ed., vol. 8, "Metallography, Structures and Phase Diagrams," Metals Park, Ohio, 1973, p. 301.)

precipitate. This alloy is used widely for springs, nonsparking tools, and parts requiring good strength, plus high thermal and electrical conductivity. Other precipitation-hardening alloys contain silicon.

5.8 Cast copper alloys

The cast alloys offer a wider range of structures because high ductility is not required for working. Of great importance are the high-lead alloys for bearings and the high-tin alloys for gears.

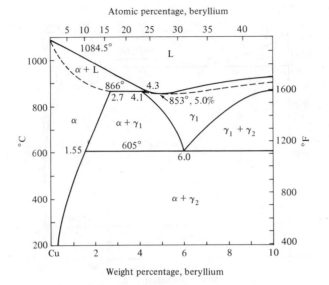

Atomic percentage, beryllium

Weight percentage, beryllium

Fig. 5.5 *Copper-beryllium phase diagram.*

(American Society for Metals Handbook, 8th ed., vol. 8, "Metallography, Structures and Phase Diagrams," Metals Park, Ohio, 1973, p. 271.)

Copper itself is used extensively in castings requiring good electrical and thermal conductivity. It is remarkable that although copper melts at 1084°C, *water-cooled* copper tuyères and lances can be used in the processing of steel where temperatures reach 1750°C in the atmosphere above the liquid metal. As in the case of the wrought alloys, strong solid solutions are formed with zinc, nickel, and aluminum.

The lead alloys are of special interest because liquid copper can dissolve lead in unlimited amounts (Fig. 5.6). On cooling the lead precipitates as metallic lead because it is insoluble in solid copper. Alloys such as 85% Cu, 5% Sn, 5% Zn, 5% Pb are valuable for bearings because of the lubricating effects of the lead droplets (Fig. 5.7a).

The alloys with tin contain a hard intermetallic compound δ ($Cu_{31}Sn_8$). The presence of δ in a ductile α matrix gives an excellent gear bronze because of a good mating surface against hardened steel gears. The photomicrograph of Fig. 5.7b shows that the δ phase will crack only after considerable deformation of the surrounding α.

The two-phase alloys include aluminum bronze, in which the aluminum exceeds the solid solubility in α and a hard γ phase is formed on cooling (Fig. 5.2). Under stress the failure occurs through γ regions (Fig. 5.7c). The alloy is also used in the heat-treated condition, obtained by heating to the β region and quenching. The β transforms on cooling to a structure called "martensite"

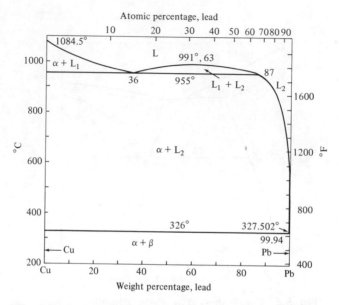

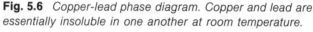

Fig. 5.6 *Copper-lead phase diagram. Copper and lead are essentially insoluble in one another at room temperature.*

(American Society for Metals Handbook, 8th ed., vol. 8, "Metallography, Structures and Phase Diagrams," Metals Park, Ohio, 1973, p. 296.)

(Fig. 5.7*d*), which is not shown on the equilibrium diagram. The nature of this transformation was discussed in Chap. 4. Other polyphase alloys include cast beryllium-copper, copper-silicon, and manganese bronze.

5.9 Nickel alloys in general

Nickel is an element somewhat similar to iron in strength, but its alloys have exceptional resistance to corrosion and elevated temperatures, as well as important magnetic properties. In this section we will discuss only the corrosion-resistant alloys, reserving discussion of the complex nickel-chromium-iron alloys, together with iron-base heat-resistant alloys, for Chap. 6.

The atomic structure of nickel is related to both copper and iron: $1s^2$, $2s^2, 2p^6, 3s^2, 3p^6, 3d^8, 4s^2$. As in copper, there is little difference in energy between the $3d$ and $4s$ electrons, and the metal is relatively noble and corrosion-resistant. The unit cell is FCC, and the lattice parameter a_0 is close to that of copper: $3.52\,\text{Å}$ vs. $3.62\,\text{Å}$.

In the wrought alloys the principal materials are Monel and Inconel (Table 5.5). The nickel-copper alloy Monel is an extension of the copper-nickel alloys to the high-nickel side of the phase diagram (Fig. 5.8).

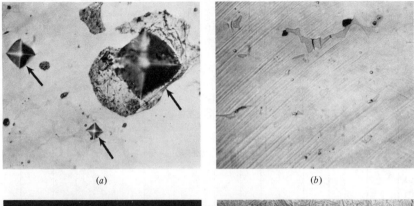

Fig. 5.7 *Typical microstructures of cast copper-base alloys. (a) 85% Cu, 5% Sn, 5% Pb, 5% Zn alloy. Lead phase (dark gray), VHN 16; matrix (white), VHN 98; δ phase (light gray), VHN 293. 500X, chromate etch. (b) 88% Cu, 8% Sn, 4% Zn. Islands of δ phase in the FCC copper matrix. The specimen was stressed after polishing. The ductile matrix shows slip, while the brittle δ phase cracked. 500X, chromate etch. (c) 89% Cu, 10% Al, 1% Fe (as cast). Gray γ_1 phase in FCC copper matrix. During stress the ductile matrix showed slip, but failure occurred through the γ_1. 500X, chromate etch. (d) Alloy of (c) heated to 1650°F for 1 hr and water-quenched, forming a martensitic structure. The specimen was not stressed. 500X, chromate etch.*

The rules of solid solubility suggest that copper and nickel should form an extensive series of solid solutions (Chap. 3), and this is the case. It is interesting to note that although nickel is stronger than copper, the intermediate nickel-copper alloy (Monel) has a higher yield strength than nickel. This is another illustration of solid solution strengthening. It shows that the element added in solid solution will harden and strengthen the solvent metal even though the solute element is soft itself. Both Monel and Inconel may be age-hardened, as shown by the alloys Duranickel 301 and Monel K500.

Table 5.5 TYPICAL PROPERTIES OF NICKEL ALLOYS

Alloy Number	Chemical Analysis, percent*	Condition	Tensile Strength,† psi × 10³	Yield Strength,† psi × 10³	Percent Elongation	Hardness	Typical Use
		Single-phase Wrought Alloys					
Nickel 200	99.5 Ni	Annealed	65	22	47	BHN 75	Corrosion-resistant parts
		Cold-worked	120	92	8	BHN 230	
Monel 400	66 Ni, 32 Cu	Annealed	72	35	42	BHN 110	Corrosion-resistant parts
		Cold-worked	120	110	8	BHN 241	
Inconel 600	78 Ni, 15 Cr, 7 Fe	Annealed	100	50	35	BHN 170	Corrosion-resistant parts
		Cold-worked	150	125	15	BHN 290	
		Polyphase Wrought Alloys					
Dura nickel 301	94 Ni, 4.5 Al, 0.5 Ti	Annealed	105	42	40	90 R_B	Corrosion-resistant parts
		Cold-worked, age-hardened	200	180	8	40 R_C	High-strength parts
Monel K500	65 Ni, 2.8 Al, 0.5 Ti, 30 Cu	Annealed	97	52	35	85 R_B	Corrosion-resistant parts
		Cold-worked, age-hardened	185	155	7	34 R_C	High-strength parts
		Single-phase Cast Alloys					
Nickel 210	95 Ni, 0.8 C	As cast	52	25	22	BHN 100	Condensers
Monel 411	64 Ni, 32 Cu, 1.5 Si	As cast	77	38	35	BHN 135	Paper mill equipment
Inconel 610	68 Ni, 15 Cr, 2 Nb, 10 Fe	As cast	82	38	20	BHN 190	Dairy equipment
		Polyphase Cast Alloys					
Monel 505	63 Ni, 29 Cu, 4 Si	Aged	127	97	3	BHN 340	Valve seats
Inconel 705	68 Ni, 9 Fe, 6 Si, 15 Cr	Aged	110	95	3	BHN 340	Exhaust manifolds

* These represent the important elements. Other elements may be present.
† Multiply psi by 6.9 × 10⁻³ to obtain MN/m² or by 7.03 × 10⁻⁴ to obtain kg/mm².
Note: Nickel-base superalloys are discussed with other superalloys in Chap. 6.

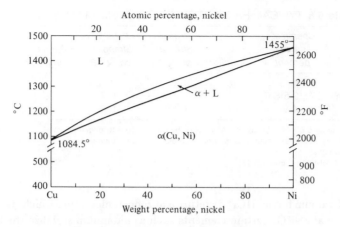

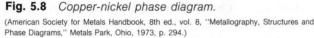

Fig. 5.8 *Copper-nickel phase diagram.*
(American Society for Metals Handbook, 8th ed., vol. 8, "Metallography, Structures and Phase Diagrams," Metals Park, Ohio, 1973, p. 294.)

5.10   Cast nickel alloys

The cast alloys are parallel to the wrought alloys with two exceptions, Monel 505 and Inconel 705 (Table 5.5). These materials contain 4 to 6 percent silicon, which produces a hard intermetallic compound Ni_3Si. This dispersion hardens the cast alloy and the material is further hardened by aging. Since the elongation is only 3 percent, these alloys are not available in wrought form.

Titanium, Zinc, and Other Alloys

5.11   Titanium alloys

Titanium alloys have been developed only recently because their great chemical reactivity makes them difficult to melt, cast into ingots, and hot-work. However, it has been found that despite their general reactivity, titanium alloys are very corrosion-resistant in certain cases, as discussed in Chap. 12. Also, the relatively low density and high strength of titanium make it a candidate for aircraft applications. This is especially important in cases where the skin heating at high speed results in loss of strength in aluminum and magnesium alloys while the titanium alloys can withstand the elevated temperatures.

The atomic structure of titanium is $1s^2$, $2s^2$, $2p^6$, $3s^2$, $3p^6$, $3d^2$, $4s^2$. In contrast to copper and nickel, titanium has only four $(3d + 4s)$ electrons, and these have a stable shell of eight beneath. Therefore, these electrons are held rather loosely, and the element is quite active. For example, there is no known material for a crucible which will not react with liquid titanium, so it is melted in a water-cooled copper crucible, thus producing a shell of solid titanium between the liquid and the copper.

Table 5.6 TYPICAL PROPERTIES OF TITANIUM ALLOYS (WROUGHT)

Material	Chemical Analysis, percent	Structure	Tensile Strength,* psi $\times 10^3$	Yield Strength,* psi $\times 10^3$	Percent Elongation	BHN
Ti	99 Ti†	α	38 to 100	22 to 85	17 to 30	115 to 220
Ti, Al, Sn	5 Al, 2.5 Sn‡	α	125	117	18	360
Ti, Al, V	6 Al, 4 V§	α-β	170	155	8	380

*Multiply psi by 6.9×10^{-3} to obtain MN/m² or by 7.03×10^{-4} to obtain kg/mm².
† Variations in properties depending on whether cold-worked or annealed.
‡ Annealed sheet.
§ Solution heat-treated and aged.

Titanium has an HCP structure α at room temperature which transforms to BCC β at 880°C. Certain elements such as aluminum stabilize the α phase giving the "all-α" alloys. Chromium, iron, molybdenum, and vanadium lower the transformation temperature, stabilizing β and leading to α-β or all-β alloys. Typical properties are given in Table 5.6.

5.12 Zinc alloys

The principal use of zinc in structures is for die castings. These zinc alloys are ideally suited for the process because of the low melting point and lack of corrosion of steel crucibles or dies. As a result, a number of automotive parts, toys, and building hardware are made of zinc alloys (Table 5.7).

5.13 The less common metals, precious metals

We are entering a period of tremendous growth in the use of the rarer metals. The era was ushered in with the discovery that in piston-driven aircraft each part is literally worth its weight in gold. In other words, the removal of one pound of dead weight makes it possible to carry one more pound of cargo each trip, and over the life of the plane this extra revenue would no doubt equal the value of a pound of gold. In recent times, with the advent of space travel, the value of a pound of weight is orders of magnitude greater. As a result, parts made of beryllium, rhenium alloys, molybdenum alloys, platinum, gold, silver, and indium are used freely in commerce. A common example is the use of expendable fine wire platinum-platinum, 10 percent rhodium thermocouples

Table 5.7 TYPICAL PROPERTIES OF ZINC ALLOYS (DIE-CAST)

Alloy Number	Chemical Analysis, percent	Nominal Tensile Strength,* psi $\times 10^3$	Percent Elongation	BHN
AG40A	4 Al, 0.05 Mg	41	10	82
AC41A	4 Al, 1 Cu, 0.05 Mg	48	7	91

*Multiply psi by 6.9×10^{-3} to obtain MN/m² or by 7.03×10^{-4} to obtain kg/mm².

to check each heat of steel. Years ago a thermocouple of this type was a carefully guarded laboratory tool. While a detailed discussion of the properties of the less common metals is beyond the scope of this text, we should mention that the following metals are readily available:

Refractory metals for high temperature service: molybdenum, tantalum, tungsten, rhenium
Metals for nuclear reactor use: zirconium, hafnium
Precious metals: gold, silver, platinum, palladium, rhodium, ruthenium, osmium, iridium
Tin and its alloys
Rare earths

Data such as the analyses and physical properties of these materials are available in the American Society for Metals handbooks and *The Materials Selector* (see References).

EXAMPLE 5.3 The discussion in the chapter suggests that most commercial alloys have compositions close to the extremities of any phase diagram. In other words, compositions such as 50%X, 50%Y are seldom used. Suggest why this is the case.

ANSWER Alloys are classified as either single-phase or polyphase. Single-phase alloys contain relatively small amounts of other elements to prevent second phases from forming, except in systems such as the copper-nickel system where there is complete solid solubility (Fig. 5.8). Here we stay near the copper end because of the higher cost of nickel unless we absolutely require a particular property of nickel, such as high-temperature resistance.
 In polyphase alloys the center portion of the phase diagram usually shows hard and brittle compounds, which, although wear- and abrasion-resistant, are difficult to maintain in a desirable distribution to maximize strength.

SUMMARY

As promised in the introductory comments, we devoted our efforts in this chapter to explaining how the properties of a wide range of nonferrous alloys can be understood in terms of their structures.

In the aluminum alloys the wrought materials can be divided into two groups: the single-phase materials, which are strengthened by work hardening, and the polyphase alloys, which are age-hardened and may also be work-hardened. The cast alloys are strengthened by age hardening or by controlling the composition to produce large amounts of a hard dispersed phase such as silicon.

The magnesium alloys are similar in many ways to the aluminum alloys, but the single-phase alloys are not as ductile because of limited slip in the hexagonal structure.

In the copper alloys great ductility is obtained in the pure metal and in the single-phase alloys. The polyphase alloys may be age-hardened or strengthened with a martensite reaction.

The nickel and other alloys follow similar reasoning.

DEFINITIONS

Nonferrous alloys   Alloys that do not have iron as the base element, such as the alloys of aluminum, copper, magnesium, nickel, zinc, and titanium.

Ferrous alloys   Alloys having iron as the base element, such as wrought iron, steel, and cast iron.

Castings   Parts that are made by pouring liquid metal into a mold of the proper dimensions. The mold may be made of sand bonded with clay or resin (sand castings) or metal (die or permanent mold castings).

Wrought products   Parts that are produced by working a solid metal either hot or cold; for example, by hot and cold forging, rolling, extrusion, spinning, stamping, or wire drawing.

Aluminum alloys   HX indicates the amount of cold work and annealing for single-phase alloys; see text. TX indicates the combination of age hardening and cold work for two-phase alloys; see text.

Copper alloys
  Brass   An alloy of copper and zinc.
  Bronze   An alloy of copper and a specified metal, such as tin bronze, aluminum bronze, silicon bronze.
  Cupronickel   An alloy of copper and 10 to 30 percent nickel.

Nickel alloys
  Monel   An alloy of nickel and 30 percent copper.
  Inconel   An alloy of nickel and 15 percent chromium.

Magnesium alloys   Described by the same H and T designations used for aluminum alloys, but with a different code for composition.

Titanium alloys   The α-β alloys are a combination of HCP and BCC structures.

PROBLEMS

5.1   Referring to Table 5.2, explain why the tensile and yield strengths of 3003 in the H18 condition are higher than those for 1060 in the same cold-worked condition. Why is the elongation much lower than in the annealed condition?

5.2 The two most important coherent precipitates in age-hardening aluminum alloys are $CuAl_2$ and Mg_2Si. The solubility of Mg_2Si varies from 1.8 percent at the eutectic temperature to less than 0.1 percent at room temperature. Explain why 5050 is listed as a single-phase alloy while 6061 can be age-hardened (Table 5.2).

5.3 Calculate the parameter called the "strength-to-weight ratio" or "specific strength" for several of the higher-strength aluminum alloys of Table 5.2, and compare these to high-strength steel with yield strength of 250,000 psi $(1.725 \times 10^3 \, MN/m^2)$. The value is equal to yield strength divided by density (in pounds per cubic inch). What is the significance of this value? Take 2.7 as the specific gravity of aluminum and 7.8 as that of steel. Density = specific gravity $\times$ 0.0361 lb/in.^{3}.

5.4 At room temperature the slip plane for magnesium is (0001), while at elevated temperatures the $(11\bar{2}2)$ and $(10\bar{1}1)$ planes are operative in the unit cell. Sketch these planes in the HCP cell.

5.5 Considering only the silicon, calculate the amount of silicon-rich β produced at freezing in aluminum 356. Why is this alloy not available in wrought form? (Assume magnesium has no effect on the aluminum-silicon equilibrium diagram.)

5.6 The density of magnesium is 0.064 lb/in.^{3}. What yield strength would be required to enable a magnesium alloy to compete with the best aluminum alloy on a strength-to-weight basis? (See Prob. 5.3.)

5.7 The engine block of the Chevrolet Vega contains 16 percent silicon in aluminum. What is the percentage of area of the cylinder wall containing the β silicon phase? The specific gravity of silicon is 2.33 and that of aluminum is 2.70.

5.8 Copper-nickel alloys are used above 330°C, but copper alloys containing lead are not. Explain this, using phase diagrams.

5.9 Construct fraction charts for the phases you would expect for alloys of copper containing 20 percent lead and 40 percent lead slowly cooled. Why is the lead content of commercial castings always below about 30 percent?

5.10 The range of tensile strength for a 1 percent beryllium-copper alloy is 125,000 to 160,000 psi $(0.863 \text{ to } 1.105 \times 10^3 \, MN/m^2)$ but 190,000 to 215,000 psi $(1.31 \text{ to } 1.48 \times 10^3 \, MN/m^2)$ for a 2 percent beryllium-copper alloy. Explain this quantitatively on the basis of the phases present.

5.11 Why is the yield strength of wrought Inconel greater than that of wrought Monel 400 although the alloy content is lower?

5.12 Calculate the weight of magnesium in a cubic mile of seawater. (There are 1.27 g of magnesium per kilogram of typical seawater; the specific gravity of seawater is 1.01.)

1μ

6

STEEL, SUPERALLOYS, CAST IRON, DUCTILE IRON, MALLEABLE IRON

IN this chapter we will take up the second principal group of metallic materials, those based on the element iron and a few related high-nickel alloys called the "superalloys." Only a few structures are involved, the two crystal forms of iron BCC and FCC that we discussed previously and a hard compound, iron carbide (Fe_3C). By altering the amount, shape, and distribution of the carbide, we can obtain a wide range of properties.

The electron micrograph shows another, special type of precipitation in a nickel-base superalloy. The matrix is FCC, but the precipitate is $Ni_3(Al,Ti)$. This gives an exceptional combination of high-temperature strength and oxidation resistance required for gas turbine blades.

The large photomicrograph shows Ni_3Al (γ') precipitate in a nickel alloy (γ) matrix (approximately 10,000X). The small insert shows dislocations cutting through the γ' precipitate (approximately 50,000X).

6.1 Introduction to iron alloys

Iron and steel are so widely used and inexpensive that at first they seem without glamour compared to the colorful copper alloys and the light alloys with unusual aircraft and space travel applications. After a little study we shall see, however, that some of the most advanced materials—such as special stainless steels for corrosion resistance, alloys for high-temperature service in gas turbines, and ductile iron for castings—are all based on the metal iron. The widest range of microstructures and properties is found in the family of iron alloys ranging from high-strength steels to the lower-strength but elegantly castable gray cast iron.

Over 90 percent of the tonnage (1973) of metallic materials is based on the element iron. In the family of iron-base alloys the principal difference is in carbon content.

Material	Carbon, percent	Annual tonnage, in millions
Wrought iron	0 to 0.05	1
Steel (wrought and cast)	0.05 to 2	120
White cast iron	2 to 4.5	1
Malleable iron	2.5 to 3	1
Gray iron	3 to 4	16
Ductile (nodular) iron	3.5 to 4.5	2

To illustrate the importance of iron alloys, let us consider the makeup of a typical automobile:

Body (doors, sides, frame, fenders): low-carbon steel
Chassis: welded steel
Engine: 95 percent cast and ductile iron
Gears, drive shaft, axle, torsion bar, springs: alloy steel
Brake disks or drums: cast iron
Wheels: steel
Axle housings, differential housing: malleable iron or ductile iron
Bumpers: steel

We see, therefore, that most parts that bear loads or transmit power are iron alloys. We will discuss the family of steels first, then the other alloys. Then after analyzing the high-alloy steels, we will take up other heat-resistant alloys including the nickel- and cobalt-base superalloys. Following this, we will consider cast iron, malleable iron, and ductile iron.

6.2 The iron–iron carbide diagram and its phases

To understand the basic differences among iron alloys and the control of properties, the iron–iron carbide diagram is essential. The diagram that is commonly used is shown in Fig. 6.1. We note first that the carbon scale only goes up to 6.67 percent carbon, where we encounter iron carbide, Fe_3C. This should cause us no concern, since we have already used a part of a diagram, the portion of the copper-aluminum diagram extending to $CuAl_2$ (θ), in our discussion of age hardening.

The first step is to examine the individual phases. Beginning with pure iron, we see three solid phases, alpha (α), gamma (γ), and delta (δ), shown at the left side. If we begin with α at 68°F (20°C) and heat slowly, we can expect α to tranform to γ at 1670°F (910°C), γ to δ at 2540°F (1393°C), and δ finally to melt at 2802°F (1538°C).

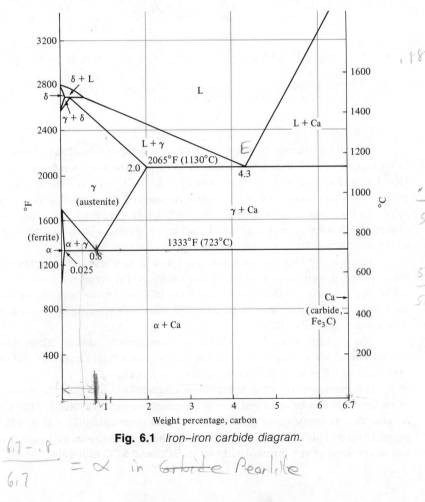

Fig. 6.1 *Iron–iron carbide diagram.*

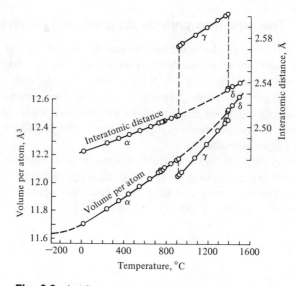

Fig. 6.2 *Lattice parameter of iron as a function of temperature.*

(C. S. Barrett and T. B. Massalski, "Structure of Metals," 3d ed., McGraw-Hill Book Company, New York, 1966, Fig. 10.5, p. 232.)

We will look in vain for β iron; it was lost in antiquity. Naturally, because of its importance the iron–iron carbide diagram was the first to receive attention. Early investigators noted that near 1400°F (760°C) iron lost its ferromagnetism. This was considered to be a phase change and the symbol β was given to iron in the range 1400 to 1670°F (760 to 910°C). It was found, however, that this magnetic effect is not a phase change but is due to a shift in alignment of the atoms, a point to be considered further in Chap. 14.

This left metallurgists with α, γ, and δ. The picture was further simplified when α and δ were found to be of the same crystal structure from x-ray diffraction evidence. Both these phases have BCC structures and the lattice parameter a of δ is the same as that for α if allowance is made for expansion with temperature (Fig. 6.2).

The γ phase is FCC and therefore more densely packed. There is a contraction of about 1 percent in volume in the $\alpha \rightarrow \gamma$ transformation and an expansion of 0.5 percent in volume in the $\gamma \rightarrow \delta$ change.

It is important to notice in the phase diagram (Fig. 6.1) that the amount of carbon that can be dissolved in γ is 2 percent maximum (2065°F, 1130°C), a value that is many times greater than the maximum solubility in α, 0.025 percent carbon (1333°F, 723°C). The basis for this may be shown by a comparison of the sizes of the interstitial holes in BCC and FCC unit cells.

EXAMPLE 6.1 Compare the sizes of the largest interstitial holes that can be placed in the BCC and FCC structures found in iron. How does the size of the carbon atom compare? The atomic radius of iron is 1.27 Å in FCC and 1.24 Å in BCC.

ANSWER A horizontal plane through the center of an FCC structure contains the largest interstitial hole (the same hole can be found in the face of the FCC).

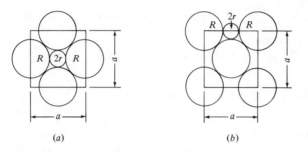

(a)   (b)

$$a = \frac{4R}{\sqrt{2}} = 2R + 2r$$

or $r = 0.414R = 0.414 \times 1.27 = 0.52$ Å

In the BCC structure the largest interstitial hole occurs at the coordinates $\frac{1}{2}$, 0, $\frac{3}{4}$.

$$(r + R)^2 = (\tfrac{1}{2}a)^2 + (\tfrac{1}{4}a)^2$$

$$r^2 + 2rR + R^2 = \tfrac{5}{16}a^2 = \frac{5}{16}\left(\frac{4R}{\sqrt{3}}\right)^2$$

$$r^2 + 2rR + R^2 = \frac{5R^2}{3}$$

$$r + R = R\sqrt{\tfrac{5}{3}}$$

or $r = 0.291R = 0.291 \times 1.24 = 0.36$ Å

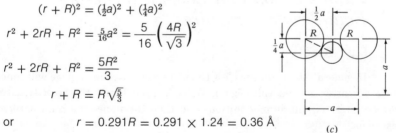

(c)

Since the atomic radius of carbon is 0.77 Å, it has a much better chance of interstitial solution in FCC iron. (Note that the radius of the iron atom is slightly different in the BCC and FCC configurations.)

In addition to the solid solutions of iron and carbon, the other solid phase in the diagram is iron carbide, Fe_3C, also called "cementite." This curious name arose from an ancient process in which carbon was diffused into iron by making up a package of layers of iron and carbon compounds and then heating in a furnace. The phase cementite resulted from this so-called "cementation" process. It is evident from the diagram that cementite will be encountered at 20°C in any alloy from 0.006 to 6.67 percent carbon. In sharp contrast to the BCC and FCC iron structures which are both ductile (over 40 percent elongation) and soft (BHN 100 to 150), cementite is brittle (0 percent plastic elongation) and hard [BHN over 700 (VHN 1200)].

We can summarize the phases and their several names as follows:

Temperature range, pure Fe	Phase	Crystal structure	a, Å	BHN	Percent elongation
Room temperature to 1670°F (910°C)	α, ferrite	BCC	2.86	~150	~40
1670 to 2540°F (910 to 1393°C)	γ, austenite	FCC	3.60	~150	~40
2540 to 2802°F (1393 to 1538°C)	δ, delta ferrite	BCC	2.89	~150	~40
	Iron carbide (cementite, Fe_3C)	orthorhombic	—	~700	~0

Plain Carbon and Low-alloy Steels: Equilibrium Structures

6.3 Steel

In general, steel contains less than 2 percent carbon, and we will concentrate our attention on this region. We will discuss first the steel structures formed under equilibrium conditions and then the nonequilibrium hardening reactions.

6.4 Hypoeutectoid steels

We will consider first the hypoeutectoid steels (up to 0.8 percent carbon) (Fig. 6.3), then the hypereutectoid steels (0.8 to 2 percent carbon).† The low-carbon steels are by far the most important of the group, mainly because of their high ductility, both hot and cold, which enables them to be readily

---

†Hypoeutectoid means below eutectoid composition, that is, <0.8 percent carbon. Hypereutectoid means above eutectoid composition, that is, >0.8 percent carbon.

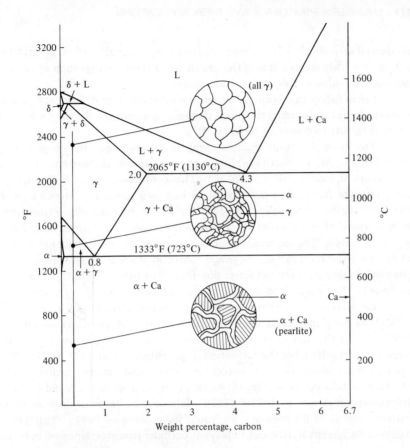

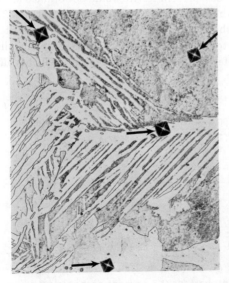

Fig. 6.3 *Changes in micro-structure of a hypoeutectoid steel upon cooling. The photograph is an actual photomicrograph, 500X, 2 percent nital etch. Ferrite (white), VHN 171. Pearlite (gray), VHN 215.*

fabricated into shapes of excellent toughness and strength. All structural and automobile body steels fall into this group. Most steel castings such as railroad car parts are also in this composition range.

Let us follow the cooling of a typical steel of this type (0.3 percent carbon) from the liquid state, using the phase diagram as a map and drawing a fraction chart (Fig. 6.4) as discussed in Chap. 4.

The most important changes occur during cooling in the range 1600 to 1300°F (871 to 704°C). Ferrite (α) begins to precipitate at the austenite (γ) grain boundaries at 1560°F (849°C) (Fig. 6.3). This continues to 1334°F (724°C) and the γ changes from 0.3 to 0.8 percent carbon as a result (tie line changes in the $\alpha + \gamma$ region upon cooling). At 1333°F (723°C) the remaining γ transforms to a mixture of $\alpha +$ carbide by a constant temperature eutectoid reaction, $\gamma \rightarrow \alpha +$ carbide. The important point is that an interleaved or lamellar mixture of ductile α and hard carbide is formed. This composite is called "pearlite" and possesses higher strength yet lower ductility than ferrite. Note that the name is given to the unique *lamellar mixture* of the two phases $\alpha +$ carbide.

Let us return to our discussion of the fraction chart (Fig. 6.4). At 1334°F (724°C) we have approximately 63 percent ferrite and 37 percent austenite. We shall refer to this ferrite as primary ferrite so as not to confuse it with the ferrite that results from the formation of pearlite. Now at 1333°F (723°C) all the remaining austenite (of 0.8 percent carbon) transforms to pearlite, and we therefore obtain eutectoid ferrite. As an example, if we had started out with 100 percent austenite at 1333°F (723°C) (this would have required an initial carbon content of 0.8 percent), we would have obtained (6.67 − 0.8)/(6.67 − 0.025) or 88 percent ferrite and 12 percent carbide. However, since we only had 37 percent austenite, 0.88 × 37 = 32.5 percent ferrite would be obtained on slow cooling. The *total* ferrite would therefore be 95.5 percent (63 + 32.5) at 1332°F (722°C). The fact that Fig. 6.4 shows no change in the amounts of ferrite or carbide when cooled from 1332 to 68°F (722 to 20°C) indicates that a slight decrease in carbon solubility in ferrite (0.025 to <0.01) has a negligible effect on the calculation.

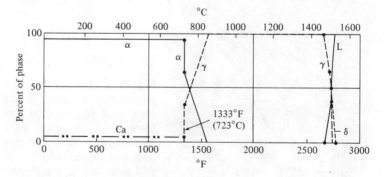

Fig. 6.4 *Phase fraction chart for 0.3 percent carbon steel.*

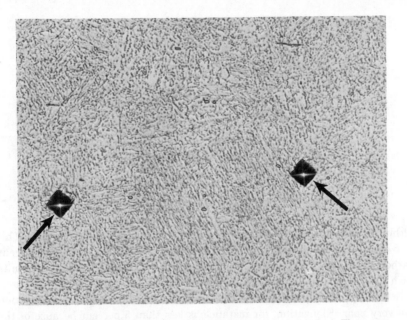

Fig. 6.5 *Spheroidized structure in 0.5 percent carbon steel. There are very fine spheroids at iron carbide in the α matrix, VHN 170. 500X, 2 percent nital etch.*

It should be emphasized that in order to achieve equilibrium, slow cooling must be used. A more rapid cooling such as air cooling would result in a finer pearlite and is called "normalizing." A slow cooling such as furnace cooling through the eutectoid is called "process annealing" and results in a slightly lower strength yet higher ductility than obtained in normalizing. Still another variation in grain size and pearlite can be obtained by reheating or holding for extended periods just below the eutectoid (approximately 1300°F, 704°C). The carbides in the pearlite change from the normal plates to spheres (Fig. 6.5). This microstructure is called "spheroidite" or "spheroidized pearlite" and has lower hardness but higher ductility and toughness.

6.5 Hypereutectoid steels

High-carbon steels naturally contain more of the hard carbide phase and are useful where higher strength, hardness, and wear resistance are needed, as in a knife blade, other cutting tools, and bearings. Let us follow the cooling of a 1 percent carbon steel on the phase diagram and plot the results as a fraction chart (Fig. 6.6).

The important temperature range for this material is 1500 to 1300°F (815 to 704°C). It is vital to observe that a brittle phase, iron carbide, precipitates at the austenitic grain boundaries from 1450 to 1333°F (788 to 723°C) (Fig. 6.7),

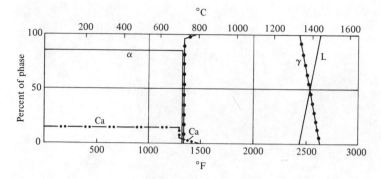

Fig. 6.6 *Phase fraction chart for 1 percent carbon steel.*

although this primary carbide amounts to only approximately 3.5 percent. This is in contrast to the ductile primary ferrite which precipitated in the 0.3 percent carbon steel. The eutectoid reaction is exactly the same in both cases; austenite with 0.8 percent carbon in solid solution forms the α + carbide in pearlite.

The properties of this high-carbon material in the slow-cooled condition are very poor. Elongation, for instance, is less than 5 percent because of the brittle carbide network. However, if we heat to 1500°F (815°C) to dissolve the carbide in the austenite, cool rapidly to below 1300°F (704°C) to allow transformation to a fine mixture of α plus carbide, and then reheat to 1300°F (704°C), the coarse-grain boundary carbide does not have time to form on cooling and a better carbide shape and distribution are obtained with higher ductility. This material is called "spheroidite." In general, the toughness of these hypereutectoid steels is still lower than that of the hypoeutectoid type because of the greater amount of carbide.

6.6 Specifications

We can now discuss a few specifications and uses of these materials (Table 6.1). Wrought iron and ingot iron both show maximum ductility. Wrought iron is essentially pure iron with slag fibers rolled into the structure. The use of these materials in pipe and architecture is well known, although present-day wrought iron is usually a low-carbon steel. Next we have the so-called "plain carbon" steels, as distinguished from alloy steels. All contain about 0.5 percent manganese, however. The code symbol for these steels is 10xx, where xx designates the percentage of carbon; for example, 1020 steel contains approximately 0.20 percent carbon. As we increase the carbon, the percentage of pearlite rises and the strength increases from 44,000 psi (299 MN/m^2) at 0.02 percent carbon to 112,000 psi (774 MN/m^2) at 0.8 percent carbon. The cold-drawn material shows still higher strength and hardness because of work hardening. However, above 0.5 percent carbon it is necessary to spheroidize the steel prior to cold

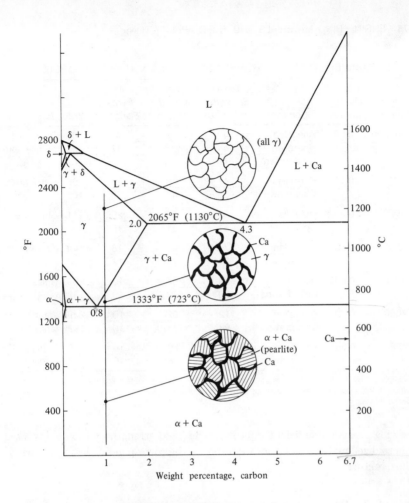

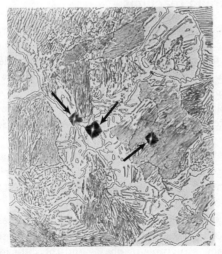

Fig. 6.7 *Changes in microstructure of a hypereutectoid steel upon cooling. The photograph is an actual photomicrograph, 500X, 2 percent nital etch. Pearlite (gray), VHN 235. Some ferrite, VHN 160, precipitated next to carbide (white).*

Table 6.1 TYPICAL PROPERTIES OF PLAIN CARBON STEELS

Steel Number	Carbon Content, percent	Condition	Tensile Strength,* psi × 10³	Yield Strength,* psi × 10³	Percent Elongation	BHN	Typical Use
Ingot iron†	0.02	Annealed	42	19	48	69	Pipe,
		Hot-rolled	44	23	47	83	architecture
		Cold-drawn	73	69	12	142	
1010	0.10	Hot-rolled	47	26	28	95	Car fenders
		Cold-drawn	53	44	20	105	
1020	0.20	Hot-rolled	55	30	25	111	Structures
		Cold-drawn	61	51	15	121	
1040	0.40	Hot-rolled	76	42	18	149	Crankshaft
		Cold-drawn	85	71	12	170	
1060	0.60	Hot-rolled	98	54	12	201	Chisel
		Cold-drawn‡	90	70	10	183	
1080	0.80	Hot-rolled	112	62	10	229	Wear-resis-
		Cold-drawn‡	98	75	10	192	tant parts
1095	0.95	Hot-rolled	120	66	10	248	Cutting
		Cold-drawn‡	99	76	10	197	blades

* Multiply psi by 6.9×10^{-3} to obtain MN/m^2 or by 7.03×10^{-4} to obtain kg/mm^2.
† Wrought iron has mechanical properties similar to ingot iron.
‡ Spheroidized, then cold-drawn.

drawing to attain sufficient ductility. The yield strength of the cold-worked spheroidized material is higher than that of the corresponding hot-rolled pearlitic structure.

6.7 Steel—nonequilibrium reactions

If we were limited to the equilibrium structures and the plain carbon steels of the iron–iron carbide diagram, a great many critical tools and components could not be made. Imagine our problems without hardened drills, files, lathe tools, gears, chisels, plows, ball bearings, rolls, razor blades, knives, saws, dies, and hundreds of other parts in which we need hardness and strength! As an example of the hardening operation, let us consider the changes in properties in SAE 1080 steel which can be accomplished by nonequilibrium cooling. See the table on page 179.

We must emphasize immediately that it is not the *act* of quenching but the *effect* of quenching on the structure that produces the change. For example, if we performed the same treatment on a piece of pure iron or on an 18% Cr, 8% Ni stainless steel, there would be no change in hardness.

Before proceeding with our discussion of how this change in hardness takes place, we should recall that all reactions take time because of the requirements for the nucleation of new structures or phases and the diffusion

Condition	Tensile strength,† psi $\times 10^3$	Yield strength,† psi $\times 10^3$	Percent elongation	BHN
Slow-cooled from 1550°F (843°C)	112	62	10	192
Water-quenched from 1550°F (843°C)	>200	200	1	680

of atoms to allow growth. We will see in the next section that the rate of transformation of austenite to ferrite and carbide, as in pearlite, is a result of nucleation and growth.

6.8 Austenite transformation

The key to understanding the variation in hardness is a knowledge of the austenite transformation. Let us consider first the simplest reactions, those that can occur with eutectoid (0.8 percent carbon) austenite. Later we will take up the effects of analysis. To this point we have discussed only the transformation of austenite under equilibrium conditions to a relatively coarse platelike mixture of carbide and ferrite called pearlite.

What if we avoid the reaction at 1333°F (723°C) by cooling the austenite rapidly from 1500 to 1200 or 800°F (815 to 649 or 427°C) or lower and allowing it to react at these temperatures?

Let us perform an experiment to determine these effects. First we machine a thin specimen of 0.8 percent carbon steel with two holes which serve as reference points for measurement (Fig. 6.8). Second we hang the specimen from a hook *A* on a tube of fused silica. The tube itself is rigidly held from platform *B*. Next we place the hook of an inner tube into the lower specimen hole *C*. The specimen is now in light tension, supporting the inner tube which slides smoothly in the outer tube. Finally we fasten a dial gage onto the top of the inner tube with the point bearing on the support of the outer tube.

This instrument is called a "dilatometer" because it measures change in length or dilation of the specimen. Whether the specimen expands or contracts, the movement is shown faithfully by the dial. Also, the use of fused silica, a very low-expansion material, reduces any errors that might be caused by expansion of the tubes.

Next let us change the structure of the specimen to *austenite* by heating to 1600°F (871°C). This is done simply by immersing the lower part of the dilatometer with the specimen in a pot of liquid lead at 1600°F (871°C). The fused silica is inert and has excellent thermal shock resistance.

---

†Multiply by 6.9×10^{-3} to obtain MN/m^2 or by 7.03×10^{-4} to obtain kg/mm^2.

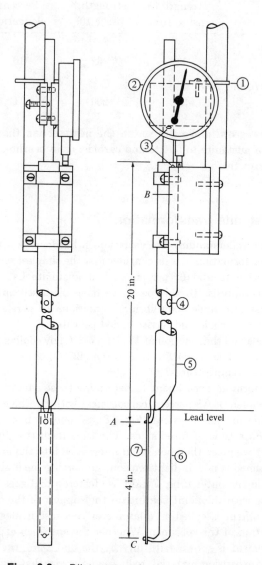

Fig. 6.8 *Dilatometer for measuring transformations. 1 = dial guide, 2 = dial gage, 3 = quartz lugs, 4 = inner-tube guides (quartz, two sets), 5 = outer quartz tube, 6 = inner quartz tube, 7 = specimen ($4\frac{1}{2}$ by $\frac{1}{2}$ by $\frac{1}{32}$ in.).*

6.9 Pearlite formation

To observe transformation at 1300°F (704°C), we quickly shift another pot of lead at 1300°F (704°C) into the place of the 1600°F (871°C) pot. We record the changes in length of the specimen as time passes and obtain the graph shown in Fig. 6.9a.

By interpreting the graph we find the microstructural changes in the specimen. The first stage (1) is a contraction as the austenite cools. The magnitude of this contraction corresponds exactly with the value expected using the coefficient of expansion (or contraction) of austenite. The specimen is at constant temperature during period (2), and since no change in microstructure is taking place, no length change occurs. At a later time (3) we notice that the length of the specimen is increasing although it is in a bath at constant temperature. This must be due to the change in structure, austenite → pearlite.

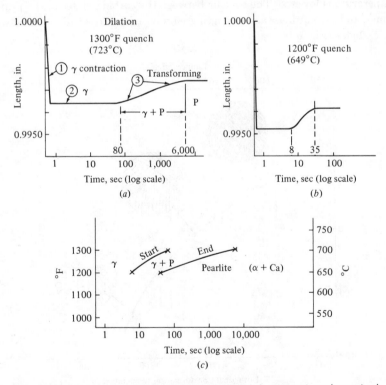

Fig. 6.9 (a) Length change on quenching 0.8 percent carbon steel to 1300°F (723°C). (b) Length change on quenching 0.8 percent carbon steel to 1200°F (649°C). (c) TTT curve [derived from data from (a) and (b)].

The length change should not surprise us if we recall the data showing the volume change when FCC iron transforms to BCC. The graph tells us, therefore, the time at which transformation begins and ends at this temperature for this steel: 80 and 6,000 sec. After the transformation there is no change in dimensions as long as the temperature remains constant.

    If we repeat the experiment using a temperature of 1200°F (649°C) for transformation, we obtain different transformation times: 8 and 35 sec (Fig. 6.9b).

    We now make a graph showing the time to beginning and end of transformation as a function of temperature. A logarithmic scale is used to accommodate the wide variation in time (Fig. 6.9c). This graph can be called either an isothermal transformation curve or a TTT (time-temperature-transformation) curve.

    If we measure the hardness of the steel at room temperature after transformation, we find that the hardness increases as the transformation temperature is lowered. The spacing between the carbide plates in the pearlite is finer, and the hardness and strength are proportional to the distance between the carbides (Fig. 6.10).

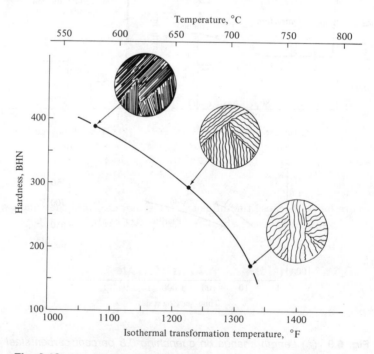

Fig. 6.10 *Hardness of 100 percent pearlite (0.80 percent carbon) as related to the isothermal transformation temperature. Lower temperatures result in finer pearlite.*

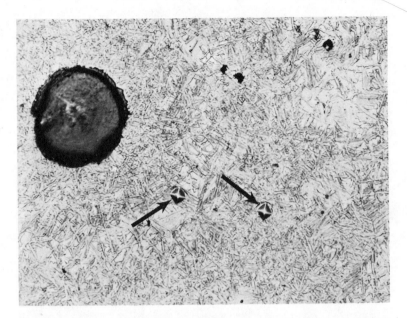

Fig. 6.11 *Bainite produced by isothermal transformation at 600°F (316°C), VHN 400, from 0.7% C austenite. 500X, 2 percent nital etch. The large spherical particle is graphite (see Sec. 6.36).*

6.10 Bainite formation

Below 1000°F (538°C) the transformed structure changes in appearance from the alternating plates of pearlite to a feathery or acicular structure called "bainite" (Fig. 6.11). The hardness continues to increase because the carbide is becoming increasingly finer, and as a result the distance over which slip can take place in the ferrite is becoming shorter. The transformation start and end times are longer because the diffusion rate is slower. From these data we can construct more of the isothermal transformation curve (Fig. 6.12).

6.11 Transformation to martensite

Below 420°F (215°C) we encounter a type of transformation quite different from the isothermal transformations we have just discussed and the structure is different (Fig. 6.13). As we decrease the temperature below 420°F (215°C), a fraction of the new structure, called "martensite," forms *instantaneously* with each decrease in temperature. If we halt the cooling and hold the sample at say 300°F (149°C), no further martensite is produced until we resume cooling. The sample is completely transformed when we reach 180°F (82°C). We call

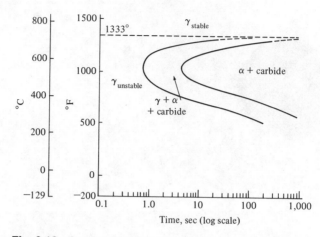

Fig. 6.12 *Construction of isothermal transformation curve (1080 steel).*

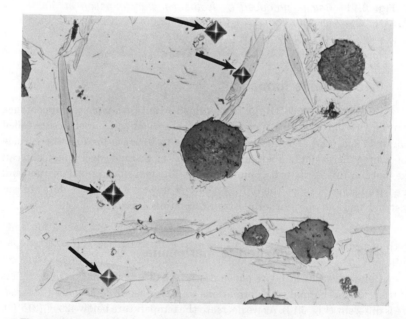

Fig. 6.13 *Martensite produced in high-carbon, high-alloy matrix, VHN 470. The matrix is austenite, VHN 205. The spheres are graphite (see the section on ductile iron). 500X, 2 percent nital etch.*

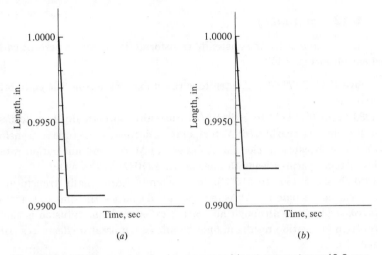

Fig. 6.14 *Dilation curves for lower quenching temperatures (0.8 percent carbon, 1080 steel). (a) Quenched to 420°F (216°C), M_s. (b) Quenched to 300°F (149°C), γ + martensite.*

the temperature at which transformation begins the "martensite start temperature, M_s," and the temperature at which it ends the "martensite finish temperature, M_f."

This effect is shown clearly by the dilatometer curves (Fig. 6.14). If we quench to just above the M_s, we obtain a normal transformation curve to bainite. If we quench to below the M_s, the decrease in length *on quenching* is less than it would be if we had quenched to 420°F (215°C). Therefore, some expansion due to transformation took place *during* cooling. We may now complete our isothermal transformation curve for 1080 steel to include the M_s and M_f, as shown in Fig. 6.15.

The importance of martensite is that it is the hardest structure formed from austenite. X-ray diffraction evidence indicates that the structure is a distorted BCC (tetragonal) and that the distortion is caused by the atoms of carbon which are trapped, as shown in Fig. 6.16.

A mechanism that explains the difference between the nucleation and growth of pearlite and bainite and the rapid transformation to martensite is as follows. In the case of pearlite and bainite, nucleation of the two new phases ferrite and carbide has to occur, and diffusion and long-range movement of atoms take place. In the case of martensite the structure can be formed by short shearing movements. In Fig. 6.17 we see the similarity between martensite and the parent austenite. The end product can be formed by movements that result in an expansion of the a direction in the tetragonal cell, which is sketched within the FCC lattice of the austenite. Note that the carbon atoms are trapped along the sides, so that the c axis is finally longer than a. At the maximum carbon level the c/a ratio is 1.08, as shown in Fig. 6.16.

6.12 Summary

Let us summarize the austenite transformation of an 0.8 percent carbon steel shown in Fig. 6.15.

1. Above 1333°C (723°C): Austenite is the stable phase from the equilibrium diagram.
2. 1333 to 1050°F (723 to 566°C): (a) Austenite isothermally transforms to α + carbide as pearlite. (b) As temperature decreases, the time to transform *decreases* because nucleation is easier. (c) More rapid nucleation rate at lower temperatures leads to finer pearlite (BHN 170 to 400).
3. 1050 to 420°F (566 to 215°C): (a) Austenite isothermally transforms to α + carbide as bainite. (b) As temperature decreases, the time to transform *increases* because although nucleation rates are high, diffusion is slower. (c) High nucleation results in finer bainite as temperature decreases (BHN 400 to 580).
4. 420 to 180°F (215 to 82°C), M_s to M_f: Austenite transforms to martensite upon cooling. The hardness of martensite is BHN 680.

It is especially important to note that once the austenite has transformed to one of the above structures, pearlite for example, it cannot be transformed to another unless the sample is reheated to the equilibrium austenite temperature [above 1333°F (723°C)]. It is possible, however, for a sample to transform to a mixture of transformation products by spending a short interval in the pearlite and bainite transformation ranges and then cooling through the martensite range.

EXAMPLE 6.2 Draw schematic time-temperature diagrams to obtain the following microstructures in 1080 steel. Indicate the microstructures on the diagram at each stage of the treatment.

(a) Coarse pearlite
(b) Fifty percent pearlite and 50 percent bainite
(c) Eighty percent martensite and 20 percent pearlite

ANSWER

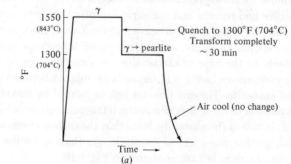

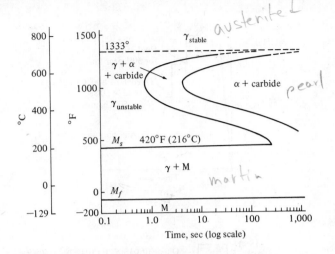

Fig. 6.15 *Isothermal transformation curve and austenite-martensite transformation (0.8 percent carbon 1080 steel).*

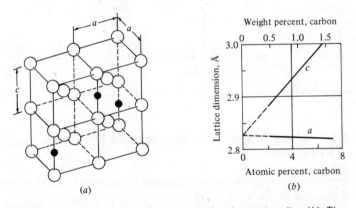

Fig. 6.16 *(a) Location of carbon atoms in martensite. (b) The carbon atoms expand the lattice, giving rise to strain and higher hardness.*

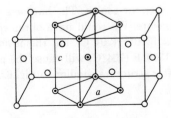

Fig. 6.17 *Relation of unit cell of martensite to parent austenite. Two unit cells of austenite (FCC) and the resultant BCT lattice of martensite are shown.*

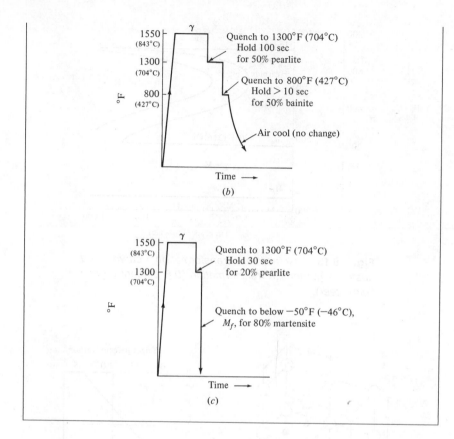

(b)

(c)

6.13 Uses of the time-temperature-transformation (TTT) curve

Now that we have discussed austenite transformation, let us illustrate some of the uses of the TTT curve.

Suppose we wish to machine a knife blade from a piece of 1080 steel and then harden it. Let us say further that we find that the hardness of the piece of steel which we have on hand is BHN 370 because of its previous history. This is too hard to machine conveniently. We therefore heat the steel to the austenite range, 1450°F (788°C), cool it to 1250°F (677°C) in the furnace, allow it to transform in the furnace for 1 hr, and then air-cool it. The hardness is BHN 250 because of the resultant *coarse* pearlite. We could have obtained a still softer pearlite by using a longer time at 1300°F (704°C), but this is adequate. Also, holding a long period of time at 1300°F (704°C) will cause the plates of carbide to spheroidize, giving a still softer structure, similar to Fig. 6.5 except at a higher carbon level.

Now we machine the blade and place it in the furnace at 1450°F (788°C) again. (It is usual to hang the part on a wire or use some support to prevent sagging of thin sections at the high temperature where the steel is soft.) After austenite is obtained (in less than $\frac{1}{2}$ hr), the blade is quenched in rapidly circulating oil at 68°F (20°C). The quenching must be sufficiently rapid to avoid transformation to either pearlite or bainite so that martensite is produced.

6.14 Tempering

If we test the knife blade, we find it is at the hardness we wanted, BHN 680 or 64 R_C. However, at the same time it is quite brittle and will break with slight bending, and the edge will chip with impact. We recall that martensite is a highly stressed supersaturated solid solution of carbon in a distorted ferrite. Therefore, we heat just enough to cause the body-centered tetragonal structure to collapse to the BCC and to allow the carbon atoms to migrate and form some very fine iron carbide crystals. Admittedly the hardness may decrease slightly, but the ductility should improve.

We find the relations shown in Fig. 6.18. The martensite transforms to ferrite plus find carbide; and the higher the tempering temperature, the coarser the carbide, the lower the hardness, and the greater the ductility.† Furthermore, the sample may be quenched from the tempering temperature without changing the effect on hardness.

There are many examples in which a balance between hardness and ductility is desired. After quenching a typical claw hammer head, the striking face is tempered to 60 R_C but the claws are tempered to 48 R_C because greater toughness is needed in this region. This is accomplished by selective heating with a flame or induction coil.

---

† In some steels certain tempering temperatures are avoided because of secondary reactions in the structure which cause temper embrittlement.

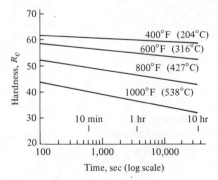

Fig. 6.18 *Relation of hardness of tempered martensite to tempering temperature and time. Since tempering is a diffusion phenomenon, the time-temperature dependence might be anticipated (1080 steel).*

(L. H. Van Vlack, "Elements of Materials Science," 2d ed., Addison-Wesley Publishing Company, Inc., Reading, Mass., 1964, Fig. 10.29, p. 294.)

EXAMPLE 6.3 What are the different microstructures that are mixtures of ferrite and carbide, and how are they obtained?

ANSWER

1. *Pearlite:* normal equilibrium cooling of austenite or isothermal holding of austenite above the nose of a TTT diagram.
2. *Bainite:* isothermal holding of austenite below the nose of a TTT diagram. (Bainite may also be obtained by cooling if two noses are present, with the bainite nose out ahead of the pearlite nose. See Sec. 6.18 and Fig. 6.29.)
3. *Spheroidite:* spheroidization of pearlite by extended holding at approximately 1300°F (704°C) or lower.†
4. *Tempered martensite:* reheating of the distorted body-centered tetragonal martensite to precipitate the carbon as carbide and allow the crystal structure to return to the BCC ferrite plus carbide.

6.15 Marquenching

One of the hazards of quenching is the possibility of distorting and cracking the part. Let us consider these effects in a simple knife blade (Fig. 6.19). To simplify the discussion, let us assume that the thin portion A of the blade will cool first and transform to martensite while the thicker section B remains austenite (above M_s), at say 500°F (260°C). In passing through the martensite transformation, B will expand, causing the knife to bow severely. Even in a round bar the same effect will be found (Fig. 6.20).

Let us consider that the region A will cool before B and transform to martensite. Portion B is still austenite, soft and hot, and follows all these

---

†Bainite and martensite can also be spheroidized.

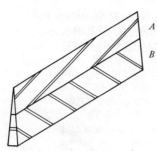

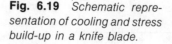

Fig. 6.19 *Schematic representation of cooling and stress build-up in a knife blade.*

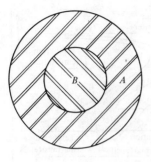

Fig. 6.20 *Schematic representation of cooling and stress build-up in a round bar.*

changes in length. Later it is B's turn to transform. The expansion cracks the surface layer A because it is brittle.

The solution to both these problems is to have the austenite-martensite transformation take place in A and B at the same time. This is done by *marquenching* as follows:

1. Austenitize (heat at 1450°F, 788°C).
2. Quench in hot oil, liquid metal, or molten salt at just *above* the M_s. Equalize the temperature throughout the part.
3. Air-cool through the M_s-to-M_f range.

The quenching in liquid is necessary to avoid pearlite or bainite formation. Then the part is removed from the hot bath *before* bainite begins to form and is allowed to transform throughout to martensite. Tempering is still used *after* cooling to below the M_f (Fig. 6.21).

6.16 Austempering

This is another process for avoiding distortion and cracking. Here the part is simply transformed to bainite at the desired level of the bainite temperature range. The part is first austenitized, quenched in a salt bath above the M_s, allowed to transform, and then air-cooled. Naturally the as-quenched hardness is lower than if martensite were formed. However, since martensite is usually tempered to a lower hardness level, the results are comparable. It is often not practical to use austempering with some alloy steels because of the long time required to form bainite, as discussed later. On the other hand, the advantage of austempering is that the extra tempering step is avoided and in general the distortion is less. We can compare conventional quenching and tempering, marquenching, and austempering with reference to the TTT curves, as shown in Fig. 6.21.

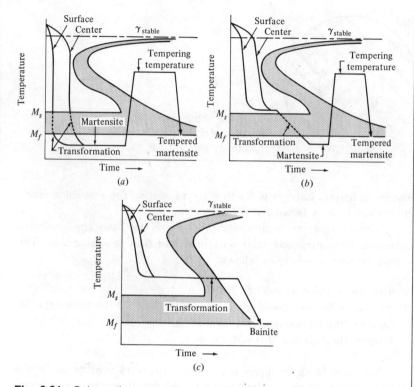

Fig. 6.21 *Schematic comparison of different quenching methods, with cooling curves superimposed on TTT diagrams. (a) Conventional process. (b) Martempering. (c) Austempering.*

(American Society for Metals Handbook, 8th ed., vol. 2: "Heat Treating, Cleaning and Finishing," Metals Park, Ohio, 1964.)

6.17 Effects of carbon on austenite transformation and transformation products

Up to this point we have considered only the simple eutectoid steel 1080. The heat treatment of hypoeutectoid and hypereutectoid steels is also important, and the use of alloys permits the successful hardening of many complex designs that could not be produced otherwise.

To attain maximum hardness in a hypoeutectoid steel, the first step is to austenitize completely. Therefore, the austenitizing temperature for 1040 steel is 150°F (83°C) above that for 1080 steel (Fig. 6.22). The steel may then be hardened by quenching to martensite at rapid cooling rates.

If the steel is cooled slowly from 1500°F (815°C), ferrite will form from 1480 to 1333°F (804 to 723°C), and the austenite will change from 0.4 to 0.8 percent carbon. Then at 1333°F (723°C) pearlite will form. Now let us see what happens with isothermal transformation. If a sample is quenched from 1500 to 1400°F (815 to 760°C), we encounter start and end times for the formation of the amount of α at equilibrium with austenite at this temperature (Fig. 6.23). However, until 1333°F (723°C) we have isothermal transformation only to ferrite. If we quench from 1500°F (815°C) to below 1333°F (723°C) to as low as 1000°F (538°C), we find first a period of ferrite formation, then pearlite.

We notice too that the time for transformation in the pearlite range has been shortened compared to that for 1080 steel. The practical consequence is that martensite or bainite can only be obtained in thin sections, which are rapidly quenched.

If the steel is transformed below 1000°F (538°C), separate formation of ferrite does not take place. Therefore the TTT curve shows only the start and end of bainite, just as for 0.8 percent carbon.

The hardness of the transformation products at any temperature is lower than that of 0.8 percent carbon steel, because the amount of the hard carbide phase is smaller.

If the steel is quenched rapidly enough to avoid prior transformation to pearlite or bainite, martensite will form on crossing the M_s down to the M_f. It is important to note that as the carbon content decreases, the M_s increases (Fig. 6.24) and the hardness of the martensite decreases (Fig. 6.25).

Now let us consider the hypereutectoid steels, which usually fall in the range 0.8 to 1.2 percent carbon. If we cool a 1.2 percent carbon steel

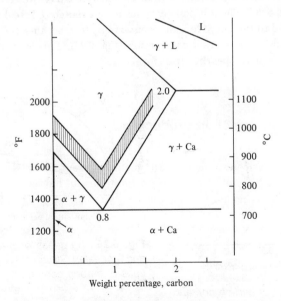

Fig. 6.22 *Austenitizing temperatures for different carbon contents (cross-hatched area).*

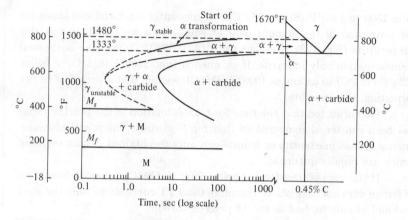

Fig. 6.23 *TTT curve for 0.45 percent carbon steel. There is an additional region, compared to 1080 steel, above the nose of the curve. A portion of the iron–iron carbide diagram is also included to show why primary α occurs.*

(L. H. Van Vlack, "Elements of Materials Science," 2d ed., Addison-Wesley Publishing Company, Inc., Reading, Mass., 1964, p. 292.)

slowly from the austenite field, we will form hypereutectoid carbide first. Then at 1333°F (723°C) the 0.8 percent carbon austenite will transform to pearlite. Therefore, the TTT curve for a hypereutectoid steel shows curves for the start and end of the carbide formation and then curves for the pearlite formation (Fig. 6.26). The dilation curve for a hypereutectoid steel is interesting compared to the curve for a hypoeutectoid steel, because the precipitation of carbide alone gives a contraction (Fig. 6.27). This is followed by an expansion when the pearlite transforms.

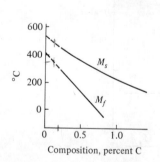

Fig. 6.24 *Effect of carbon content on the M_s and M_f.*

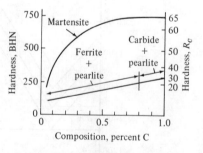

Fig. 6.25 *Hardness of martensite and annealed structures as functions of carbon content.*

(L. H. Van Vlack, "Elements of Materials Science," 2d ed., Addison-Wesley Publishing Company, Inc., Reading, Mass., 1964, Fig. 10.21, p. 289.)

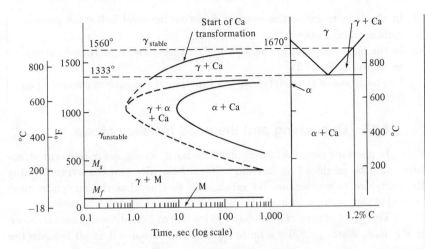

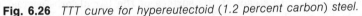

Fig. 6.26 *TTT curve for hypereutectoid (1.2 percent carbon) steel.*

The differences in the TTT curve for the hypereutectoid steels are a longer time to transformation, a lower M_s than for 0.8 percent carbon steel, and the addition of the curve for carbide precipitation.

We can summarize the effects of carbon in the TTT curve as follows:

1. Above 1000°F (538°C) the formation of ferrite in hypoeutectoid steel and of carbide in hypereutectoid steel precedes the transformation to pearlite.

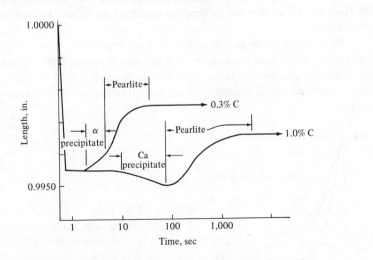

Fig. 6.27 *Dilation curves for isothermal transformation of 0.3 percent carbon and 1 percent carbon steels. Transformation temperature is 1300°F (704°C).*

2. In the bainite range the product is softer for steel below 0.8 percent carbon, harder above.

3. In the martensite range the M_s and M_f decrease with increased carbon, as shown in Fig. 6.24.

4. The hardness of the martensite is a function of the carbon content (Fig. 6.25).

6.18   Quenching and the need for alloy steels

In previous sections we found that a hard, strong, wear-resistant structure could be produced by heating to form austenite and then transforming the austenite to martensite. We established further that the austenite must be cooled fast enough to avoid transformation to pearlite or bainite.

The effect of type of quenchant on cooling of 1-in.-diameter bars is shown in Fig. 6.28. Water provides a more rapid quench than salt or oil because the conversion of water to steam at the surface of the bar absorbs a great deal of heat (high latent heat of vaporization). On the other hand, quenches in oil and salt lead to less distortion because of lower thermal gradients.

We may encounter many cases in which the section is thick and martensite cannot be obtained with even a water quench. In other parts a water quench cannot be used because of cracking or distortion. The remedy is to increase the time for austenite transformation to pearlite, so that a slower cooling rate will still avoid this reaction. This is done by the addition of alloys such as nickel, molybdenum, chromium, and manganese to the steel.

The basic effects of alloys are demonstrated by the TTT curves of Fig. 6.29. Note the great increase in time for pearlite formation compared to plain carbon steels.

These alloy steels give flexibility in heat treatment in two ways. First, an intricate part of thin section can be more slowly cooled from the austenitic field, lessening the susceptibility to cracking. Second, a part of thick section which would cool slowly when quenched can still be produced with a martensitic structure.

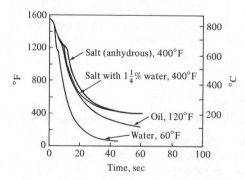

Fig. 6.28 *Effect of various quenchants on cooling rates of 1040 steel 1-in.-diameter by 4-in. bars.*

(American Society for Metals Handbook, 8th ed., vol. 2: "Heat Treating, Cleaning and Finishing," Metals Park, Ohio, 1964.)

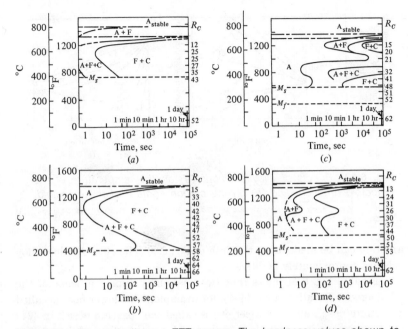

Fig. 6.29 *Effect of alloy on TTT curves. The hardness values shown to the right of each steel are those to be obtained after complete isothermal transformation at the temperatures indicated. (a) TTT diagram for 1034 steel, 0.34% C. (b) TTT diagram for 1090 steel, 0.9% C. (c) TTT diagram for 4340 steel, 0.4% C, 1.7% Ni, 0.8% Cr, 0.25% Mo. (d) TTT diagram for 5140 steel, 0.4% C, 0.8% Cr. Previous symbols used for A, F, and C were γ, α, and Ca.*

(American Society for Metals Handbook, 8th ed., vol. 2: "Heat Treating, Cleaning and Finishing," Metals Park, Ohio, 1964, Figs. 4 and 5, p. 38.)

Since alloy steels are more expensive than plain carbon steels, some of the effects of alloys acting in combination are worth noting. In general, a triple-alloy steel, such as NiCrMo 8640, will provide a better TTT curve, i.e. longer pearlite formation time, than a single alloy steel of the same cost. The alloys also affect the M_s, and an empirical formula provides an estimate of this point. (Note that all the alloys lower the M_s.)

$$M_s = 930°F - [540 \times (\text{percent C}) + 60 \times (\text{percent Mn})$$
$$+ 40 \times (\text{percent Cr}) + 30 \times (\text{percent Ni}) + 20 \times (\text{percent Mo})]$$

It should be emphasized that the alloys do not change the hardness of the pearlite, bainite, or martensite, but merely increase the transformation time for pearlite and bainite and lower the temperature range for martensite.

In alloy steels the eutectoid temperature and the composition are somewhat different from 1333°F (723°C) and 0.8 percent carbon, and the

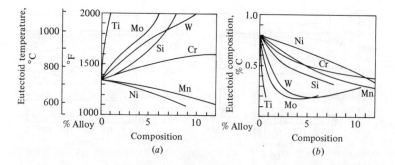

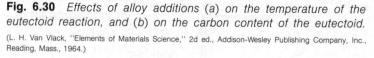

Fig. 6.30 *Effects of alloy additions (a) on the temperature of the eutectoid reaction, and (b) on the carbon content of the eutectoid.*

(L. H. Van Vlack, "Elements of Materials Science," 2d ed., Addison-Wesley Publishing Company, Inc., Reading, Mass., 1964.)

correction can be made using the data of Fig. 6.30. The effects at low alloy levels are approximately additive.

Finally, Fig. 6.29 shows that two noses may appear in some of the TTT diagrams for alloy steels (4340, for example). The lower nose is called the "bainite nose," and the upper one is called the "pearlite nose." In 4340 steel, for example, it is possible to cool at such a rate as to miss the pearlite nose yet encounter the bainite nose and therefore obtain bainite by continuous cooling instead of by the isothermal treatment discussed previously.

EXAMPLE 6.4 Calculate the eutectoid temperature, eutectoid carbon, and M_s for a steel of the following alloy composition: 2% Ni, 1% Cr, 0.5% C.

ANSWER From Fig. 6.30 we find:

Alloying element	Change in eutectoid temperature	Change in eutectoid carbon, percent
2% Ni	−30°F	−0.05
1% Cr	+50°F	−0.10
	+20°F	−0.15

Approximate eutectoid temperature = 1353°F (1333°F + 20°F)

(734°C)

Approximate eutectoid carbon = 0.65% (0.80 − 0.15%)

$M_s = 930 - 540 \times 0.5 - 40 \times 1 - 30 \times 2 = 560°F$ (293°C)

Hardenability; Properties of Plain Carbon and Low-alloy Steels

6.19 General; method for choosing the proper steel for a given part

In the previous sections we established the importance of the time and temperature of austenite transformation on the hardness of the structures produced—pearlite, bainite, and martensite. The basic effect of the carbon and alloy content was shown by TTT curves. With this background we can now come to grips with the important engineering problem of how to select the steel for a given part and heat treatment.

Let us suppose we have a simple shaft 2 in. (5.08 cm) in diameter, and we wish to obtain a hardness of 50 R_C to at least $\frac{1}{2}$ in. (1.27 cm) beneath the surface. We wish to use an agitated oil quench for hardening, having experienced cracking in tests with a water quench. The question is which steel should be used. This leads us to the concept of hardenability evaluation.

6.20 Hardenability

Many years ago steel producers developed extensive handbooks showing the hardness profiles that could be developed in different steels with different quenching conditions in different section thicknesses. After a little thought we see that with the variables of composition, section thickness, and quenching medium, literally thousands of curves are needed. In addition, the question of how to check a given heat of steel for its *ability to harden* developed. True, the determination of the TTT diagram would be useful, but it is time-consuming.

To meet this problem of evaluating the millions of tons of steel used by the automotive industry, Walter Jominy and his associates developed the "hardenability bar." The idea behind this test is to produce in one bar a wide variety of known cooling rates. Then by measuring the hardness along the bar, the hardnesses obtainable with different cooling rates from the austenitizing temperature are known.

The method of making the test is shown in Fig. 6.31. A bar of the steel to be tested is machined to give a cylinder 4 in. (10.16 cm) long and 1 in. (2.54 cm) in diameter with an upper lip. The bar is then austenitized in standard fashion in a furnace and placed in the fixture. The water is quickly turned on and provides a smooth film striking only the end of the bar. The quenched end cools rapidly, and the regions away from the end cool at rates proportional to their distances from the quenched portion. After

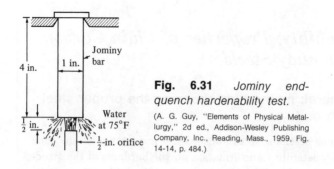

Fig. 6.31 *Jominy end-quench hardenability test.*

(A. G. Guy, "Elements of Physical Metallurgy," 2d ed., Addison-Wesley Publishing Company, Inc., Reading, Mass., 1959, Fig. 14-14, p. 484.)

the bar is cool, it is removed from the fixture, a flat is ground along one side, and R_C hardness readings are taken every $\frac{1}{16}$ in. (0.159 cm). A typical plot is shown in Fig. 6.32. The greatest hardness is at the quenched end where martensite is formed, and the lower hardness farther away is due to softer transformation products.

The most significant point of this plot is not the hardness at a given distance but the hardness at a given cooling rate. (The cooling rates have been measured at the different points and are quite constant for different steels.) We must bear in mind that a little region in the bar does not know it is in a Jominy end-quench test; it only feels the effect of a given cooling rate. For this reason a position on the Jominy bar will give the hardness that would be obtained at a point in an oil- or water-quenched bar with the *same cooling rate*. In Fig. 6.33 the relationships between given Jominy

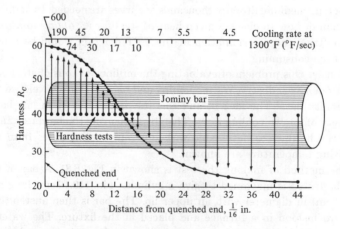

Fig. 6.32 *Typical hardness distribution in Jominy bars.*

(A. G. Guy, "Elements of Physical Metallurgy," 2d ed., Addison-Wesley Publishing Company, Inc., Reading, Mass., 1959, Fig. 14-14, p. 484.)

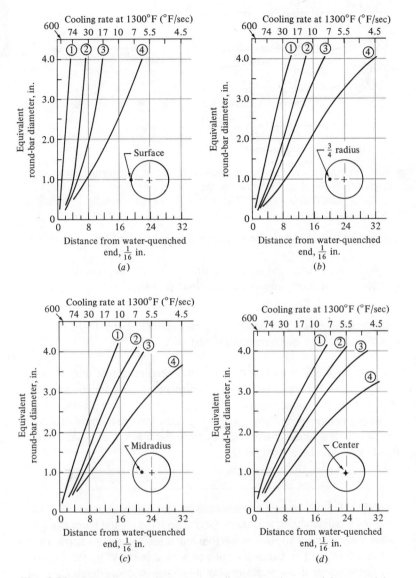

Fig. 6.33 *Relationships between cooling rates in round bars and in Jominy locations. 1 = still water; 2 = mildly agitated oil; 3 = still oil; 4 = mildly agitated molten salt.*

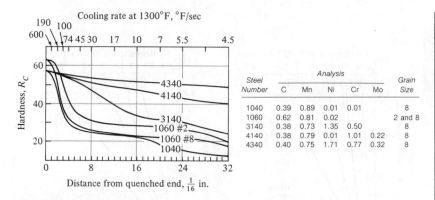

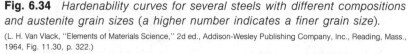

| Steel | Analysis | | | | | Grain |
Number	C	Mn	Ni	Cr	Mo	Size
1040	0.39	0.89	0.01	0.01		8
1060	0.62	0.81	0.02			2 and 8
3140	0.38	0.73	1.35	0.50		8
4140	0.38	0.79	0.01	1.01	0.22	8
4340	0.40	0.75	1.71	0.77	0.32	8

Fig. 6.34 *Hardenability curves for several steels with different compositions and austenite grain sizes (a higher number indicates a finer grain size).*

(L. H. Van Vlack, "Elements of Materials Science," 2d ed., Addison-Wesley Publishing Company, Inc., Reading, Mass., 1964, Fig. 11.30, p. 322.)

positions and different locations in actual bars quenched in several media are given.

With these data and a few Jominy curves for representative steels (Fig. 6.34), we can now answer our previous problem.

EXAMPLE 6.5 How do we obtain 50 R_C minimum $\frac{1}{2}$ in. beneath the surface of a 2-in.-diameter bar quenched in still oil?

ANSWER We read from Fig. 6.33 that the cooling rate at this position (midradius) is 20°F/sec and that this corresponds to a Jominy position of $\frac{11}{16}$ in. from the quenched end of the bar. We then draw a vertical line at this position on the Jominy curve of Fig. 6.34. We see that of this group only 4340 steel would provide satisfactory hardness, that is, above 50 R_C (4140 would be almost 50 R_C).

EXAMPLE 6.6 We have a complex shape that has been made of 1040 steel. After austenitizing and quenching in agitated oil, the hardness $\frac{1}{8}$ in. below the surface is 30 R_C. We find that this is too soft for our application and that 50 R_C at $\frac{1}{8}$ in. below the surface is necessary. What steel would you recommend if the austenitizing treatment and quenching are to remain the same?

ANSWER From the Jominy curves for 1040 steel (Fig. 6.34), 30 R_C appears at $\frac{1}{4}$ in. from the water-quenched end. This means that the material at $\frac{1}{4}$ in. from the water-quenched end of a Jominy test cools at the *same* rate ($74°F/sec$) as the material in our complex shape quenched in oil *at $\frac{1}{8}$ in. below the surface*. Therefore, we look at the hardness obtained for other steels at the same cooling rate ($\frac{1}{4}$ in. station or $74°F/sec$) and see that 4340, 4140, and 3140 steels will all give over 50 R_C under these quenching conditions.

6.21 Analysis and properties of typical low-alloy steels

To simplify the specification of these steels, the Society of Automotive Engineers (SAE) and the American Iron and Steel Institute (AISI) issued a joint specification of SAE-AISI steels (Table 6.2). The key is as follows. The last two digits give the carbon content, as in plain carbon steels; that is, 4340 steel contains 0.40 percent carbon. In rare cases, such as 52100 ball-bearing steel, five digits are used and the last three represent the carbon, which in this example is 1.00 percent. It should be realized that the chemical analysis of any given steel will vary somewhat from that given in Table 6.2 because of manufacturing variables.

Table 6.2 SAE-AISI CHEMICAL SPECIFICATIONS

			Nominal Chemical Analysis, percent								
Type	Name	Example	C	Mn	P Maximum	S Maximum	Si	Ni	Cr	Mo	Other
10xx	Plain carbon	1020	0.2	0.4	0.04	0.05					
11xx	Free machining	1111	0.1	0.7	0.09 average	0.12 average					
13xx	Mn	1330	0.3	1.7	0.04	0.04	0.3				
3xxx	NiCr	3140	0.4	0.8	0.04	0.04	0.3	1.0	0.6		
40xx	Mo	4042	0.4	0.8	0.04	0.04	0.2			0.25	
41xx	CrMo	4140	0.4	0.8	0.04	0.04	0.3		1.0	0.20	
46xx	NiMo	4620	0.2	0.6	0.04	0.04	0.3	1.8		0.25	
47xx	NiCrMo	4720	0.2	0.6	0.04	0.04	0.3	1.0	0.4	0.20	
48xx	NiMo	4820	0.2	0.6	0.04	0.04	0.3	3.5		0.25	
50xx	Cr	5015	0.1	0.4	0.04	0.04	0.3		0.4		
52xx	Cr	52100	1.0	0.4	0.02	0.02	0.3		1.4		
61xx	CrV	6120	0.2	0.8	0.04	0.04	0.3		0.8		0.10 V
81xx	NiCrMo	8115	0.15	0.8	0.04	0.04	0.3	0.3	0.4	0.10	
86xx	NiCrMo	8650	0.5	0.8	0.04	0.04	0.3	0.5	0.5	0.20	
87xx	NiCrMo	8720	0.2	0.8	0.04	0.04	0.3	0.5	0.5	0.25	
88xx	NiCrMo	8822	0.2	0.8	0.04	0.04	0.3	0.5	0.5	0.35	
92xx	Si	9260	0.6	0.8	0.04	0.04	2.0				
93xx	NiCrMo	9310	0.1	0.6	0.02	0.02	0.3	3.0	1.2	0.10	
94xx	NiCrMo	94B30*	0.3	0.8	0.04	0.04	0.3	0.4	0.4	0.10	0.0005 B
98xx	NiCrMo	9840	0.4	0.8	0.04	0.04	0.3	1.0	0.8	0.25	

* "B" refers to the presence of boron.

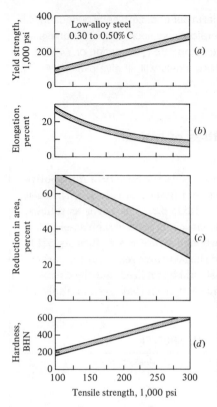

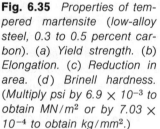

Fig. 6.35 *Properties of tempered martensite (low-alloy steel, 0.3 to 0.5 percent carbon). (a) Yield strength. (b) Elongation. (c) Reduction in area. (d) Brinell hardness. (Multiply psi by 6.9 × 10⁻³ to obtain MN/m² or by 7.03 × 10⁻⁴ to obtain kg/mm².)*

(American Society for Metals Handbook, 8th ed., vol. 1: "Properties and Selection of Metals," Metals Park, Ohio, 1961, Fig. 3, p. 109.)

The first two digits indicate alloy content as shown, including the plain carbon steels for completeness. The combinations of strength, ductility, and hardness obtainable with these steels are shown in Fig. 6.35. Therefore, from a simple hardness test after quenching and tempering it is possible to determine the other properties. In general, it is the carbon that determines the hardness of the martensite in the quenched structure. The alloy content merely serves as a means to obtain the desired structure at a given cooling rate.

Cast Steel. In the preceding discussions we have generally referred to specifications for wrought steel, that is, for material that is cast and then rolled or forged to shape. However, many important parts are made as steel castings. In this way many shapes can easily be obtained that would be difficult to fabricate. To illustrate this point, let us consider several applications.

Electrical machinery such as generators and motors require a cast low-carbon steel (0.05 to 0.15 percent carbon) to obtain optimum magnetic properties.

Structural parts for the railroad industry such as side frames, use steel in the range 0.2 to 0.5 percent carbon to attain a good combination of strength and ductility.

Railroad car wheels and mining equipment components which require better wear resistance use eutectoid (0.8 percent carbon) steel.

Other parts that are to be hardened, such as large gears, use low-alloy steels similar to the SAE-AISI grades already discussed.

6.22 Tempering of alloy and noneutectoid steels

We have already discussed the tempering of 0.8 percent carbon steel in Sec. 6.14, and the other steels follow the same principles.

In general terms, tempering is a reheating operation which leads to precipitation and spheroidization of carbide. There are notable side effects, such as the change in martensite from a stressed tetragonal structure to BCC and often the transformation of retained austenite (untransformed during cooling in high-carbon alloys), but the basic effect on carbide is the most important.

Although we normally refer to the tempering of martensite, the term is also used for the softening of bainitic and even pearlitic structures. If samples of martensite, bainite, and pearlite of the same steel are heated at 1200°F (649°C), for example, all will contain spheroidite (spheroidal carbide plus ferrite) after sufficient time has passed.

Some quantitative relations among alloy content, tempering temperature, and hardness are shown in Fig. 6.36a. The alloy steels generally retain their hardness more than the plain carbon steels. The steels shown in Fig. 6.36a are all 0.45 percent carbon, but the effect of other carbon contents can be determined by using Fig. 6.36b.

EXAMPLE 6.7 Find the tempered hardness of a 4330 steel after 1 hr at 600°F (315°C).

ANSWER We read the hardness for 4345 at 50 R_C on Fig. 6.36a, and then from Fig. 6.36b we find that a subtraction of 6 R_C is needed, giving 44 R_C as the final reading. (*Note:* These corrections should be applied only to steels of similar alloy content.)

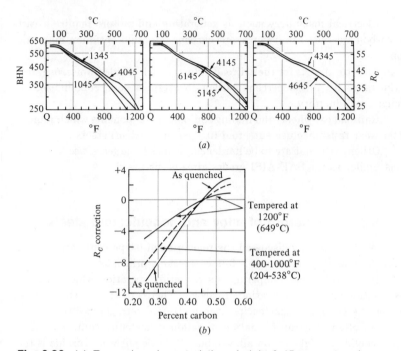

Fig. 6.36 (a) *Tempering characteristics of eight 0.45 percent carbon alloy steels. Duration of tempering = 1 hr. (b) Effect of carbon content on hardness of quenched and tempered steel. R_C units to be added or subtracted from the value for 0.45 percent carbon for different temperatures are given.*

(American Society for Metals Handbook, 8th ed., vol. 1: "Properties and Selection of Metals," Metals Park, Ohio, 1961, Fig. 2, p. 109.)

High-alloy Steels and Superalloys

6.23 General

Up to this point we have considered the use of only relatively small amounts of alloy for the purpose of changing the TTT curve and hardenability. There is another group of steels in which well over 5 percent alloy is added for various special purposes. These high-alloy steels may be divided into the groups shown in the table on page 207.

We should note that a number of these steels are called "austenitic," so that one of the effects of the alloy must be to produce a structure that is austenite at room temperature. This can be explained only by a radical alteration of the iron–iron carbide diagram we have studied. The phase relationships in the high-alloy steels must therefore be our first consideration.

Type	Common chemistry, percent
Stainless steel	
Austenitic	18 Cr, 8 Ni, balance Fe
Ferritic	16 Cr, 0.1 C
Martensitic	17 Cr, 1 C
Precipitation-hardened stainless steel	17 Cr, 7 Ni, 1 Al
Maraging steel	18 Ni, 7 Co
Tool steel; high-speed steel	18 W, 4 Cr, 1 V
Manganese steel, austenitic	12 Mn

6.24   Phase diagrams of the high-alloy steels

The most important effect of alloys is the change in the austenite field. It is possible to classify all alloys on the basis of whether they contract or expand the γ field. Examples are shown in Fig. 6.37. Note that nickel widens the field until, at 32 percent nickel, γ is encountered at room temperature under *equilibrium* conditions. Conversely, chromium narrows the field until at 13 percent chromium we have only α as the solid phase at all temperatures. Although the interstitial elements such as carbon and boron do not produce completely open austenitic or ferritic fields, they expand or contract the γ field. A summary of the effects of various elements is given on a periodic basis in Fig. 6.38. It is interesting that all the group VIII elements plus manganese expand the γ field while the other transition elements contract it.

We are now in a position to discuss the structures encountered in the special steel groups.

6.25   The stainless steels

We have already pointed out in the introduction that the stainless steels fall into three principal groups—ferritic, marstensitic, and austenitic—named according to the predominating structure.

Ferritic Steels. We see from the iron-chromium diagram (Fig. 6.37) that the formation of austenite is completely suppressed in a pure iron-chromium alloy above 13 percent chromium. However, if carbon is present, the chromium will form chromium carbide, and in this case 1 percent of carbon will combine with 17 percent of chromium. This carbide will precipitate and deprive the iron matrix of a corresponding amount of chromium. For this reason the commercial steel 430 listed in Table 6.3 contains 16 percent chromium to retain the ferritic structure. The advantage of this material over unalloyed ferrite is its corrosion resistance, as discussed in Chap. 12, and it is used in automotive trim and kitchen utensils in the cold-worked condition.

Martensitic Steels. When the carbon content of the high-chromium steels is high, as in 440C (Table 6.3), we can heat to the γ field and quench to produce martensite. Also with lower chromium, as in type 410, we can produce martensite with only 0.15 percent carbon. Because of the

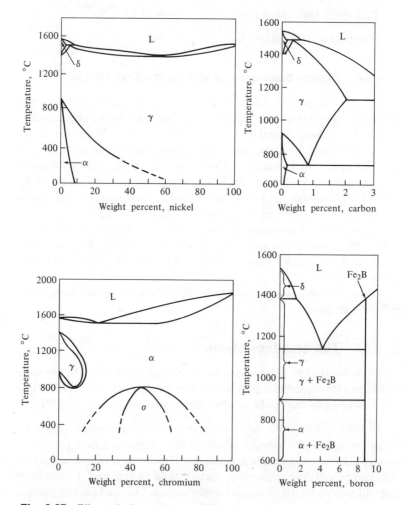

Fig. 6.37 *Effect of alloys of the γ field of iron. Some elements narrow the field while others widen it.*

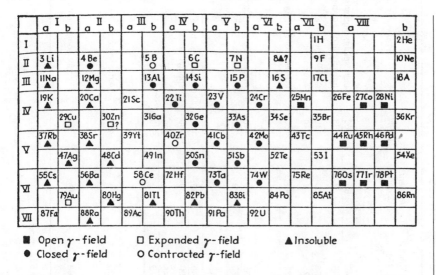

■ Open γ-field   □ Expanded γ-field   ▲ Insoluble
● Closed γ-field   O Contracted γ-field

Fig. 6.38 *Periodic table showing the effects of elements on the γ field.*
(C. Barrett, "Structure of Metals," 2d ed., McGraw-Hill Book Company, New York, 1952, Fig. 16, p. 251.)

high alloy content these steels have high hardenability and in some cases it is only necessary to cool in air to form martensite, easily avoiding pearlite and bainite formation. These steels are excellent for cutlery and dies.

We can summarize the carbon-chromium relationships for the ferritic and martensitic types as follows:

When [percent Cr − (17 × percent by weight C)] > 13, ferrite only is present. When [percent Cr − (17 × percent by weight C)] < 13, austenite can be formed and quenched to martensite.

Austenitic Steels. In these steels we find substantial amounts of nickel as well as chromium, as in the familiar 18% Cr, 8% Ni types. The nickel acts to enlarge the austenite field to such an extent that it is stable at room temperature. To illustrate this effect we can use a horizontal section of the iron-nickel-chromium ternary diagram (Fig. 6.39) of the type discussed in Chap. 4. This section represents the phases present at 650°C. (It is simpler to use this temperature to avoid the complications of transformation of the austenite in the low-alloy steels which occur at lower temperature.) We note that the composition 18% Cr, 8% Ni is close to the boundary between the $\alpha + \gamma$ and γ fields. If we want to be sure to have an all-γ structure at room temperature, we use a type 304 composition (Table 6.3), with 10 percent nickel. However, there is an advantage to type 301 steel in which some ferrite is encountered. This steel is more responsive to cold working, as shown by

Table 6.3 TYPICAL PROPERTIES OF STAINLESS STEELS

Number	Chemical Analysis, percent	Condition	Tensile Strength,[*] psi × 10³	Yield Strength,[*] psi × 10³	Percent Elongation	BHN	Typical Use
		Austenitic Steels					
301	17 Cr, 7 Ni	Annealed	110	40	60	160	Lightweight, high-strength transportation equipment
		Cold-worked	185	140	9	388	
304	19 Cr, 10 Ni	Annealed	85	35	60	149	General chemical equipment
		Cold-worked	110	75	12	240	
347	18 Cr, 11 Ni[†]	Annealed	90	35	45	160	Welded construction
		Ferritic Steels					
430	16 Cr, <0.1 C	Annealed	80	55	25	140	Automobile trim, kitchen equipment
		Cold-worked	90	80	20	200	
		Martensitic Steels					
410	12 Cr, 0.15 C	Annealed	70	40	30	155	General purpose springs, rules
		Quenched and tempered	140	100	20	300	
440C	17 Cr, 1 C	Annealed	110	65	14	230	Instruments, cutlery, valves
		Quenched and tempered	285	275	2	580	
		Precipitation-hardened Steels					
17-7PH	17 Cr, 7 Ni, 1 Al	Hardened	235	220	6	400	Airframe parts
		Maraged Steels					
Maraging steel	18 Ni, 7 Co[‡]	Maraged	275	268	11	500	Aircraft components

[*] Multiply psi by 6.9 × 10⁻³ to obtain MN/m² or by 7.03 × 10⁻⁴ to obtain kg/mm².

[†] Contains niobium (columbium) to prevent weld embrittlement. Percent Nb = 10 × percent C. Percent C is about 0.08 in usual grades.

[‡] 0.025% C, 0.1% Mn, 0.1% Si, 0.22% Ti, 0.003% B.

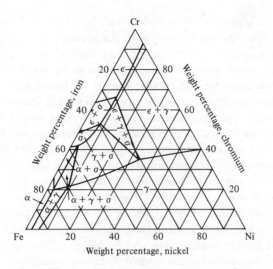

Fig. 6.39 *Iron-nickel-chromium ternary phase diagram at 650°C.*

(American Society for Metals Handbook, 8th ed., vol. 1: "Properties and Selection of Metals," Metals Park, Ohio, 1961.

the yield strength compared to type 304. Type 347 is used where there will be welding, to avoid the combination of carbon with chromium which leads to poor corrosion resistance. In type 347 the carbon combines with the niobium, instead of with chromium, preventing the corrosion due to chromium depletion of the matrix, as discussed in Chap. 12.

6.26 Precipitation hardening and maraging steels

Where maximum strength and some corrosion resistance are required, 17-7PH and maraging 18 percent nickel steels are used. In the 17-7PH type a solution treatment at 1925°F (1050°C) is followed by aging at 900°F (482°C) to produce a precipitate. In the maraging steels the material is austenitized at 1500°F (815°C) to dissolve precipitated phases, and then relatively soft, low-carbon martensite is formed on cooling. The material can be worked or machined in this condition. To strengthen the material, the alloy is heated to 900°F (482°C) for 3 hr, which causes aging. Thus the hardening is due to both martensite and aging precipitates. There is little distortion.

6.27 Tool steels

The tool steels encompass a wide range of compositions. It is possible to produce a hard cutting edge in a high-carbon plain carbon steel. The advantages of the high-alloy steels such as 18% W, 4% Cr, 1% V are twofold. Higher hardenability makes it possible to use slower cooling rates during quenching which protect tools from cracking. Second, the highly alloyed martensite plus excess alloy carbides retain their hardness at the high temperatures generated by fast cutting speeds; i.e. they resist tempering.

The various grades of tool steels are represented in Table 6.4 in order of increasing cost. The plain carbon W1 grade is heat-treated, as is the simple iron-carbon alloy we have already discussed. The final hardness depends on the carbon content and the tempering temperature. It is worth noting that the same hardness can be attained in this steel as in the highly alloyed types. The next grade O1 can be oil-hardened in simple sections because of the alloy and shows less distortion. The D1 grade with 12 percent chromium resembles the martensitic stainless steels we just discussed at a high-carbon level. The H grade is different from all the others because of the lower carbon level. It is used for hot-working dies and is therefore exposed to thermal shock. A higher-carbon steel would crack under these conditions. To prevent softening of the martensite, a considerable amount of alloy is used: 5% Cr, 1.5% Mo, 0.4% V. The M1 and T1 steels can be considered together as high-speed steels. The T1 type with 18 percent tungsten was developed first, and then with the shortage of tungsten in World War II the molybdenum modification was developed.

The most significant feature of the heat treatment of these steels is the need for a very high austenitizing temperature. The phase diagram for the T1 type (Fig. 6.40) explains this point. In the early research in these steels the hardness obtained after quenching was disappointingly low. This

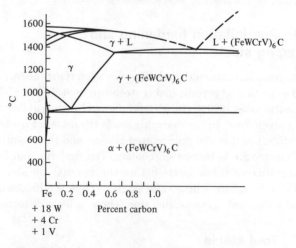

Fig. 6.40 *Phase diagram for T1 type tool steels.*

Table 6.4 TYPICAL CHARACTERISTICS OF TOOL STEELS

Number	Chemical Analysis, percent*					Austenitizing Temperature, °F (°C)	Quenching Medium	Tempering Temperature, °F (°C)	R_C	Typical Use
	C	Cr	Mo	W	V					
W1	0.6 / 1.4					1550 (843) / 1400 (760)	Water	350 (77) / 650 (343)	64 / 50	Tools and dies used below 350°F (177°C)
O1	0.9	0.5		0.5		1500 (815)	Oil	350 (177) / 500 (260)	62 / 57	Tools and dies requiring less distortion than W1
D1	1.0	12.0	1.0			1800 (982)	Air	400 (204) / 1000 (538)	61 / 54	Wear-resistant, low-distortion tools
H11	0.35	5.0	1.5		0.4	1850 (1010)	Air	1000 (538) / 1200 (649)	54 / 38	Hot-working dies
M1	0.8	4.0	8.0	1.5	1.0	2200 (1204)	Air	1000 (538) / 1100 (593)	65 / 60	High-speed tools
T1	0.7	4.0		18.0	1.0	2350 (1288)	Air	1000 (538) / 1100 (593)	65 / 60	High-speed tools

* Balance Fe.

was because with an austenitizing temperature of 900°C (1652°F) the carbon content of the γ is only 0.3 percent, and therefore a low-carbon martensite is obtained at about 40 R_C. However, when the austenitizing temperature is raised to 1250°C (2270°F), the carbon content of the austenite is 0.6 percent, which gives martensite of 64 R_C on quenching.

The tempering of high-speed steels has received extensive study because an effect called "secondary" hardening is encountered. The hardness rises slightly after tempering at 1000°F (538°C). The as-quenched structure contains some retained austenite, and upon heating to 1000°F (538°C) some alloy carbides precipitate. The M_s of the retained austenite is raised because of this effect (carbides take alloy out of solution), and the structure transforms to martensite on cooling from the tempering temperature. Therefore, although the original martensite was softened at 1000°F (538°C), the new martensite raises the overall reading. The toughness of the structure is often improved by retempering.

6.28 Hadfield's manganese steel

We recall that manganese, as well as nickel, produces an enlargement of the γ field. We take advantage of this in a very tough abrasion-resistant austenitic steel containing 12 percent manganese and 1 percent carbon. The as-cast structure contains carbide, austenite, and some pearlite, but after heating to 2000°F (1093°C) and water-quenching, the structure is completely austenitic. A water quench may be used even with complex castings because there are no stresses from transformation of austenite to martensite. Typical properties are:

> Tensile strength: 120,000 psi†
> Yield strength: 50,000 psi†
> Percent elongation: 45
> BHN: 160

The steel is widely used in mining and earth-moving equipment and in railroad track work.

6.29 Superalloys; general

The name "superalloys" arises from the fact that these materials exhibit far greater strength at high temperatures (1500 to 2000°F) (815 to 1093°C) than conventional alloys. Some of these alloys are even called "exotic" because of the high cost of some of the elements that are used. The basic reason for the development of this interesting field is that the efficiency of a gas turbine

---

†Multiply psi by 6.9×10^{-3} to obtain MN/m² or by 7.03×10^{-4} to obtain kg/mm².

increases with the temperature at which the rotor can be operated. However, the problem of strength at elevated temperatures is not merely in the rotor itself but in the vanes, the combustion chamber, and even the compressor section.

There are two principal problems in the development of materials for all parts operating at high temperatures—oxidation resistance and strength. Oxidation resistance is essentially a problem in gas corrosion and will be taken up in Chap. 12. At this point we will consider only the problem of strength.

We find very quickly in testing for strength at high temperatures that simply running a hot tensile test (i.e. testing the conventional 0.505-in.-diameter specimen in a furnace) does not give the required information. To illustrate this point, let us suppose that the conventional test gives a yield strength of 50,000 psi ($345 \, MN/m^2$). Let us load the specimen to 40,000 psi ($276 \, MN/m^2$) and observe it over a period of time at the elevated temperature. First we note that the sample continues to elongate although it is at constant load. This continued elongation at constant load is called "creep." Second, we find that the specimen will rupture after a certain time, say 20 hr. These are two new important factors in design: the gradual extension or creep and the lower rupture strength as a function of time of testing. Therefore, if we design a rotor for a gas turbine merely on the basis of yield strength during a tensile test, the blades will either stretch excessively and contact the housing or break after a period of time.

6.30 Creep and stress rupture test data

The type of data obtained in creep testing is shown in Fig. 6.41. Bars are equalized at temperature in a test furnace, and then a fixed load is applied through a lever system. The extension is plotted as a function of time. Note that increased creep rates (slopes) and shorter rupture times are obtained by increasing either the load or the temperature.

Three stages of creep are recognized: stage I in which initial yielding is fairly rapid, stage II in which a straight line is obtained, and stage III in which yielding is again rapid and failure occurs. From these data it is possible to plot points for the stress required to produce various amounts of deformation and to rupture as a function of time (Fig. 6.42).

The test temperature has a pronounced effect on the creep and rupture properties of all materials and will be discussed again in Chap. 11.

6.31 High-temperature properties of typical superalloys

At first glance the names and analyses of superalloys seem random and complex (Table 6.5). There are, however, certain common features which correlate analysis, structure, and performance.

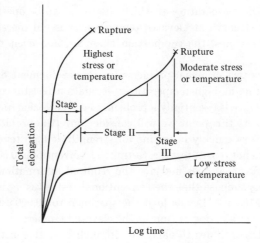

Fig. 6.41 *Typical creep curves.*

The first simplification is that all the important alloys have an FCC structure based on a combination of nickel, iron, and chromium. Therefore, they may be viewed as a simple extension of the austenitic field of the ternary diagram (Fig. 6.39) to higher nickel compositions. The chromium, while not added to obtain austenite, is essential to provide oxidation resistance. Thus we have accounted for the iron, nickel, and chromium content. Cobalt is used to replace part of the iron or nickel and dissolves in solid solution, augmenting the strength of the material.

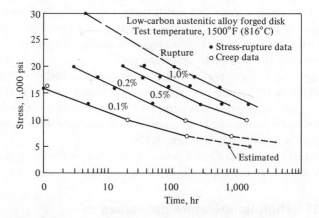

Fig. 6.42 *Stress deformation data for a low-carbon austenitic alloy.*

(American Society for Metals Handbook, 8th ed., vol. 1: "Properties and Selection of Metals," Metals Park, Ohio, 1961.)

Table 6.5 SUPERALLOYS: IRON, COBALT, AND NICKEL-BASE ALLOYS

Code and Name	Chemical Analysis, percent								Stress to Rupture in 100 hr,* psi × 10³, at Indicated Temperature, °F							Availability	
	Fe	Ni	Cr	Co	Al	Ti	Mo	Other	1200	1350	1400	1500	1600	1700	1800	Cast	Wrought
HF (18/8 type)	Balance	10	20					0.3 C	30		14		6			X	X
Incoloy 901	Balance	43	13		0.2	3.0	6	0.01 B		50							X
Inconel 718	18	52	19		0.6	0.8	3	5.0 Nb			45					X	X
Waspalloy	2	Balance	20	13	1.3	3.0	4	†				40	25			X	X
Inco 713	2	Balance	13		6.0	0.8	9	2.0 Nb					44			X	
Udimet 500	4	Balance	18	18	3.0	3.0	4	†				45	32	19		X	X
Udimet 700	1	Balance	15	18	4.5	3.5	5	†					42			X	X
In 100	1	Balance	10	15	5.5	5.0	3						56			X	
René 41	5	Balance	19	10	1.5	3.0	10					45	28	17		X	X

* Multiply psi by 6.9 × 10⁻³ to obtain MN/m² or by 7.03 × 10⁻⁴ to obtain kg/mm².
† Plus small amounts of boron and/or zirconium.

Aluminum and titanium have a special use in providing a very fine precipitate called γ' (read "gamma prime"). The name is derived from the fact that the structure is close to the γ solid solution. On the other hand, γ' has the formula Ni_3Al, and some of the aluminum can be replaced with titanium with only small changes in a_0. The γ' structure has long-range order with aluminum or titanium atoms occupying the corners of the unit cell and nickel atoms in the face centers. The precipitate is in the form of cubes, and the magnification of the electron microscope is necessary to resolve it (see the frontispiece for Chap. 6). The presence of this precipitate is essential to high strength in these alloys.

Molybdenum adds to the strength of the matrix and also forms complex carbides which reduce the creep rate. Small amounts of zirconium and boron are also added for their effects on the fine carbides. Usually the carbon content is about 0.1 to 0.2 percent, but the carbides are carefully controlled.

In general, this field of alloys is changing rapidly because in the development of no other alloy are the rewards higher for a material with just moderately better performance. As an example, some new alloys are being used with an addition of 0.5 percent hafnium. In another development turbine blades are being solidified directionally using an induction field to develop the desired temperature gradient in the mold. It should be noted, moreover, that practically all grades are cast but only a limited number are wrought. The reason is that with increased creep strength the hot working becomes difficult and die life is short. Also, some of the alloys have low ductility which is adequate for service but not for rolling. The investment casting process is the most widely used method and is capable of producing intricate shapes.

White Iron, Gray Iron, Ductile Iron, and Malleable Iron

6.32 General; importance of high-carbon alloys

Gray cast iron is one of the oldest alloys and ductile iron, which is being used more and more, is one of the newest. To illustrate the importance of these alloys, let us review the parts in a typical automotive engine. Ninety-five percent of the weight is gray iron and ductile iron. Typical gray iron parts are the block, head, camshaft, piston rings, tappets, and manifolds. The crankshaft and rocker arms are ductile iron. Malleable iron and ductile iron are used in many other parts of the automobile, such as the differential housings.

The annual production of castings in this group of high-carbon iron alloys is about 20 million tons and is second only to steel in the field of materials.

Nevertheless, the structures and properties of these alloys are not well understood because of a few fundamental differences from steel. To use these alloys properly and to understand their heat treatment, it is important to focus first on the portion of the structure that is different from steel:

White iron: massive carbide
Gray iron: flake graphite
Ductile iron: spheroidal graphite $\Big\}$ + steel matrix (ferrite,
Malleable iron: clumplike graphite   pearlite, martensite, etc.)
           (temper carbon)

Examples of these microstructures are shown in Fig. 6.43.

We are already familiar with the control of austenite transformation to ferrite and other products, so let us study first the factors affecting the massive carbide and graphite structures.

6.33 Relationships between white iron, gray iron, ductile iron, and malleable iron

White Iron. We see from the fraction chart of white iron (Fig. 6.44) that carbide precipitates during three important periods (3 percent carbon iron):

1. 2065°F (1130°C): the eutectic reaction, $L \rightarrow \gamma$ + carbide
2. 2065 to 1333°F (1130 to 723°C): from the eutectic to the eutectoid, $\gamma \rightarrow \gamma$† + carbide
3. 1333°F (723°C): the eutectoid reaction, $\gamma \rightarrow \alpha$ + carbide (as pearlite)

We should note further that in reaction 1 large massive carbides are formed from the liquid. In reaction 2 the carbide crystallizes onto the existing massive carbide, and in reaction 3 pearlite is formed, which results in the microstructure shown in Fig. 6.43a. The final product, therefore, contains a high percentage of massive carbide and is hard and brittle.

Gray Iron. To explain gray cast iron, we need to understand that iron carbide is not basically a stable phase and that with abnormally slow cooling (or in the presence of certain alloys such as silicon), graphite (pure carbon) and iron will crystallize (Fig. 6.43). Furthermore, if we heat iron carbide for extended periods, it will decompose: iron carbide → iron + graphite.

In other words, the true equilibrium diagram is the iron-graphite system, which is superimposed on the iron–iron carbide diagram in Fig. 6.45.

---

† The analysis of the austenite changes from 2 percent carbon to 0.8 percent carbon with decreasing temperature.

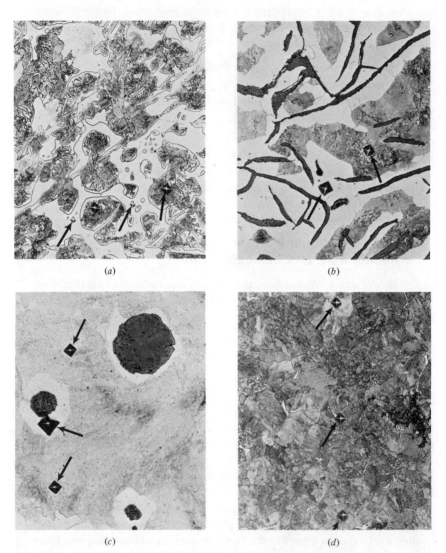

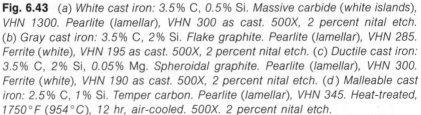

Fig. 6.43 (a) White cast iron: 3.5% C, 0.5% Si. Massive carbide (white islands), VHN 1300. Pearlite (lamellar), VHN 300 as cast. 500X, 2 percent nital etch. (b) Gray cast iron: 3.5% C, 2% Si. Flake graphite. Pearlite (lamellar), VHN 285. Ferrite (white), VHN 195 as cast. 500X, 2 percent nital etch. (c) Ductile cast iron: 3.5% C, 2% Si, 0.05% Mg. Spheroidal graphite. Pearlite (lamellar), VHN 300. Ferrite (white), VHN 190 as cast. 500X, 2 percent nital etch. (d) Malleable cast iron: 2.5% C, 1% Si. Temper carbon. Pearlite (lamellar), VHN 345. Heat-treated, 1750°F (954°C), 12 hr, air-cooled. 500X. 2 percent nital etch.

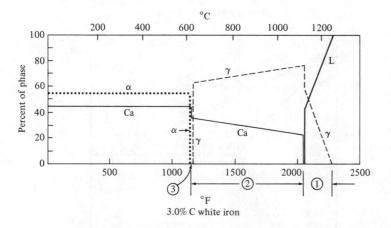

Fig. 6.44 *Fraction chart for white cast iron.*

It is not necessary to learn a new diagram because for all practical purposes all we need to do is to substitute graphite for carbide in the two-phase fields, as shown in Fig. 6.45, and move the right-hand vertical line to 100 percent carbon. There are, however, slight differences in eutectic and eutectoid carbon contents and temperatures.

Using this diagram, let us follow the formation of graphite in a ferritic gray cast iron with 3 percent carbon (i.e. ferrite matrix with flake graphite). To produce graphite in place of carbide, we will either cool very very slowly or add silicon, which also promotes graphite formation. We obtain the fraction chart shown in Fig. 6.46.

Now we can contrast the formation of graphite with that of carbide in white cast iron of eutectic composition (4.3 percent carbon).

Temperature °F(°C)	Gray iron†	White iron	Comments
2065 (1130)	$L \rightarrow \gamma$ + graphite	$L \rightarrow \gamma$ + carbide	
2065 to 1333 (1130 to 723)	$\gamma \rightarrow \gamma$ + graphite	$\gamma \rightarrow \gamma$ + carbide	Graphite (carbide) precipitates from γ on existing graphite (carbide). Carbon solubility is decreasing.
1333 (723)	$\gamma \rightarrow \alpha$ + graphite	$\gamma \rightarrow \alpha$ + carbide (as pearlite)	The analysis of the austenite is 0.8 percent carbon.

†We will continue to use the iron–iron carbide equilibrium temperatures in this intro-ductory section for simplicity. Later we will modify these remarks in discussing the actual commercial alloys containing, for example, silicon.

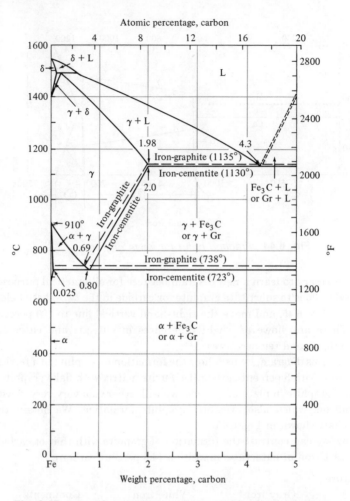

Fig. 6.45 *Iron–iron carbide and iron-graphite phase diagrams.*

We see therefore that in place of the carbide precipitation previously encountered in white cast iron we obtain graphite in the gray cast iron. (We shall modify this position somewhat when we discuss other gray cast iron microstructures. It is possible, for example, by fast-cooling gray iron from 1500°F (815°C) to obtain pearlite or the other austenite transformation products we found in the austenite transformation of steel.)

The final structure in ferritic gray cast iron is therefore flake graphite plus ferrite. The flakes formed at 2065°F (1130°C) grew during cooling to 1333°F (723°C) and at the eutectoid transformation $\gamma \rightarrow \alpha + $ Gr.

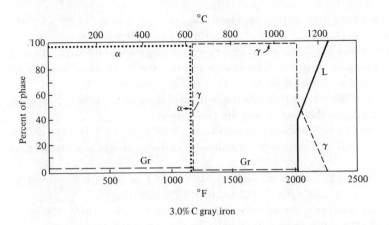

Fig. 6.46 *Fraction chart for gray cast iron.*

Finally it should be pointed out that "gray" and "white" not only refer to the microstructure but also describe the appearance of the fracture of the cast irons. Therefore, a fracture test of cast irons will at least tell whether the high-carbon phase is carbide or graphite.

Ductile Iron. While gray iron is excellent for parts such as engine blocks because the graphite flakes give good machinability, a spheroidal graphite shape is better for strength. In 1949 Millis and Gagnebin discovered that a small (0.05 percent) amount of magnesium dissolved in the liquid iron would cause the formation of spheroidal graphite instead of flake graphite from the melt. Then the graphite that precipitates from 2065 to 1333°F (1130 to 723°C) and at the eutectoid continues to crystallize in the spheroidal form.

The final structure in ferritic ductile iron is therefore *spheroidal* graphite plus ferrite. The phase amounts and analyses are the same as for ferritic gray cast iron, but the different shape of the graphite gives twice the tensile strength and 20 times the ductility.

Malleable Iron. The difference between malleable iron and gray or ductile iron is that the *as-cast* structure of malleable iron is the same as that of white cast iron and contains no graphite. In other words, the cooling rate and composition produce white cast iron in the original casting. The cold casting is then heated in a malleablizing furnace at 1750°F (954°C) to give the iron carbide an opportunity to change to the equilibrium structure of iron plus graphite. At this point a fair question to ask is: Why not produce the graphite during the original cooling and avoid the heat treatment expense? The answer is that malleable iron was developed many years before ductile iron. It was found that malleable iron was stronger and more ductile than *gray cast* iron because during the heat treatment graphite was formed in clumplike nodules (something like spheroidal graphite); this graphite is called "temper carbon." Since the discovery of ductile iron many malleable iron producers have changed to the new material, although malleable iron is still used for many thin-section castings.

It is interesting to follow the structural changes in the heat treatment cycle for malleable iron. There are two stages:

STAGE 1. Graphitization. Heat at 1750°F (954°C) for 12 hr. The structure of austenite plus carbide changes to austenite plus temper carbon.

STAGE 2. Cool from 1750 to 1400°F (954 to 760°C). Slow-cool at 10°F/hr (5.5°C/hr) to 1200°F (649°C).

Whereas austenite would normally transform to α + carbide, the slow-cooling rate above gives α + graphite, in the eutectoid temperature range. The graphite crystallizes on the existing temper carbon formed at the higher temperatures. The final structure is then ferrite plus temper carbon.

EXAMPLE 6.8 Why is it difficult to produce malleable cast iron in large sections?

ANSWER It should be remembered that malleable cast iron must first be cast white (carbide as the high-carbon constituent). In large sections the cooling rate is slow enough so that graphite can form at the high temperatures during solidification. Since this graphite is in flake form, the desirable properties of the temper carbon structure cannot be obtained. Although it is possible to stabilize the carbide by using a lower percentage of silicon or adding carbide-stabilizing elements such as chromium, the time required for stage 1 would become prohibitively long and uneconomical to use.

Now let us consider the variations of properties in these different families as we change the structure of the steel matrix (Table 6.6). Since we have austenite at 1600°F (871°C), we can produce any of the transformation products found in steel: ferrite, pearlite, bainite, martensite, and even retained austenite. These variations will now be discussed for each type of cast iron.

6.34   White cast iron

The largest tonnage of white cast iron is made with a pearlitic matrix, that is, a structure of pearlite and massive carbide. If alloys are added to suppress the pearlite transformation, martensitic white irons are obtained.

Table 6.6 MINIMUM PROPERTIES OF WHITE IRON, GRAY IRON, DUCTILE IRON, MALLEABLE IRON, AND SPECIAL ALLOYS

Name and Number	Chemical Analysis, percent	Condition	Tensile Strength,[a] psi × 10³	Yield Strength,[a] psi × 10³	Percent Elongation	BHN	Typical Use
White cast iron, unalloyed	3.5 C, 0.5 Si	As cast	40	40	0	500	Wear-resistant parts
Gray Iron							
Ferritic class 25	3.5 C, 2.5 Si	As cast	25	20	0.4	150	Pipe, sanitary ware
Pearlitic class 40	3.2 C, 2 Si	As cast	40	35	0.4	220	Machine tools, blocks
Quenched martensitic	3.2 C, 2 Si[b]	Quenched	80	80	0	500	Wearing surfaces
Quenched bainitic	3.2 C, 2 Si[c]	Quenched	70	70	0	300	Camshafts
Ductile Iron							
Ferritic (60-40-18)	3.5 C, 2.5 Si	Annealed	60	40	18	170	Heavy-duty pipe
Pearlitic (80-55-06)	3.5 C, 2.2 Si	As cast	80	55	6	190	Crankshafts
Quenched (120-90-02)	3.5 C, 2.2 Si	Quenched and tempered	120	90	2	270	High-strength machine parts
Malleable Iron							
Ferritic (35018)	2.2 C, 1 Si	Annealed	53	35	18	130	Hardware, fittings
Pearlitic (45010)	2.2 C, 1 Si	Annealed	65	45	10	180	Couplings
Quenched (80002)	2.2 C, 1 Si	Quenched and tempered	100	80	2	250	High-strength yokes
Special Alloy Irons							
Austenitic gray	20 Ni, 2 Cr[d]	As cast	30	30	2	150	Exhaust manifolds
Austenitic ductile	20 Ni[d]	As cast	60	30	20	160	Pump casings
High-silicon gray	15 Si, 1 C	As cast	15	15	0	470	Furnace grates
Martensitic	4 Ni, 2.5 Cr[e]	As cast	40	40	0	600	Wear-resistant parts, liners
white	20 Cr, 2 Mo, 1 Ni[f]	Heat-treated	80	80	0	600	

[a] Multiply psi by 6.9 × 10⁻³ to obtain MN/m² or by 7.03 × 10⁻⁴ to obtain kg/mm².
[b] +1% Ni, 1% Cr, 0.4% Mo
[c] +1% Ni, 1% Mo
[d] +3% C, 2% Si
[e] +3.2% C, 0.8% Si
[f] +2.7% C, 0.8% Si

Examples are:

Analysis, percent

Material	C	Si	Cr	Ni	Mo	BHN
Typical pearlitic white cast iron	3.2	0.5	1.0			450
Martensitic white iron	3.2	0.5	2.5	4.0		600
Martensitic high chromium iron	2.5	0.5	20.0	1.0	2.0	600

The high-alloy content pushes the nose of the TTT diagram far enough to the right so that martensite can even be obtained in as-cast sections. These alloys are used chiefly for abrasion resistance in grinding mills as liners and balls, in cement manufacture and mining equipment, and in rolls for finishing steel.

6.35 Gray cast iron

As in the case of white cast iron, it is possible to vary the matrix of gray cast iron. The range of structures in actual use is wider than that of white iron and includes ferrite, pearlite, bainite, martensite, and even austenite. In all cases the structure depends on the alloy content and cooling rate of the austenite.

Pearlite is easily obtained by cooling fast enough through the eutectoid range to avoid the transformation of austenite to α + graphite. An equally important factor is the percentage of silicon and other elements which affect the stability of carbide. For example, a pearlitic gray iron motor block is made with 3.2 percent carbon and 2 percent silicon and is air-cooled through the eutectoid after pouring. If the block were shaken out at 1400°F (760°C) and cooled slowly in a furnace at 10°F/hr (5.5°C/hr), the structure would be ferritic. The presence of small amounts of tin or copper results in a pearlite matrix even with cooling in the sand molds.

Bainite can be obtained either by an isothermal heat treatment or by casting an alloy combination which produces a TTT curve with a pronounced bainite nose. For example, to produce a bainite structure in a 1-in.-diameter crankshaft, the following analysis is used: 3.2% C, 2% Si, 0.7% Mn, 1% Ni, 1% Mo. The BHN is 300. The hardness is lower than that of a bainitic steel because of the presence of graphite.

Martensite is usually obtained by heat treating—austenitizing plus oil quenching. A typical analysis for an automotive camshaft is: 3.2% C, 0.7% Mn, 2% Si, 1% Ni, 1% Cr, 0.4% Mo. The BHN is 550. A section of an induction-

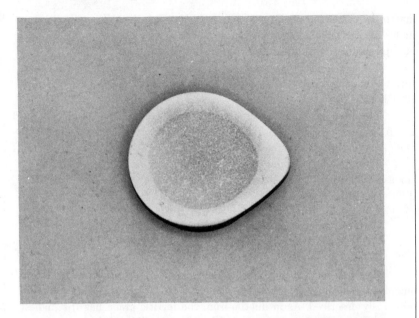

Fig. 6.47 *Cross-section of a cam lobe from an automobile camshaft (gray iron). The camshaft was heated on the surface by induction, then quenched. The hardened layer shows as a light etching structure. Actual size, 5 percent nital etch.*

hardened shaft is shown in Fig. 6.47. In this case, because of the selective heating, only the surface layers were austenitized before quenching.

Austenite is obtained by adding enough alloy to avoid pearlite trans-formation *and* to lower the M_s below room temperature. This can be done with a 20% Ni, 2% Cr alloy or a 14% Ni, 6% Cu, 2% Cr alloy. Examples are exhaust manifolds in which a stable heat-resistant structure is desired and pump parts for which the corrosion resistance of a high-nickel chromium iron is desired.

6.36 Ductile iron

We see from Table 6.6 that the tensile strength of the common grades of gray iron varies from 25,000 to 40,000 psi (173 to 276 MN/m^2) and the elongation is very low. By contrast, the ductile irons have strengths up to 120,000 psi (828 MN/m^2) and high elongation.

The commonest grade is the as-cast pearlitic type 80-55-06. The analysis is controlled to avoid massive carbide, but in addition a pearlitic matrix is formed on cooling. In other words, the silicon is maintained at under 2.5 percent and the manganese is above 0.4 percent. Small amounts of copper and tin are also used to promote pearlite.

The ferritic grade is obtained by slow cooling from the casting temperature or by reheating to 1600°F (871°C) and holding, or slow cooling near 1300°F (704°C).

The quenched and tempered grade is heat-treated in a manner similar to a 0.8 percent carbon steel. It is austenitized at 1600°F (871°C), quenched, and tempered to the desired hardness.

The austenitic grade has higher strength and toughness than the corresponding gray iron. This of course, could have been predicted from the differences in graphite shape.

6.37 Malleable iron

The principal grades of malleable iron are ferritic, pearlitic, and quenched. The ferritic grade is obtained by the two-stage graphitization previously described. As in ductile iron, the ferritic grade has the highest ductility (Table 6.6).

The pearlitic grade is made by avoiding the stage 2 graphitization. In other words, after stage 1 the casting is air-cooled and the austenite transforms to pearlite. If the pearlite is too hard, it is subsequently tempered (spheroidized).

The quenched grade is also called pearlitic, but this is a misnomer. A martensitic structure is produced by quenching from austenite and then tempering to spheroidite.

SUMMARY

The ferrous or iron-base alloys can be divided into three principal groups:

1. Plain carbon and low-alloy steels
2. High-alloy steels (over 10 percent alloy)
3. Cast iron, ductile iron, malleable iron

The equilibrium structures of the first group, the low-alloy steels, can be predicted from the section of the iron–iron carbide diagram below 2 percent carbon. Only three important phases are involved: ferrite (α, BCC with up to 0.02 percent carbon in solid solution), austenite (γ, FCC with up to 2 percent carbon in solid solution), and iron carbide, Fe_3C (6.67 percent carbon). The majority of plain carbon steels, such as constructional and automobile body grades, consist simply of ferrite with increasing amounts of carbide as the carbon content is increased. They are specified in the hot-rolled or cold-rolled condition. When the carbon content is above 0.3 percent in unalloyed steel, it is possible to harden thin sections by heating to produce austenite, followed by rapid quenching in water to form martensite. However, for most heat-treated parts low-alloy steels are used because the hard martensitic structure can be obtained with a less drastic quench and in thicker sections. The TTT curve and the Jominy hardenability bar provide accurate measures of the kinetics of austenite transformation and the relationship between cooling rate and martensite formation. The control of other microstructures, such as pearlite and bainite, employs the same data.

In the second group the addition of large amounts of alloy alters substantially the iron–iron carbide diagram. One group of elements, such as nickel, gives the open γ (austenite) field, leading to alloys that are austenitic at room temperature, such as autenitic stainless steels. By contrast, the elements that give a closed γ field, such as chromium, tend to produce an α structure at all temperatures, as in the ferritic stainless steels. An intermediate condition occurs in the martensitic stainless steels. The quantitative relationships are indicated by ternary diagrams. Tool steels and other specialty grades depend on similar principles.

In the third group, the cast irons and related materials all contain over 2 percent carbon. Some of this may be present as graphite, and to analyze these relationships, the iron-graphite diagram is used. Each of these materials can be considered as a steel matrix (pearlite, bainite, etc.) plus carbide or graphite. In white cast iron no graphite is present, and the product is hard and brittle because of the presence of massive particles of iron carbide. Gray cast iron is soft and machinable because the iron carbide is replaced by graphite flakes, but the ductility is low because of the notch effect of the graphite. Ductile iron exhibits better ductility than gray cast iron because the shape of the graphite is spheroidal. Malleable iron also is ductile because of the temper carbon produced during heat treatment of white cast iron.

DEFINITIONS

Wrought iron Iron with very low carbon content (0.02 percent) which is easily worked while hot into intricate forgings.

Steel An iron-carbon alloy with 0.02 to 2 percent carbon. The most common range is 0.05 to 1.1 percent carbon. Steel can be hot- or cold-worked.

White cast iron An iron-carbon alloy with 2 to 4 percent carbon. The most common range is 3 to 3.5 percent carbon. Because of the presence of large amounts of brittle iron carbide the material is not hot-worked or machined.

Gray cast iron An iron-carbon alloy with 2 to 4 percent carbon and graphitizing elements such as silicon. Most of the carbon is in the form of graphite flakes, so the material is easily machined but low in ductility.

Ductile iron An iron-carbon alloy with 3.5 to 4 percent carbon and graphitizing elements such as silicon. It differs from gray cast iron in that the graphite is in the form of spheres and the ductility is high. The spheroidal graphite is produced by the addition of 0.05 percent magnesium just before the liquid metal is poured into castings.

Malleable iron An iron-carbon alloy with 2 to 3 percent carbon. Castings are produced of white cast iron and then are heat-treated at 1750°F (954°C). The massive iron carbide is converted to nodules of graphite plus iron called "temper carbon." The properties are similar to ductile iron.

Ferrite α iron, BCC, with a maximum of 0.02 percent carbon in solid solution.

Austenite γ iron, FCC, with a maximum of 2 percent carbon in solid solution.

Carbide In the context of iron-base alloys carbide means iron carbide, Fe_3C. It is also called "cementite."

Eutectoid composition In the iron-carbon system this is 0.8 percent carbon. The eutectoid reaction is γ (austenite) $\rightarrow$ α (ferrite) + iron carbide at 1333°F (723°C). The α + iron carbide mixture is called "pearlite."

Spheroidite α plus spheroidized carbide, produced by heating pearlite, bainite, or martensite at elevated temperatures.

Hypoeutectoid steel Steel with carbon content below 0.8 percent (plain carbon).

Hypereutectoid steel Steel with carbon content 0.8 to 2 percent (plain carbon).

Pearlite A mixture of α and carbide phases in parallel plates, produced by the transformation of austenite between 1333°F (723°C) and 1000°F (538°C).

Bainite A mixture of α and very fine carbides showing a needlelike structure, produced by transformation of austenite between 1000°F (538°C) and about 500°F (260°C).

Martensite A nonequilibrium phase consisting of body-centered tetragonal iron with carbon in supersaturated solid solution.

Tempered martensite Martensite that has been heated to produce BCC iron and a fine dispersion of iron carbide.

Austenitizing Heating to the austenite temperature range (which will vary for the particular steel) to produce austenite.

M_s The temperature at which austenite-to-martensite transformation starts.

M_f The temperature at which austenite-to-martensite transformation finishes.

Marquenching A hardening process involving austenitizing then quenching in a liquid bath at about the M_s, followed by air cooling through the martensite transformation range.

Austempering A hardening process involving austenitizing then quenching in a liquid bath above the M_s *and holding* in the bath to transform to bainite, then cooling.

Dilatometer An instrument for measuring the length of a sample during heating and cooling.

TTT curve A time-temperature-transformation curve which indicates the time required for austenite of a given composition to transform to α plus carbide at different temperatures (isothermally).

Nose of TTT curve The temperature range in which isothermal transformation is fastest.

Hardenability A general term indicating for a given steel the ease of avoiding pearlite or bainite transformation so that martensite can be produced. High hardenability does not mean the same as high hardness. The maximum hardness attainable is a function of the carbon content.

Hardenability curve, Jominy curve A graph showing the hardness attained for a given steel using the Jominy end-quench test.

Jominy bar, hardenability bar A 1-in.-(2.54-cm-) diameter by 4-in.-(10.16-cm-) long bar that is austenitized, water-quenched on one end, and then tested for hardness along its length.

Stainless steel A steel that is highly alloyed with chromium and often nickel to improve corrosion resistance.

Maraging steel A highly alloyed steel that is hardened both by martensite and by age hardening.

Superalloys Alloys used for high-temperature service. FCC alloys with nickel, cobalt, iron, and chromium are usually strengthened with an $Ni_3(Al,Ti)$ precipitate.

Stress rupture The value of stress needed to cause rupture in a specified time at a given temperature.

Creep Plastic elongation as a function of time in stressed material which is also related to temperature.

PROBLEMS

6.1 What would be the maximum solubility in percent by weight if all the $\frac{1}{2}, \frac{1}{2}, \frac{1}{2}$ positions in FCC iron were occupied by carbon? Why is this not attained? (If 2 percent by weight carbon is soluble in the FCC iron, what is the number of unit cells of iron per carbon atom?)

6.2 Draw fraction charts showing the amount of each phase present as a function of temperature for the following alloys:

a. SAE-AISI 1010 steel (0.1 percent carbon)
b. SAE-AISI 1080 steel (0.8 percent carbon)
c. 2.5 percent carbon white cast iron

6.3 If a sample of 3.3 percent carbon white cast iron were heat-treated so that only ferrite and graphite were present, what would be the percent by *volume* of graphite in the sample? (The specific gravity of iron is 7.87 and that of graphite is 2.25.)

6.4 Using the TTT curve for 1080 steel, give the microstructure present in a thin [0.010-in. (0.0254-cm)] strip after each step of the following treatments. The original material is hot-rolled. (See Fig. 6.15.)

a. 1600°F (871°C), 1 hr.
   Quench in lead at 1300°F (704°C); hold 20 min.
   Quench in lead at 900°F (482°C); hold 1 sec.
   Water-quench.
b. 1300°F (704°C), 6 hr.
   Water-quench.
c. 1600°F (871°C), 1 hr.
   Quench in salt at 800°F (427°C); hold $\frac{1}{2}$ hr.
   Heat rapidly to 1100°F (593°C); hold 1 hr.
   Water-quench.
d. 1600°F (871°C), 1 hr.
   Quench in salt at 500°F (260°C); hold 1 min.
   Air-cool.
   What is the name of this treatment?
e. 1600°F (871°C), 1 hr.
   Quench in salt at 600°F (315°C); hold 1 hr.
   Air-cool.
   What is the name of this treatment?

6.5 Draw schematic time-temperature diagrams to obtain the following structures in an SAE 1045 steel, giving the structure at each step. (Percentages are approximate.) (See Example 6.2.)

a. 45 percent ferrite, 55 percent pearlite
b. 20 percent ferrite, 80 percent martensite
c. 100 percent martensite
d. 100 percent spheroidite
e. 90 percent bainite, 10 percent ferrite
f. Approximate the M_s for the martensite in part b.

6.6   Draw schematic time-temperature diagrams to obtain the following structures in a 1.2 percent carbon steel. Give the structure after each step.

   a. Spheroidized carbide + pearlite
   b. Grain boundary carbide + pearlite
   c. Spheroidized carbide + tempered martensite

6.7   Although a single specimen of steel will give as a hardenability curve a single line, results of the tests of commercial steel will fall within a hardenability band such as that shown for the 86*xx* series steels in Fig. 6.48.

   a. Why is the hardness band at the $\frac{2}{16}$-in.-station higher in 8650H than in 8630H? What is the microstructure in each case?
   b. What will be the microstructure of 8620H at the $\frac{18}{16}$ position compared with 8640H? Are there several structures which might be encountered in the 8640H at the $\frac{18}{16}$ position?
   c. What is the principal element responsible for the bandwidth at the $\frac{2}{16}$ position in all cases?

6.8   A 1-in.-diameter shaft is made from a sample of 8630H with hardenability that falls in the middle of the band (Fig. 6.48). The shaft is austenitized at 1600°F (871°C) and oil-quenched. A hardness survey shows 40 R_C $\frac{1}{8}$ in. (0.317 cm) beneath the surface. A hardness of 35 R_C minimum is desired at this position, but minimum carbon (that is, the minimum curve from those steels in Fig. 6.48) is recommended. Which 86*xx* steel should be used?

6.9   A midradius hardness of 50 R_C minimum, 62 R_C maximum is required in a 2-in. (5.08-cm) bar. Which of these steels would meet the specification and how should it be quenched (Fig. 6.48)?

6.10 The following readings are obtained from bars of an 8627 steel. Is the hardenability within specification?

	R_C, midradius	R_C, center
A 2-in. (5.08-cm) bar quenched in agitated water	45	39
A 2-in. (5.08-cm) bar quenched in agitated oil	36	35

6.11 Indicate whether the following steels (see page 236) might be hypoeutectoid or hypereutectoid. (*Hint:* The alloy content changes both the percent carbon and temperature associated with the eutectoid.)

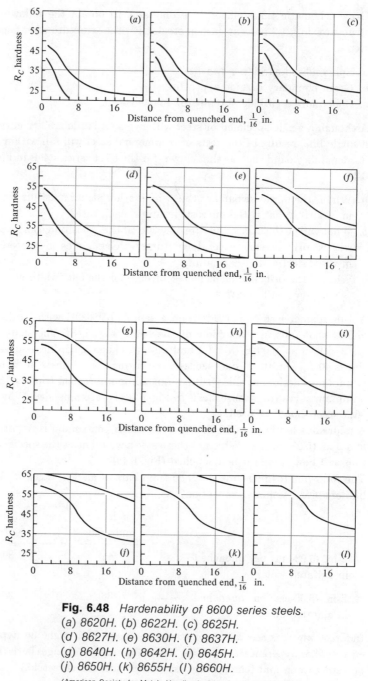

Fig. 6.48 *Hardenability of 8600 series steels.*
(a) 8620H. (b) 8622H. (c) 8625H.
(d) 8627H. (e) 8630H. (f) 8637H.
(g) 8640H. (h) 8642H. (i) 8645H.
(j) 8650H. (k) 8655H. (l) 8660H.

(American Society for Metals Handbook, 8th ed., vol. 1: "Properties and Selection of Metals," Metals Park, Ohio, 1961.)

Table for Fig. 6.48

| SAE-AISI Steel | Chemical Analysis, percent | | | | | | Normalizing Tempera- ture, °F | Austenitizing Tempera- ture, °F |
	C	Mn	Si	Ni	Cr	Mo		
8620H	0.17 to 0.23	0.60 to 0.95	0.20 to 0.35	0.35 to 0.75	0.35 to 0.65	0.15 to 0.25	1700	1700
8622H	0.19 to 0.25	0.60 to 0.95	0.20 to 0.35	0.35 to 0.75	0.35 to 0.65	0.15 to 0.25	1700	1700
8625H	0.22 to 0.28	0.60 to 0.95	0.20 to 0.35	0.35 to 0.75	0.35 to 0.65	0.15 to 0.25	1650	1600
8627H	0.24 to 0.30	0.60 to 0.95	0.20 to 0.35	0.35 to 0.75	0.35 to 0.65	0.15 to 0.25	1650	1600
8630H	0.27 to 0.33	0.60 to 0.95	0.20 to 0.35	0.35 to 0.75	0.35 to 0.65	0.15 to 0.25	1650	1600
8637H	0.34 to 0.41	0.70 to 1.05	0.20 to 0.35	0.35 to 0.75	0.35 to 0.65	0.15 to 0.25	1600	1550
8640H	0.37 to 0.44	0.70 to 1.05	0.20 to 0.35	0.35 to 0.75	0.35 to 0.65	0.15 to 0.25	1600	1550
8642H	0.39 to 0.46	0.70 to 1.05	0.20 to 0.35	0.35 to 0.75	0.35 to 0.65	0.15 to 0.25	1600	1550
8645H	0.42 to 0.49	0.70 to 1.05	0.20 to 0.35	0.35 to 0.75	0.35 to 0.65	0.15 to 0.25	1600	1550
8650H	0.47 to 0.54	0.70 to 1.05	0.20 to 0.35	0.35 to 0.75	0.35 to 0.65	0.15 to 0.25	1600	1550
8655H	0.50 to 0.60	0.70 to 1.05	0.20 to 0.35	0.35 to 0.75	0.35 to 0.65	0.15 to 0.25	1600	1550
8660H	0.55 to 0.65	0.70 to 1.05	0.20 to 0.35	0.35 to 0.75	0.35 to 0.65	0.15 to 0.25	1600	1550

     a. 0.6% C, 1% Mo, 1% Cr
     b. 0.6% C, 1% Ni, 1% Cr

6.12 Of the eight steels shown in Fig. 6.36, which one would make the best drill bit? Explain.

6.13 Given a stainless steel with 15 percent chromium, what is the maximum carbon content it can contain and still remain ferritic?

6.14 Hadfield's manganese steel (12% Mn, 1% C) when quenched will be all austenite, yet when slowly cooled may contain extensive amounts of martensite. Explain what might take place. (*Hint:* The alloy content of the austenite determines the M_s.)

6.15 Which steel(s) of Table 6.3 would you specify for the following applications? Explain the deficiencies of the others.

     a. A kitchen knife of intricate blade shape for preparing grapefruit.
     b. A blade for carving which would hold its edge best.
     c. Automobile trim.
     d. Corrosion-resistant aircraft parts with the best strength-to-weight ratio.
     e. A cast alloy for exhaust valves contains 15% Ni, 15% Cr, 1% C. From the point of view of structure why is this more resistant to deformation than one of the austenitic steels listed?

6.16 Calculate the alloy cost per pound of the tool steels given in Table 6.4 using $0.30/lb for chromium, $1.30/lb for molybdenum, $5/lb for tungsten, and $4/lb for vanadium (1974 prices).

6.17 Why are the austenitizing temperatures for the M1 and T1 types much higher than for other grades of low-alloy steel?

6.18 Why are higher tempering temperatures used for the M1 and T1 types than those for low-alloy steels?

6.19 From the creep data given in Fig. 6.42, calculate the following:

     a. Minimum creep rate (percent strain per hour) at a stress of 15,000 psi (103 MN/m^2). (It may be necessary to replot the data. Use only the straight-line portion of the curve.)
     b. Stress to rupture in 1,000 hr.
     c. Total creep in 1,000 hr.

6.20 Which of the following data would you use for the applications given below for operation at elevated temperatures (over 1000°F, 538°C)?

     1. Stress rupture
     2. Second-stage creep (minimum creep rate per hour)
     3. Total elongation vs. time
     4. Hot tensile test
     5. Room temperature tensile and elongation values

In some cases the failure is determined by excessive deformation, in others by time to rupture. Example: In a superheater tube considerable distortion is allowable but rupture cannot be tolerated. By contrast, in a valve stem the part would not function properly after deformation.

Applications:

a. Bolts
b. Steam valves
c. Oil still tubes
d. Turbine rotors
e. Turbine casings, steam turbine blading

6.21 As mentioned in the text, nickel combines with aluminum and titanium to form the compound Ni_3X, where X is aluminum or titanium or both. From the analysis given in Table 6.5 for Udimet 700, calculate the percent by weight of this compound present, assuming all aluminum and titanium are in the compound. Now convert to percent by volume. At what content of aluminum weight percent or titanium weight percent would the structure be 100 percent γ'?

6.22 Draw fraction charts for a 3 percent carbon iron (showing percentage of phases as a function of temperature) that is cooled under different conditions.

Case A. Cooled at an intermediate rate to produce pearlite plus flake graphite
Case B. White cast iron reheated to 1750°F (954°C), held 12 hr, and air-cooled
Case C. White cast iron reheated to 1750°F (954°C), held 12 hr, slowly cooled to 1300°F (704°C), held 12 hr, and air-cooled
Case D. Case C reheated to equilibrium at 1600°F (871°C), and water-quenched

6.23 When appreciable silicon is present (over 0.5 percent), the effect on the iron-carbon diagram is important. In Fig. 6.49 the phase diagram for 2 percent silicon is shown.

a. According to the phase rule, why can $\alpha + \gamma$ + graphite (three phases) exist over a range of temperature?
b. The eutectic composition is no longer 4.3 percent carbon but follows the formula eutectic = 4.3 − 0.3% Si. Verify this for the diagram shown.
c. What is the carbon content of a pearlitic region in a 2 percent silicon cast iron?
d. How does the austenitizing temperature prior to quenching compare with a 0.8 percent carbon steel?

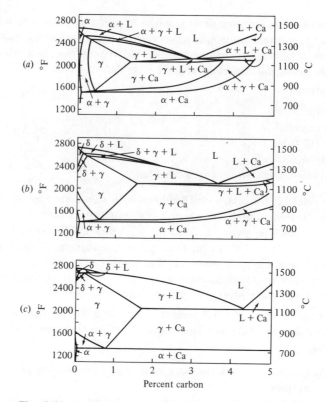

Fig. 6.49 *Iron–iron carbide diagram at different silicon levels. (a) 4 percent silicon. (b) 2 percent silicon. (c) 0 percent silicon.*

(After Greiner, et al.)

6.24 Which material(s) could be used to fill the following specifications, and what analyses and heat treatments should be used? (See Table 6.6.)

Case A. Automobile crankshaft: maximum section 2 in. (5.08 cm), BHN 220, endurance limit in fatigue 20,000 psi (138 MN/m²). This is discussed later and can be taken as 0.25 × tensile strength.

Case B. Engine head: sections $\frac{1}{4}$ to 1 in. (0.635 to 2.54 cm) in diameter, BHN 200, good machinability, tensile strength 30,000 psi (207 MN/m²).

Case C. Tappet: 1 in. (2.54 cm) in diameter by 2 in. (5.08 cm) long. One end to be BHN 400, free from graphite to depth of $\frac{1}{4}$ in. (0.635 cm). Balance of casting BHN 250 maximum, machinable.

Case D. Differential housing: sections $\frac{1}{4}$ to 1 in. (0.635 to 2.54 cm) in diameter, tensile strength 50,000 psi (345 MN/m²), good toughness to resist road conditions, good machinability.

Case E. Camshaft: good machinability. Cam lobes 50 R_C, structure: tempered martensite, flake graphite, 5 percent massive carbide.

Case F. Disk brake: good machinability, structure: flake graphite plus pearlite, BHN 200.

Case G. Cylinder block: flake graphite plus pearlite. (A plain gray iron 3.2% C, 0.5% Mn, 2% Si contains 15% ferrite, balance pearlite, and is not satisfactory because of the effect of wear on ferrite.)

Case H. Grinding mill roll: BHN 550 minimum on surface, no graphite.

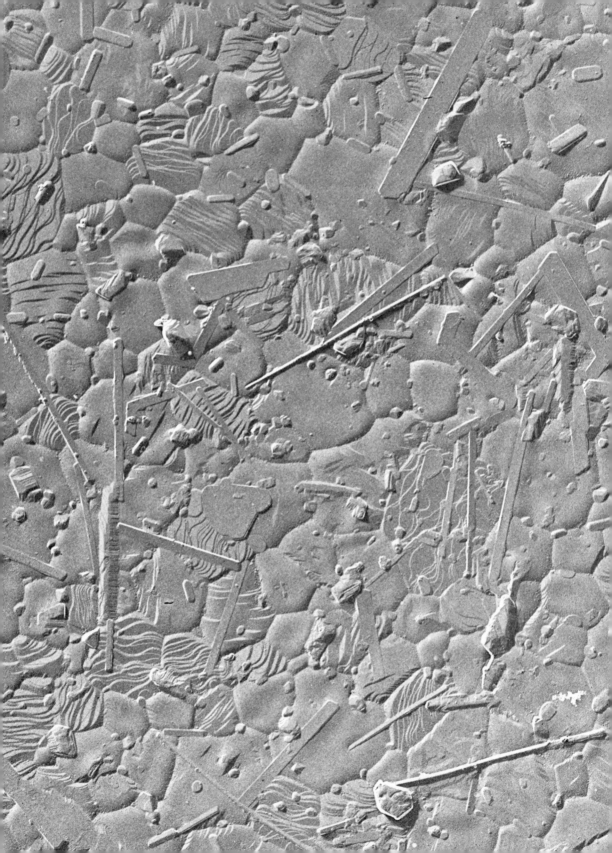

7

CERAMIC STRUCTURES AND
THEIR PROPERTIES

*THE microstructure of a newly invented ceramic called "Pyro-
ceram" is shown in the photomicrograph. This ceramic is typical
of the important advances taking place in the very old field of
ceramics.*

*There is still some argument as to whether glass should
be considered a ceramic, but the new field of glass ceramics of
which Pyroceram is a typical product provides an answer.*

*In this photomicrograph, taken with an electron micro-
scope at 25,000X, the long bladelike crystals, which are* TiO_2
*(rutile), have assisted in the nucleation of the spodumeme from
a glassy matrix (see text). (Corning Glass Laboratory, Dr.
J. H. Munier.)*

*In this chapter we will study the ceramic structures in
the same way we studied the metals. First we consider the types
of bonding: ionic, covalent, and van der Waals. Then we study
the important unit cells of which ceramics are composed, and
finally we consider the properties.*

7.1 Introduction

We will advance now from the metals to the second group of materials, the ceramics, and we find that there is a double reward in studying this field. First, from the point of view of engineering materials, it is vital to understand not only the traditional ceramics such as concrete, glasses, whiteware, and brick but also the newer ceramics such as laser materials, the barium titanate ferroelectrics, ceramic magnets, and light-sensitive glasses. Second, ceramics are our oldest and most beautiful materials, and it gives a deep feeling of satisfaction to know the basic materials and processing of the very earliest shapes of baked clay and shaped stone as well as the elegant porcelain forms and exotic enamels of today.

Let us first outline the field of ceramics and then review the portions of our study of metals that are of use to us in reaching a basic understanding of the ceramics.

When the word "ceramics" appears, many people think only of dinnerware and pottery, and indeed these uses are not far from the meaning of the original Greek word, "fired material." Today, however, the ceramist and the engineer who is aware of the field realize that ceramics include the following important materials:

Glasses as in containers, insulators, windows
Cements such as Portland cement, lime, gypsum
Brick and pipe for construction
Refractories such as furnace linings and insulation
Whiteware as in dishes
Vitreous enameled ware for plumbing and chemical uses
Abrasives for wheels
Molds for metal casting
Special ceramics such as ferrites for magnets, lead-zirconium-titanium oxide
(PZT) for depth sounders, ruby single crystals for lasers, photoconductive
and photochromic glasses and cermets

From this list we see that ceramics fill many conventional uses and that they are involved in many new and growing fields as well. Under the broad classification of ceramics we will include not only the oxides and silicates but also all the important inorganic nonmetallic materials such as the hard carbides and graphite.

Although at first glance there appears to be little in common between ceramics and metals, a great deal of the knowledge we have acquired concerning metallic structures can be applied to ceramics. Below we will outline step by step the reasoning we followed in arriving at our understanding of metallic materials and at each point indicate the additional or contrasting information we need to understand ceramics.

242

We began the study of metals with a review of the metallic elements and the formation of the metallic bond. By contrast, we will see that in most ceramics, instead of a single metal or metallic phases, there are usually both metallic and nonmetallic elements present with ionic or covalent bonds. Therefore, we need to consider not only the structure of the metallic atom that we have already discussed but also that of the nonmetallic atom, often oxygen, as well as the balance of charges given by the valences.

As with the metals, the unit cell is a key point in describing ceramic structures, and the cubic and hexagonal cells are still the most important. In addition, however, the difference in radii between the metallic and nonmetallic ions plays an important part in the arrangement of ions in the unit cell.

We recall that in the metals the regular arrangement of metallic atoms into densely packed planes led to the occurrence of slip under stress, giving metals their characteristic ductility. In ceramics, by contrast, both the arrangement of the atoms and the type of bonding are different. In general, we encounter brittle fracture rather than slip. However, the regular arrangement of the atoms determines the path of the fracture and the *cleavage,* the plane of fracture, is closely related to the makeup of the planes of atoms.

We shall see that the phase diagram is of great importance in understanding the formation and control of the microstructure of polyphase ceramics, just as it is for polyphase metallic materials. Also, the occurrence of nonequilibrium structures is even more prevalent in ceramics because the more complex crystal structures are more difficult to nucleate and to grow from the melt.

This leads us to a large group of ceramics, the *glasses,* with structures that are totally different from those encountered in the metals,† although they are related to some found in the plastics. For example, if we melt pure silica sand, SiO_2, and cool faster than at a certain rate, we obtain a glass called "fused silica" instead of a crystalline material. This structure is closer to a liquid than a crystalline solid, although at room temperature it is rigid and has a definite modulus of elasticity. It is made up of a giant network in space of silicon and oxygen atoms. The unit building block is a tetrahedron composed of a silicon atom in the center with four oxygen atoms at the corners, giving the formula SiO_4^{4-}. However, although the tetrahedra all have the same structure, there is no order in the way they are linked together, giving a giant random network. There is no regular cleavage in fracture, the structure softens far below the melting point of the crystalline form, and it exhibits the viscous behavior that is vital to the shaping of glass.

With this introduction, let us first consider in this chapter the special features of bonding in ceramic structures and the mechanical properties of commercial ceramics. In the next chapter we will study the processing of ceramics into useful shapes and the special problems of selected products such

---

†Except at extremely rapid cooling rates, when metal glass can be formed.

as cement and concrete, whiteware, abrasive wheels, and molds for metal castings.

The electrical, magnetic, and optical properties of ceramics are discussed in Chaps. 13, 14, and 15.

7.2 Bonding forces

In discussing the metallic bond we found that the metal atoms give up their outer (valence) electrons and take positions as positive ions in a space lattice. The electrons provide a bonding force as an electron gas. In contrast, let us now review ionic and covalent bonding in ceramics.

In the *ionic bond* of sodium chloride the single electron of the outer ring of the sodium is attracted to the outer shell of the chlorine, which contains seven electrons. We recall from chemistry that there is a strong driving force for this transfer because this added electron gives chlorine a stable outer shell of eight as well as leaving a stable inner ring for the sodium. We now have the task of determining how an assortment of equal quantities of Na^+ and Cl^- ions comes to equilibrium. By the same x-ray diffraction technique that is used for metals we find the arrangement shown in Fig. 7.1. As would be expected, the negative ions are clustered around the positive ions and, conversely, the positive around the negative. We can also say that this arrangement is obtained because ions of the same charge repel each other. The bonding force is related to how electropositive, i.e. ready to give up electrons, the sodium is and how electronegative, i.e. ready to accept electrons, the chlorine is. In general, ionic bonds are stronger than metallic bonds, as pointed out in Chap. 2. We should note finally that there are no NaCl "molecules" because each Na^+ is surrounded by and attracted equally to six equidistant Cl^- ions.†

The *covalent bond* is also encountered in ceramics and is of great importance. In solids the perfect example of the covalent bond is seen in a diamond crystal. Each carbon atom has four electrons in its outer shell and needs eight

---

†More distant ions also contribute to the bonding.

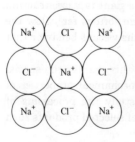

Fig. 7.1 *Two-dimensional sketch of the sodium chloride structure. Note that the center Na^+ is situated near six Cl^- (only four of which are shown, the others being above and below).*

C ⦂ C ⦂ C ⦂ C
•• •• •• ••
C ⦂ C ⦂ C ⦂ C
•• •• •• ••
C ⦂ C ⦂ C ⦂ C   **Fig. 7.2** *Covalent bonding in*
•• •• •• ••   *diamond as a two-dimensional*
C ⦂ C ⦂ C ⦂ C   *sketch. Compare to Fig. 7.10.*

for completion. It shares one of its electrons with each of four equally spaced neighbors (Fig. 7.2). Thus in the covalent case the bond is produced by electron sharing, the electrons remaining equidistant between the atoms. This bond is of the highest strength, and since there are very few free electrons or even ions that are mobile, the best electrical insulators are made of ceramics of this type.

As a matter of fact, most ceramics have bonding that is partly ionic and partly covalent. Since electrons are in motion, we can say that the bonding is divided between ionic and covalent in a substance such as silica sand, SiO_2, depending on the position of the bonding electron relative to the ions. In addition to the ionic and covalent bonds, *van der Waals forces* are active in bonding but are smaller in magnitude. As an example, the ionic and covalent bonds are strongest in the plane of a layer of clay, and weaker van der Waals forces hold adjacent layers together. Therefore, the cleavage is perfectly parallel to the layers.

7.3   Unit cells

In the metals we considered only cubic, hexagonal, and tetragonal unit cells. In ceramic structures additional types are encountered, and for completeness all the crystal systems are again shown (Fig. 7.3). However, the cubic, hexagonal, and tetragonal systems are the most important in the ceramics, just as they are in the metals.

In ceramic materials there are certain groups of crystal structures that are known by key names, such as the "sodium chloride structure" and the "calcium fluoride structure." While the chemicals of the key structure are not very important ceramic materials in themselves, they represent large structural groups. For example, the very important FeO, MgO, and CaO crystal structures are the same as the sodium chloride structure. Another way of looking at this is to point out that there are some frequently occurring chemical formulas in ceramics: AX, AX_2, A_2X, ABX_3, A_2X_3, and AB_2X_4. In each case A and B are metals and X is a nonmetal. If we know the few characteristic structures of AX compounds in general, the entire field is much simpler. Therefore, we use the term "sodium chloride type structure" to describe a whole group of AX structures. Let us examine these key structures.

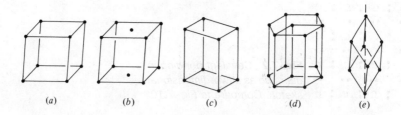

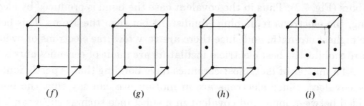

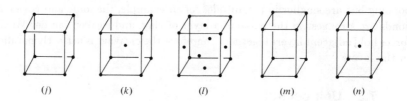

System	Axes	Axial Angles
Cubic	$a_1 = a_2 = a_3$	All angles = 90°
Tetragonal	$a_1 = a_2 \neq c$	All angles = 90°
Orthorhombic	$a \neq b \neq c$	All angles = 90°
Monoclinic	$a \neq b \neq c$	Two angles = 90°; one angle ≠ 90°
Triclinic	$a \neq b \neq c$	All angles different; none equals 90°
Hexagonal	$a_1 = a_2 = a_3 \neq c$	Angles = 90° and 120°
Rhombohedral	$a_1 = a_2 = a_3$	All angles equal, but not 90°

Fig. 7.3 *The 14 crystal lattices and their geometric relationships. The lattices continue in three dimensions.*
(a) Simple monoclinic. (b) End-centered monoclinic. (c) Triclinic. (d) Hexagonal. (e) Rhombohedral.
(f) Simple orthorhombic. (g) Body-centered orthorhombic. (h) End-centered orthorhombic. (i) Face-centered orthorhombic.
(j) Simple cubic. (k) Body-centered cubic. (l) Face-centered cubic.
(m) Simple tetragonal. (n) Body-centered tetragonal.

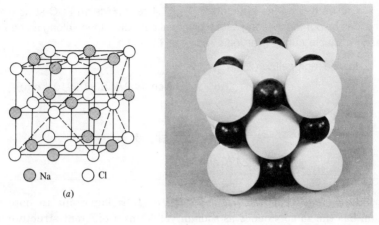

Na   Cl
(a)

(b)

Fig. 7.4 *The sodium chloride structure.* (a) *Sketch of the ion configuration.* (b) *Photograph of a hard sphere model. Note that the ions do* not *touch along what might be considered a facial diagonal. (The larger spheres are Cl⁻.)*

    The Sodium Chloride Structure. Although this structure is basically simple cubic (SC), it is more easily described as an FCC grouping of Cl⁻ ions with Na⁺ ions touching each one of a group of six Cl⁻ ions, as shown in Fig. 7.4.

EXAMPLE 7.1 Calculate the density of sodium chloride from the crystal structure (Fig. 7.4) and the atomic weights of Na^+ and Cl^-.

ANSWER We know that

$$\text{Density} = \frac{\text{mass}}{\text{volume}}$$

$$= \frac{\text{mass of a unit cell}}{\text{volume of a unit cell}}$$

    From Fig. 7.4 the larger ions (Cl⁻) seem to form an FCC-like arrangement. From Chap. 3 we know that the FCC structure has four atoms per unit cell. Therefore, sodium chloride has four Cl⁻ ions, and to maintain the 1:1 stoichiometry it must have four Na⁺ ions.

$$\frac{\text{Mass}}{\text{Unit cell}} = \frac{4\ Na^+ \times 23\ \text{g/at. wt.} + 4\ Cl^- \times 35.45\ \text{g/at. wt.}}{6.02 \times 10^{23}\ \text{atoms/at. wt.}}$$

$$= 3.88 \times 10^{-22}\ \text{g}$$

Even though the structure in Fig. 7.4 resembles an FCC structure, the ions *do not* touch along a face diagonal but rather touch along an edge, as in a simple cubic. (For ionic radii see Table 7.1.)

$$\text{Volume/unit cell} = (2 \times r_{Cl^-} + 2 \times r_{Na^+})^3$$
$$= (2 \times 1.81 \times 10^{-8}\,\text{cm} + 2 \times 0.98 \times 10^{-8}\,\text{cm})^3$$
$$= 1.737 \times 10^{-22}\,\text{cm}^3$$
$$\text{Density} = \frac{3.88 \times 10^{-22}\,\text{g}}{1.737 \times 10^{-22}\,\text{cm}^3}$$
$$= 2.24\,\text{g/cm}^3$$

The Cesium Chloride Structure. It is important to note that cesium has the same valence as sodium yet forms a different structure with chlorine (Fig. 7.5). In this case each cesium ion touches *eight* chloride ions. This is called a *coordination number* (CN) of 8 compared with a CN of 6 in sodium chloride. The basic reason for the difference is that the cesium ion is larger and can touch more Cl^- ions at the same time. By contrast, a sodium ion would "rattle around" (Fig. 7.6a, page 249) inside a cube made up of eight Cl^- ions and could touch only six at the same time, as shown in Fig. 7.4.

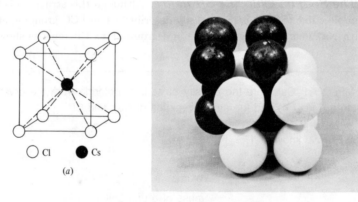

◯ Cl   ● Cs

(a)

(b)

Fig. 7.5 *The cesium chloride structure. (a) Sketch of the ion configuration. (b) Photograph of the hard sphere model. (The larger ions are the Cl^-.) The structure looks like two interpenetrating simple cubics. The ions touch along the body diagonal.*

Fig. 7.6 *(a) Stable and unstable ionic coordination configurations. Instability arises when the ion (cation in this instance) "rattles around" between the anions. (b) Radius ratios for various atom arrangements in ionic bonding. The ratios are cation/anion (r^+/r^-).*

(C. R. Barrett, W. D. Nix, and A. S. Tetelman, "Principles of Engineering Materials," Prentice-Hall, Inc., Englewood Cliffs, N.J., 1973.)

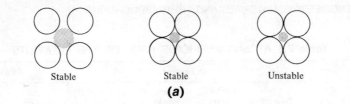

Stable	Stable	Unstable

(a)

Ratio of Cation Radius to Anion Radius	Disposition of Ions about Central Ion	CN	
1	Same as FCC structure, Chap. 2	12	
1 to 0.732	·Corners of cube	8	
0.732 to 0.414	Corners of octahedron	6	
0.414 to 0.225	Corners of tetrahedron	4	
0.225 to 0.155	Corners of triangle	3	
0.155 to 0	Single pair	2	

(b)

Table 7.1 ATOMIC AND IONIC RADII OF THE ELEMENTS

Atomic Number	Symbol	As Element				As Ion	
		Type of Structure*	Coordination Number	Inter-atomic Distances, Å†	Goldschmidt Atomic Radii, Å‡	State of Ionization	Goldschmidt Ionic Radii, Å
1	H	HCP	6, 6	—	0.46	H^-	1.54
2	He	—	—	—	—	—	—
3	Li	BCC	8	3.03	1.57	Li^+	0.78
4	Be	HCP	6, 6	2.22; 2.28	1.13	Be^{2+}	0.54
5	B	—	—	—	0.97	B^{3+}	0.2
6	C	Dia.	4	1.54	0.77	C^{4+}	<0.2
		Hex.	3	1.42	—		
7	N	Cubic	—	—	0.71	N^{5+}	0.1 to 0.2
8	O	Orthorh.	—	—	0.60	O^{2-}	1.32
9	F	—	—	—	—	F^-	1.33
10	Ne	FCC	12	3.20	1.60	—	—
11	Na	BCC	8	3.71	1.92	Na^+	0.98
12	Mg	HCP	6, 6	3.19; 3.20	1.60	Mg^{2+}	0.78
13	Al	FCC	12	2.86	1.43	Al^{3+}	0.57
14	Si	Dia.	4	2.35	1.17	Si^{4-}	1.98
						Si^{4+}	0.39
15	P	Orthorh.	3	2.18	1.09	P^{5+}	0.3 to 0.4
16	S	FC Orthorh.	—	2.12	1.04	S^{2-}	1.74
						S^{6+}	0.34
17	Cl	Orthorh.	—	2.14	1.07	Cl^-	1.81
18	Ar	FCC	12	3.84	1.92	—	—
19	K	BCC	8	4.62	2.38	K^+	1.33
20	Ca	FCC	12	3.93	1.97	Ca^{2+}	1.06
		HCP	6, 6	3.98; 3.99	2.00		
21	Sc	FCC	12	3.20	1.60	Sc^{2+}	0.83
		HCP	6, 6	3.23; 3.30	1.64		
22	Ti	HCP	6, 6	2.91; 2.95	1.47	Ti^{2+}	0.76
						Ti^{3+}	0.69
						Ti^{4+}	0.64
23	V	BCC	8	2.63	1.36	V^{3+}	0.65
						V^{4+}	0.61
						V^{5+}	~ 0.4
24	Cr	BCC (α)	8	2.49	1.28	Cr^{3+}	0.64
		HCP (β)	6, 6	2.71; 2.72	1.36	Cr^{6+}	0.3 to 0.4
25	Mn	Cubic (α)	—	2.24 to 2.96	1.12	Mn^{2+}	0.91
		Cubic (β)	—	2.36 to 2.68	1.18	Mn^{3+}	0.70
		FCT (γ)	8, 4	2.58; 2.67	~ 1.37	Mn^{4+}	0.52
26	Fe	BCC (α)	8	2.48	1.28	Fe^{2+}	0.87
		FCC (γ)	12	2.52	1.26	Fe^{3+}	0.67
27	Co	HCP (α)	6, 6	2.49, 2.51	1.25	Co^{2+}	0.82
		FCC (β)	12	2.51	1.26	Co^{3+}	0.65
28	Ni	HCP (α)	6, 6	2.49; 2.49	1.25	Ni^{2+}	0.78
		FCC (β)	12	2.49	1.25		
29	Cu	FCC	12	2.55	1.28	Cu^+	0.96
30	Zn	HCP	6, 6	2.66; 2.91	1.37	Zn^{2+}	0.83
31	Ga	Orthorh.	—	2.43 to 2.79	1.35	Ga^{3+}	0.62
32	Ge	Dia.	4	2.44	1.39	Ge^{4+}	0.44

Table 7.1 ATOMIC AND IONIC RADII OF THE ELEMENTS (*Continued*)

Atomic Number	Symbol	Type of Structure*	As Element			As Ion	
			Coordination Number	Inter-atomic Distances, Å†	Gold-schmidt Atomic Radii, Å‡	State of Ionization	Gold-schmidt Ionic Radii, Å
33	As	Rhomb.	3, 3	2.51; 3.15	1.25	As^{3+} As^{5+}	0.69 ~0.4
34	Se	Hex.	2, 4	2.32; 3.46	1.16	Se^{2-} Se^{6+}	1.91 0.3 to 0.4
35	Br	Orthorh.	—	2.38	1.19	Br^-	1.96
36	Kr	FCC	12	3.94	1.97	—	—
37	Rb	BCC	8	4.87	2.51	Rb^+	1.49
38	Sr	FCC	12	4.30	2.15	Sr^{2+}	1.27
39	Y	HCP	6, 6	3.59; 3.66	1.81	Y^{3+}	1.06
40	Zr	HCP BCC	6, 6 8	3.16; 3.22 3.12	1.60 1.61	Zr^{4+}	0.87
41	Nb	BCC	8	2.85	1.47	Nb^{4+} Nb^{5+}	0.74 0.69
42	Mo	BCC	8	2.72	1.40	Mo^{4+} Mo^{6+}	0.68 0.65
43	Tc	—	—	—	—	—	—
44	Ru	HCP	6, 6	2.64; 2.70	1.34	Ru^{4+}	0.65
45	Rh	FCC	12	2.68	1.34	Rh^{3+} Rh^{4+}	0.68 0.65
46	Pd	FCC	12	2.75	1.37	Pd^{2+}	0.50
47	Ag	FCC	12	2.88	1.44	Ag^+	1.13
48	Cd	HCP	6, 6	2.97; 3.29	1.52	Cd^{2+}	1.03
49	In	FCT	4, 8	3.24; 3.37	1.57	In^{3+}	0.92
50	Sn	Dia. Tetra.	4 4, 2	2.80 3.02; 3.18	1.58 —	Sn^{4-} Sn^{4+}	2.15 0.74
51	Sb	Rhomb.	3, 3	2.90; 3.36	1.61	Sb^{3+}	0.90
52	Te	Hex.	2, 4	2.86; 3.46	1.43	Te^{2-} Te^{4+}	2.11 0.89
53	I	Orthorh.	—	2.70	1.36	I^- I^{5+}	2.20 0.94
54	Xe	FCC	12	4.36	2.18	—	—
55	Cs	BCC	8	5.24	2.70	Cs^+	1.65
56	Ba	BCC	8	4.34	2.24	Ba^{2+}	1.43
57	La	HCP FCC	6, 6 12	3.72; 3.75 3.75	1.87 1.87	La^{3+}	1.22
58	Ce	HCP FCC	6, 6 12	3.63; 3.65 3.63	1.82 1.82	Ce^{3+} Ce^{4+}	1.18 1.02
59	Pr	Hex. FCC	6, 6 12	3.63; 3.66 3.64	1.83 1.82	Pr^{3+} Pr^{4+}	1.16 1.00
60	Nd	Hex.	6, 6	3.62; 3.65	1.82	Nd^{3+}	1.15
61	Pm	Hex.	—	—	—	Pm^{3+}	1.06
62	Sm	Rhomb.	—	—	1.81	Sm^{3+}	1.13
63	Eu	BCC	8	3.96	2.04	Eu^{3+}	1.13
64	Gd	HCP	6, 6	3.55; 3.62	1.80	Gd^{3+}	1.11
65	Tb	HCP	6, 6	3.51; 3.59	1.77	Tb^{3+} Tb^{4+}	1.09 0.89

(*Continued*)

Table 7.1 ATOMIC AND IONIC RADII OF THE ELEMENTS (*Continued*)

			As Element			As Ion	
Atomic Number	Symbol	Type of Structure*	Coordination Number	Inter-atomic Distances, Å†	Gold-schmidt Atomic Radii, Å‡	State of Ionization	Gold-schmidt Ionic Radii, Å
66	Dy	HCP	6, 6	3.50; 3.58	1.77	Dy^{3+}	1.07
67	Ho	HCP	6, 6	3.48; 3.56	1.76	Ho^{3+}	1.05
68	Er	HCP	6, 6	3.46; 3.53	1.75	Er^{3+}	1.04
69	Tm	HCP	6, 6	3.45; 3.52	1.74	Tm^{3+}	1.04
70	Yb	FCC	12	3.87	1.93	Yb^{3+}	1.00
71	Lu	HCP	6, 6	3.44; 3.51	1.73	Lu^{3+}	0.99
72	Hf	HCP	6, 6	3.13; 3.20	1.59	Hf^{4+}	0.84
73	Ta	BCC	8	2.85	1.47	Ta^{5+}	0.68
74	W	BCC (α)	8	2.74	1.41	W^{4+}	0.68
		Cubic (β)	12; 2, 4	2.82; 2.52, 2.82	1.41	W^{6+}	0.65
75	Re	HCP	6, 6	2.73; 2.76	1.38	Re^{4+}	0.72
76	Os	HCP	6, 6	2.67; 2.73	1.35	Os^{4+}	0.67
77	Ir	FCC	12	2.71	1.35	Ir^{4+}	0.66
78	Pt	FCC	12	2.77	1.38	Pt^{2+}	0.52
						Pt^{4+}	0.55
79	Au	FCC	12	2.88	1.44	Au^{+}	1.37
80	Hg	Rhomb.	6	3.00	1.55	Hg^{2+}	1.12
81	Tl	HCP	6, 6	3.40; 3.45	1.71	Tl^{+}	1.49
		BCC	8	3.36	1.73	Tl^{3+}	1.06
82	Pb	FCC	12	3.49	1.75	Pb^{4-}	2.15
						Pb^{2+}	1.32
						Pb^{4+}	0.84
83	Bi	Rhomb.	3, 3	3.11; 3.47	1.82	Bi^{3+}	1.20
84	Po	Monocl.	—	2.81	1.40	Po^{6+}	0.67
85	At	—	—	—	—	At^{7+}	0.62
86	Rn	—	—	—	—		
87	Fr	—	—	—	—	Fr^{+}	1.80
88	Ra	—	—	—	—	Ra^{+}	1.52
89	Ac	—	—	—	—	Ac^{3+}	1.18
90	Th	FCC	12	3.60	1.80	Th^{4+}	1.10
91	Pa	—	—	—	—	—	—
92	U	Orthorh.	—	2.76	1.38	U^{4+}	1.05
93	Np	—	—	—	—	—	—
94	Pu	—	—	—	—	—	—
95	Am	—	—	—	—	—	—
96	Cm	—	—	—	—	—	—

*Abbreviations: HCP = hexagonal close-packed, BCC = body-centered cubic, Dia. = diamond, Hex. = hexagonal, Orthorh. = orthorhombic, FCC = face-centered cubic, FCT = face-centered tetragonal, Rhomb. = rhombohedral, Tetra. = tetragonal, Monocl. = monoclinic.

†Some elements have two values because calculations in different directions in the unit cell give different numbers.

‡In our calculations we use the Goldschmidt radii, which compensate for the fact that the radii are measured with different coordination numbers in different unit cells.

Coordination Number (CN). At this point we interrupt our discussion of unit cells to point out the relations of CN to the ionic radii in the structure. Ionic radii are given in Table 7.1, and it is important to note that in general the negative (nonmetallic) ions have the larger radii. This is because the positive (metallic) ions give up outer electrons, and the ionic radius therefore becomes smaller than the atomic radius. However, the addition of electrons to the outer shell of the negative ions reduces the force of the nucleus on the individual electrons, and the ionic radius is greater than the atomic radius. See, for example, Fe^{2+} vs. Fe^0 and Cl^- vs. Cl^0.

To achieve a stable grouping of larger ions around a smaller ion, i.e. a given CN, the smaller ion must touch all the larger ones. The greater the ratio of the radius r of the smaller ion to the radius R of the larger (i.e. the closer the ratio is to 1), the greater the number of ions that can be grouped around the smaller ion. It is observed that in the unit cells found in nature CN can equal 2, 3, 4, 6, 8, and 12. The minimum radius ratio, r/R, for each of these groupings is illustrated in Fig. 7.6*b*.

EXAMPLE 7.2 Calculate the minimum r/R for CN = 3.

ANSWER Note that in Fig. 7.6*b* the smaller ion has just reached a radius at which it is touching all three larger ions. We can calculate r/R from geometry as follows. From the figure we see that

$$AD = r + R \qquad \tfrac{1}{2}AC = R$$

Since ABC is an equilateral triangle (each angle = 60°) and the smaller ion is at the center, α must equal 30°.

Then

$$\cos 30° = \frac{\tfrac{1}{2}AC}{AD}$$

$$0.866 = \frac{R}{r + R}$$

$$r/R = 0.155$$

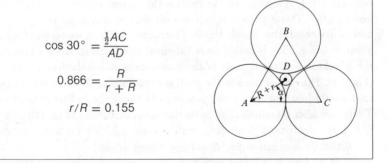

The other cases are solved in a similar way, beginning with finding $r + R$ as a function of R from the geometry of the figure.

In considering the effect of the radius ratio in this way on the shape of the unit cell, we find that in boron nitride we have a simple triangle of larger

N^{3-} ions with a B^{3+} ion in the center. If we formed a tetrahedron with four touching N^{3-} ions, the B^{3+} ion would rattle around, and therefore the configuration would be unstable ($R_{N^{3-}} = 1.17$ Å, $r_{B^{3+}} = 0.2$ Å).

As we progress to the radius ratio of silicon to oxygen, we find that the important SiO_4^{4-} complex ion is made up of a silicon atom at the center of four oxygen atoms. In another electronically balanced case we have ZnS. (In general practice the words "ion" and "atom" are used interchangeably, just as they are in discussions of metal structures. However, we must use the *ionic* radius in calculations for ceramic structures.)

Now let us look specifically at the ionic radius ratios for sodium chloride and cesium chloride, which have CN's of 6 and 8, respectively. For sodium chloride we have $0.98/1.81 = 0.54$, which is in the range 0.732 to 0.414 (CN = 6), and for cesium chloride we have $1.65/1.81 = 0.91$, which is in the range 1 to 0.732 (CN = 8). The sodium chloride coordination is described as octahedral. We should note carefully that while the Cl^- ions around the Na^+ form an octahedral figure, there are only six ions, not eight, and CN = 6.

Let us sum up the effect of CN on the unit cell structure. For a given CN the radius ratio will usually be large enough so that the smaller atom touches all the larger atoms indicated by the CN. For example, for CN = 8, r/R should be *greater* than 0.732. However, we may have a structure in which CN is lower than that permitted by the radius ratio. In this case the smaller atom is indeed touching all the larger atoms, but additional larger atoms could be packed around the smaller atom. For example, a carbon atom in diamond (Fig. 7.10) will have only four nearest neighbors because of the arrangement to provide four shared electrons.

Interstitial Sites. While we have before us the models of tetrahedral and octahedral groups of ions around a central ion for CN = 4 and CN = 6, respectively, it is a good time to discuss the terms "octahedral site" and "tetrahedral site." These are voids in the unit cell in which we can place different atoms. We recall that inside the FCC structure of iron we could place a carbon atom at $\frac{1}{2}, \frac{1}{2}, \frac{1}{2}$. This is called an octahedral site because the six nearest atoms, the iron atoms at the centers of the faces, form an octahedron around the void (Fig. 7.7). The size of the site (or void) is expressed as the diameter of the sphere that would just touch the iron atoms (0.41 times the radius of the iron atom). There are also tetrahedral sites in the same structure (Fig. 7.8). An atom at $\frac{1}{4}, \frac{1}{4}, \frac{1}{4}$ is equidistant from four iron atoms, making up a tetrahedron, and $r = 0.29R$. [r = small atom, R = large (iron) atom.]

If we look at a larger region of the space lattice, including adjacent unit cells, we find that there are octahedral sites at the center of each cube edge. On the average there are four octahedral sites per FCC unit cell, or one site for each FCC atom, and eight tetrahedral sites per cell, or two sites per atom.

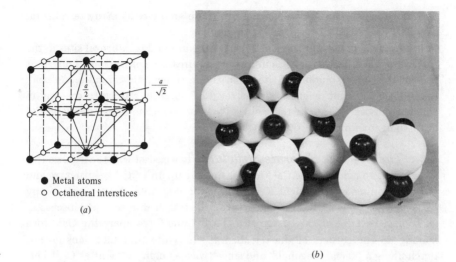

(b)

Fig. 7.7 *Octahedral sites (CN = 6). (a) Sketch of the sites in an FCC structure. (b) Photograph of a hard sphere model of an FCC structure with octahedral sites filled. The corner atoms have been removed to show the site in the center of the structure.*

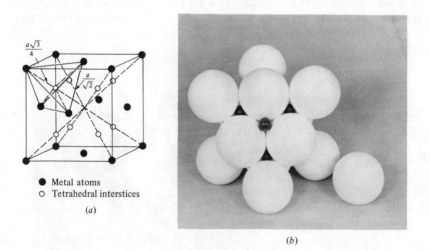

(b)

Fig. 7.8 *Tetrahedral sites (CN = 4). (a) Sketch of the sites in an FCC structure. (b) Photograph of a hard sphere model of an FCC structure showing only four tetrahedral sites. The corner atom has been removed. Note the small size of these sites compared to that of the octahedral sites (Fig. 7.7).*

Therefore, in the close-packed structures there are *twice* as many tetrahedral as octahedral sites.

In HCP structures there are also octahedral and tetrahedral sites in the same ratios of sites to atoms as in FCC structures. One of the most important uses of this understanding of sites is in the ferrimagnetic ceramics. We shall see in Chap. 14 that magnetic field strength is directly related to the way in which atoms are placed in these sites. Now let us continue our discussion of the important unit cells.

The Calcium Fluoride Structure. The easiest way to visualize this structure is to consider the Ca^{2+} ions as making up an FCC-like structure and placing a simple cube of eight F^- ions inside it (Fig. 7.9). The F^- ions occupy $\frac{1}{4}, \frac{1}{4}, \frac{1}{4}$ type positions in the "FCC structure," the tetrahedral sites just discussed. Of course, because of their large size the F^- ions force apart the Ca^{2+} ions making up the corners of the tetrahedral site. Note that Ca^{2+} ions do not touch along a "facial diagonal," and since twice as many F^- ions as Ca^{2+} ions are required (CaF_2), the F^- ion positions cannot be octahedral. This structure is also characteristic of A_2X compounds in which the cations occupy the F^- and the anions the Ca^{2+} positions.

The Diamond Structure. If we omit four of the eight F^- ions that form the cube inside the CaF_2 structure, we form a tetrahedron (Fig. 7.10). In this case the ratio of the atoms inside the cell to those forming the FCC "box"

$2\sqrt{3}r_F + 2r_F + 2r_{Ca} = B \text{ Diag of large cube}$

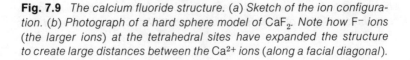

O F   ◎ Ca

(a)

(b)

Fig. 7.9 *The calcium fluoride structure. (a) Sketch of the ion configuration. (b) Photograph of a hard sphere model of CaF_2. Note how F^- ions (the larger ions) at the tetrahedral sites have expanded the structure to create large distances between the Ca^{2+} ions (along a facial diagonal).*

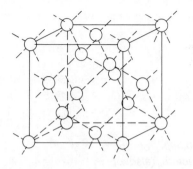

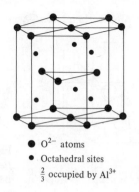

● O^{2-} atoms
● Octahedral sites
$\frac{2}{3}$ occupied by Al^{3+}

Fig. 7.10 *The diamond structure shown as a three-dimensional array of carbon atoms.*

Fig. 7.11 *Sketch of the corundum structure.*

is 1. The inner atoms are still at $\frac{1}{4}$, $\frac{1}{4}$, $\frac{1}{4}$ locations relative to the corners. This structure is found in the important semiconductor materials silicon and germanium as well as in diamond, tin, and some AX compounds.

The Corundum Structure. The mineral corundum is Al_2O_3 and, depending on small amounts of impurities, may have a red color as in the ruby or a blue color as in the sapphire. Common, clear corundum (alumina) is an important abrasive and refractory. The structure can be built up by starting with an HCP-like structure of O^{2-} ions. In this lattice the number of octahedral sites (Fig. 7.11) is equal to the number of oxygen atoms. However, since the aluminum ion is trivalent, we can have only two Al^{3+} to three O^{2-}. The result is that only two-thirds of the sites are filled.

The Spinel Structure. The basic formula for spinel is $A^{2+}B_2^{3+}O_4$, where A and B are metal ions ($MgAl_2O_4$, for example). The oxygen ions form an FCC lattice and the A and B ions can be found in tetrahedral and octahedral sites, depending on the particular spinel. This is of great importance in the magnetic ferrites, which are a special type of spinel to be taken up in detail in Chap. 14.

The Perovskite Structure. In perovskite, $CaTiO_3$, the oxygen ions are at the centers of an FCC-like unit cell, the calcium ions at the corners, and the titanium in the center (Fig. 7.12). This gives the proper formula:

$$6\ O^{2-} \times \tfrac{1}{2}\ \text{ion/face} = 3$$
$$8\ Ca^{2+} \times \tfrac{1}{8}\ \text{ion/corner} = 1$$
$$1\ Ti^{4+} \times 1\ \text{ion/center of body} = 1$$

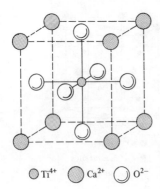

$\bigcirc$ Ti^{4+} $\bigcirc$ Ca^{2+} $\bigcirc$ O^{2-}

Fig. 7.12 *Sketch of the perovskite structure (also of barium titanate).*

This type of structure is particularly important in the similar structure of barium titanate, an unusual dielectric and ferroelectric material discussed in Chap. 13.

The Silica and Silicate Structures. These are among the most complicated structures and the details will be taken up in the next section when we discuss commercial materials. However, we should realize here that a great simplification of all these structures is possible if we note in each case the common building block, the SiO$_4^{4-}$ tetrahedron (Fig. 7.13). In other words, the silicon atoms are always bonded to four oxygen atoms. The complexity arises

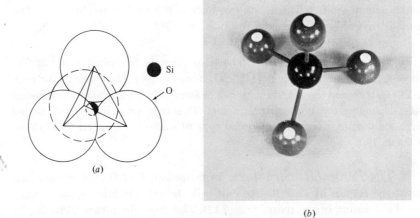

(a)

(b)

Fig. 7.13 *The silicate tetrahedron. (a) Sketch. (b) Photograph of a silicate tetrahedron model. The atoms with light-colored dots indicate unsatisfied oxygens to which other ions or molecules may attach.*

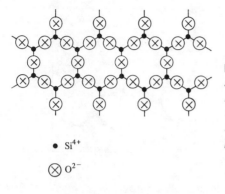

Fig. 7.14 *Schematic representation of a silica network. Note that the fourth bond of the silicon is not shown. It is normal to the plane of the paper and connects to the fourth oxygen (no foreign atoms).*

• Si^{4+}

⊗ O^{2-}

from the different attachments made to the other four unsatisfied bonds extending from the oxygen atoms outward from the tetrahedron.

In the simplest case, SiO_2, the bonds are all satisfied by adjacent silicon atoms, forming an SiO_2 network (Fig. 7.14). (It must be appreciated that the structure is really three-dimensional.) At the other extreme all four bonds are satisfied by foreign atoms such as magnesium (Fig. 7.15). This is called an "island structure" because the SiO_4^{4-} tetrahedra, as in Fig. 7.13, are isolated from each other by foreign atoms (Mg^{2+}). The resultant formula is Mg_2SiO_4. (One-half of each Mg^{2+} is shared by each O^{2-} in the SiO_4^{4-} tetrahedron.)

Glass. As mentioned in the introduction, if we take a sample of quartz sand (SiO_2), fuse it in an electric arc, and cool fairly rapidly, we will find a clear, noncrystalline material called "fused silica." It is a *glass.* The basic building block is still the SiO_4^{4-} tetrahedron we just discussed, but the tetrahedra are joined in a random network, not in a regular pattern, as shown in Fig. 7.16.

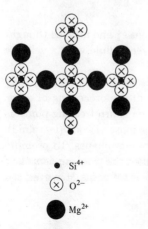

• Si^{4+}

⊗ O^{2-}

● Mg^{2+}

Fig. 7.15 *Satisfying unshared oxygen ions in the SiO_4^{4-} tetrahedron (island structure) with Mg^{2+} ions. Each O^{2-} is bonded to an Si^{4+} and an Mg^{2+}.*

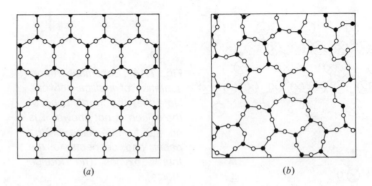

(a)   (b)

Fig. 7.16 *Two-dimensional sketch of a silica network in (a) crystalline and (b) glass (low-order) form.*

(After Zachariasen and Warren.)

One of the outstanding properties of glass is that it does not have the sharp freezing point of a crystalline solid, but as it cools, it gradually becomes more and more viscous. Since the viscosity of a glass is one of its most important properties, it should be defined.

We can consider a mass of liquid glass as being made up of layers. To move one layer past another takes a certain shearing force per unit area. The equation from physical chemistry is

$$f = \frac{nAv}{\theta}$$

where f = force
n = coefficient of viscosity
A = area between layers
v = velocity attained
θ = layer separation

The unit of the coefficient of viscosity is the poise[†] and may be thought of as the force per unit area required to move a layer of fluid with a velocity of 1 cm/sec past a parallel layer 1 cm away.

As examples of the quantitative use of the viscosity coefficient, we may consider the definitions of important temperatures in the processing of glass. As we mentioned, as a melt of glass cools, there is no sharp freezing point as there is in crystallization. The melt merely becomes more viscous (less fluid). In this condition the glass is easily formed or blown into shapes. To promote standardization in rating different glasses, this is called the *working point* and is defined as the temperature at which the viscosity is 10^4 poise. However, the

---

[†]1 poise = 0.1 Pa-sec (Pascal-second).

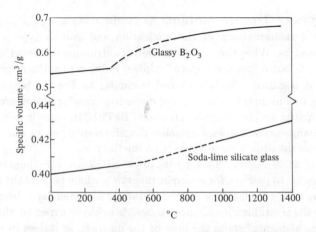

Fig. 7.17 *The specific volume of glass as related to temperature. The point at which the curve changes slope is called the "glass transition temperature."*

(After R. H. Doremus, "Glass Science," John Wiley & Sons, Inc., New York, 1973, Fig. 1, p. 115.)

material still can be deformed easily down to the *softening point,* $10^{7.6}$ poise. At this temperature a rod will still sag appreciably (of the order of 1 mm/min) of its own weight. The next point encountered on cooling is the *annealing point,* $10^{13.4}$ poise, at which residual stresses can be relieved. Finally, the *strain point,* $10^{14.6}$ poise, is reached, where the glass becomes rigid.

Another important characteristic of glass is the *glass transition temperature.* If we plot the specific volume (cm^3/g) of a glass, we obtain a graph similar to Fig. 7.17. The temperature at which the rate of contraction changes slope rather abruptly is called the "transition temperature." At this point the free movement of the molecules past each other becomes difficult and the material behaves more like a solid than a viscous liquid.

7.4 Solid solutions

Up to this point we have discussed only pure compounds, but solid solutions are encountered in ceramics as in metal systems, although additional principles affect their formation. For example, an important series of minerals occurs between pure magnesite, MgO, and iron oxide, FeO. Both have the sodium chloride structure, and we find a continuous series of solid solutions similar to the copper-nickel system in metals. However, other points such as similar valence are equally important in obtaining a continuous series of solid solutions.

Types of Transformation. As in the metals, in the ceramics we encounter transformations of the nucleation and growth type and of the diffusionless type. When the transformation of a structure involves the breaking of bonds it is called "reconstructive," while if only small shifts in position are involved, it is called "displacive" and is similar to the martensite type of transformation in metals. Examples of displacive transformation are obtained in $\alpha \rightarrow \beta$ quartz and in cubic $\rightarrow$ tetragonal $BaTiO_3$. In cases in which a large volume change accompanies displacive transformation, as in quartz, severe cracking can take place since there is no ductility.

Also, as in metals, the kinetics of reactions are very important. The transformation to pearlite, for example, takes time which permits the formation of martensite. TTT diagrams exist in ceramics also and have been used to determine the crystallization time of glasses. As might be expected, the addition of "alloying elements" shifts the nose of the diagram, as it does in metals. In general, most reactions are more sluggish in ceramics because of the necessity of moving several ions in combination to maintain electroneutrality, a problem not encountered in metals. These unusual features of ceramics will be discussed further in subsequent sections.

7.5   Defect structures, lattice vacancies

The point defects or vacancies in the lattices that we found in metals also occur in ceramic crystals. Vacancies are especially important when a foreign ion of different valence is dissolved in the standard structure, and some compensation is needed to obtain a charge balance over the crystal as a whole. A typical case is in iron oxide crystals in which both Fe^{2+} and Fe^{3+} are present. The normal FeO structure is like NaCl, but it is found that if we make up the crystal with some Fe^{3+} ions present, one Fe^{2+} is absent to balance two Fe^{3+} ions (Fig. 7.18).

The symbol $\square$ signifies an ion vacancy where an Fe^{2+} was omitted to balance the charge surplus caused by two Fe^{3+} and is called a "cation" vacancy. If, on the other hand, we substitute a cation of lower than normal valence, say K^+, then "anion" vacancies are needed to balance the charge. (In this case some O^{2-} would be left out.) Defect structures are used in voltage rectifiers, as discussed in Chap. 13.

O^{2-} Fe^{2+} O^{2-} Fe^{2+} O^{2-} Fe^{2+} O^{2-} Fe^{2+}

Fe^{2+} O^{2-} Fe^{2+} O^{2-} Fe^{2+} O^{2-} Fe^{2+} O^{2-}

O^{2-} Fe^{3+} O^{2-} Fe^{2+} O^{2-} $\square$ O^{2-} Fe^{2+}

Fe^{2+} O^{2-} $\square$ O^{2-} Fe^{3+} O^{2-} Fe^{3+} O^{2-}

O^{2-} Fe^{3+} O^{2-} Fe^{2+} O^{2-} Fe^{2+} O^{2-} Fe^{2+}

Fe^{2+} O^{2-} Fe^{2+} O^{2-} Fe^{2+} O^{2-} Fe^{2+} O^{2-}

Fig. 7.18 *The defect structure of $Fe_{<1}O$. The symbol $\square$ denotes an ion vacancy.*

EXAMPLE 7.3 Olivine sands are usually given the formula $(Mg,Fe)_2SiO_4$, which suggests complete solid solution of magnesium and iron. Justify the high solid solubility and indicate whether lithium ions might also substitute.

ANSWER Mg^{2+} (0.78 Å) and Fe^{2+} (0.87 Å) have the same charge and are very close in size, which are the criteria for a high degree of solid solubility.

Although Li^+ (0.78 Å) is also the same size, the charge is not balanced. Some solid solution is still possible, however, by the following techniques:

1. $Li^+ + Fe^{3+}$ (0.67 Å) substitute for two Mg^{2+}.
2. Two Li^+ substitute for one $Mg^{2+} + \square$.

EXAMPLE 7.4 In Fig. 7.18 (wüstite) what fraction of cation (iron) sites will be vacant if there are 10 Fe^{3+} ions to every 100 Fe^{2+} ions?

ANSWER In a problem of this sort it is necessary first to decide whether the anion or cation lattice will be perfect. In this case it is the anion (oxygen) lattice.

$$100\ Fe^{2+} + 10\ Fe^{3+} = 110 \text{ cation sites filled}$$

For every two Fe^{3+} there must be one Fe^{2+} vacancy for the charge to balance. Therefore, there are five Fe^{2+} vacancies, since we have 10 Fe^{3+} ions.

The anion sites are all filled, however, with oxygen, and there are equal numbers of anion and cation sites in FeO.

$$100\ Fe^{2+} + 10\ Fe^{3+} + 5\ \square = 115 \text{ cation sites total}$$

or

$$\text{Anion sites} = 115 \text{ (all filled)}$$
$$\text{Percent vacant cation sites} = \tfrac{5}{115} \times 100 = 4.35$$

It is not possible to select a few common properties as a basis for comparison of ceramics in the same way we did for metals. For example, the tensile test, which is a common denominator for comparing metals, is rarely performed. In many cases the insulating or optical properties are more important than the mechanical properties. It seems best, therefore, to discuss some of the properties that are important in ceramics and then take up the individual materials.

7.6 Some properties of ceramics

We will see that a number of ceramics are used as refractories. This is a broad term meaning in general materials with good resistance to heat. However, the temperature of operation of the equipment and the degree of thermal shock are important. A good refractory for a kitchen oven might contain considerable asbestos, but this combination would melt at steel-making temperatures (1600°C). For this reason the melting point or solidus temperature of ceramics is important.

Another important factor in using refractories to resist high temperatures is the thermal coefficient of expansion α, also called the "linear coefficient of expansion." If we heat a bar of length l from T_1 to T_2, the length will increase by Δl. We define α as

$$\alpha = \frac{\Delta l / l}{T_2 - T_1} \qquad \text{or} \qquad \alpha = \frac{\Delta l}{l \, \Delta T}$$

The units are therefore $(°F)^{-1}$, $(°C)^{-1}$, or $(K)^{-1}$, and the numerical values of °C and K will be the same. With a high coefficient the surface layers will expand greatly relative to the unheated interior. Since ceramics are brittle, these heated layers will flake off. This is called "spalling."

EXAMPLE 7.5 Calculate the elastic strain and then the stress developed by rapidly heating the surface layer of a fire clay brick from 100 to 600°F (38 to 315°C).

ANSWER Assume $\alpha = 2.5 \times 10^{-6}(°F)^{-1}$ $[4.5 \times 10^{-6}(°C)^{-1}]$ and the modulus of elasticity $E = 15 \times 10^6$ psi $(1.035 \times 10^5 \text{ MN/m}^2)$, which are reasonably constant in this temperature range.

If the layer is unrestrained, the expansion ε is

$$\varepsilon = 2.5 \times 10^{-6}(°F)^{-1} \times 500°F \quad [4.5 \times 10^{-6}(°C)^{-1} \times 277°C]$$
$$= 12.5 \times 10^{-4}$$

If the layer is completely restrained by surrounding material and no heat transfer occurs between layers, then

$$\sigma = E\varepsilon = 18{,}750 \text{ psi} \qquad (130 \text{ MN/m}^2) \qquad \text{(a maximum value)}$$

In reality the heated layer tends to expand, developing a shearing stress between the backing layers, and a thin layer spalls off the surface.

It should be added that in a noncubic crystal the expansion coefficient is different in different directions (Table 7.2).

Table 7.2 COEFFICIENT OF EXPANSION
IN SOME CERAMICS

Crystal	Coefficient of Expansion, $(°C)^{-1} \times 10^{-6}$	
	Normal to c Axis	Parallel to c Axis
Alumina, Al_2O_3	8.3	9.0
Mullite, $3Al_2O_3 \cdot 2SiO_2$	4.5	5.7
$ZrSiO_4$	3.7	6.2
$CaCO_3$	−6.0	25.0
SiO_2	14.0	9.0
Graphite	1.0	27.0

In some materials the coefficient of expansion for polycrystalline material is not the mean of these values, because cracking results from the difference in expansion in different directions in various grains. This is encountered in magnesium and aluminum titanates, for example. There is a general relationship of coefficients in different families of ceramics:

Corundum type (Al_2O_3, Cr_2O_3, Fe_2O_3): 8 to $12 \times 10^{-6}(°C)^{-1}$
Zircon type ($ZrSiO_4$): $4 \times 10^{-6}(°C)^{-1}$
MgF_2 type (TiO_2): 9 to $11 \times 10^{-6}(°C)^{-1}$

The coefficient for glasses is not predictable and varies from $0.5 \times 10^{-6}(°C)^{-1}$ for fused silica to $7.6 \times 10^{-6}(°C)^{-1}$ for fused GeO_2. A major breakthrough in control of the coefficient was attained in one variety of Pyroceram. The basic formula is related to the mineral β spodumene, $Li_2O \cdot Al_2O_3 \cdot 4SiO_2$, which has a negative coefficient in the c direction, $-16.9 \times 10^{-6}(°C)^{-1}$, and a positive coefficient in the a direction, $8.11 \times 10^{-6}(°C)^{-1}$. By controlling the amount of this material in a glassy matrix, an overall coefficient close to zero can be obtained.

Another factor in determining thermal shock resistance as well as heat loss through refractories is thermal conductivity k. It is essentially this characteristic that determines the heat transmitted between the end walls of a unit volume per second. The equation for heat flux is

$$J = k\frac{\Delta T}{\Delta X}$$

(See Chap. 4 and Fig. 4.18.)

If the conductivity is high, the temperature difference between the surface and the underlying layers will be small, and therefore the tendency to spall will be reduced. The excellent resistance to thermal shock of graphite is due to its high thermal conductivity, while in fused silica it is due to the low coefficient of expansion.

Hardness and strength are important in some cases, and we should review the measurement of these properties in ceramics. Two of the methods for testing

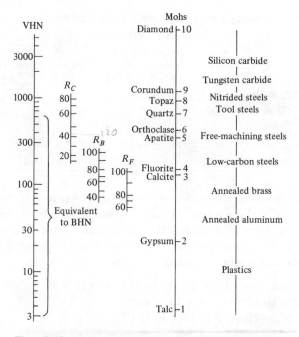

Fig. 7.19 *Interrelationship of different hardness scales. (Readings on carbides were obtained with special indenters.)*

hardness in metals, the Brinell and Rockwell tests, are not used for ceramics because the material would fracture. However, the Vickers test discussed in Chap. 3 is used. In this method a small, carefully polished diamond pyramid indenter is used with very light loads (below 100 g). Below 500 the Vickers scale is approximately the same as the Brinell scale, but above 500 the corresponding Brinell number is lower (Fig. 7.19). Before the development of the microhardness test, the Mohs scratch hardness method was used. Ten minerals ranging from diamond (10) to talc (1) were selected to cover the known spectrum of hardness (Fig. 7.19). The values were chosen on the basis that a specimen of a higher number would scratch one of a lower number; for example, diamond (10) scratches sapphire (alumina, 9). This scale is still used by mineralogists and some ceramists.

The tensile strength of a ceramic is usually difficult to determine because of the sensitivity to small cracks which are almost always present in specimens of appreciable size. Griffith demonstrated long ago that fine glass fibers had tensile strengths many times that of the bulk material because it was possible to produce the fibers relatively free from defects. In brittle materials such as most ceramics a crack propagates easily under stress because no energy is dissipated in plastic deformation ahead of the notch (as in a metal). There is even a large difference between the strength of specimens with rough surfaces which have been abraded to produce notches and specimens with smooth surfaces. We say, therefore, that the "fracture toughness" of the usual ceramic is low. As we shall see, in some cases, such as Pyroceram, the situation has been improved by the controlled growth of a very fine crystal network.

An alternative to the tensile test for brittle hard materials is a bending test which uses a small beam about 1 by 0.4 by 0.4 in. (2.54 by 1.02 by 1.02 cm), loaded at the center and supported beneath near the ends. The breaking stress at the outermost fibers can be calculated from the simple beam formula, which is based on the absence of plastic deformation of the material in the test (Fig. 7.20). This is called the "transverse rupture strength" and is roughly comparable to the tensile strength. It should also be pointed out that although the tensile strength may not be high, the compressive strength can be excellent. As we will see in the next chapter, high compressive and low tensile strengths lead

Fig. 7.20 *Bend test for ceramics and powder metal specimens. The centers of the two bottom supports are 1 in. (2.54 cm) apart. The load is applied to the top member, which places the bottom of the beam in tension. The tensile stress is given approximately by the formula $S = 3PL/(2bh^2)$, where S = transverse rupture strength, psi (kg/mm^2); P = load required to fracture, lb (kg); L = length of span, in. (mm); b = width of specimen, in. (mm); and h = thickness of specimen, in. (mm).*

to materials such as tempered glass and prestressed concrete. Furthermore, ancient civilizations did not have the advantage of many of our modern materials; however, they did know how to use ceramics. Since ceramic structures would easily fail in tension, they developed the arch, which was architecturally pleasing and from an engineering standpoint was loaded in compression.

With this introduction, let us consider a few of the more common ceramics.

7.7 Simple ceramic materials

Silica and silicates are by far the most important ceramic materials, and just as we gave a good deal of attention to iron as a base for many metallic materials, we should investigate silica carefully. Under equilibrium conditions and atmospheric pressure three allotropic forms are encountered as we cool liquid silica: cristobalite at 3110 to 2678°F (1710 to 1470°C), tridymite at 2678 to 1598°F (1470 to 870°C), and quartz below 1598°F (870°C). These phase changes are extremely sluggish, because the atomic rearrangements are far more complex than in the $\gamma \rightarrow \alpha$ transformation in iron. We have already mentioned that if we cool liquid silica at a moderately fast rate, we obtain none of the equilibrium structures but glass instead. It is also possible to retain a high temperature phase such as cristobalite at room temperature or, conversely, to melt quartz without transforming to the intermediate phases.

At this point we need to distinguish clearly between these sluggish reactions involving extensive rearrangement and simple *displacive* transformations which take place rapidly within the three crystal structures at relatively fixed temperatures, as shown in Fig. 7.21. The $\beta \rightarrow \alpha$ quartz transformation is rapid and will shatter a roof of silica brick if uncontrolled. The unit cells of the crystalline forms are fairly complex, and it is not necessary to cover these in detail. As an example, the structures of α cristobalite and

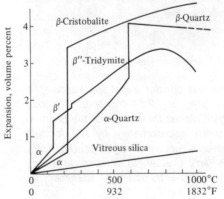

Fig. 7.21 *Volume changes with temperature in silica structures.*

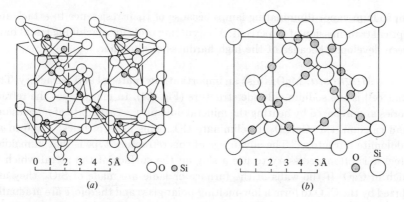

O O   ⊙ Si

O   Si

Fig. 7.22 (a) Structure of α quartz (a form of SiO₂). (b) Structure of α cristobalite (a form of SiO₂).

[Part (a) after W. L. Bragg and Gibbs. Part (b) after Wyckoff.]

α quartz are shown in Fig. 7.22. Note first that in cristobalite the silicon atoms are in the diamond structure (Fig. 7.10), and then that there are four oxygen atoms around each silicon atom in the characteristic tetrahedron.

Silica. This material, SiO_2, is used in refractory brick in furnaces where high temperatures are encountered, as in steel making (3000°F, 1649°C). The brick is made from silica, and if a small amount (2 percent) of $Ca(OH)_2$ is added, it acts to form a liquid with a lower melting point which cements the grains during firing and also catalyzes the transformation of some of the quartz to a mixture of cristobalite and tridymite. This results in better expansion characteristics than if the brick were composed of all quartz. However, even the mixture has to be heated carefully to avoid cracking from thermal shock.

Another use of quartz is for oscillating crystals of known frequency, such as those used in some radio equipment (see Chap. 13). Although a large quantity of quartz is mined, methods have been developed in the Bell Telephone Laboratories for growing excellent crystals.

Alumina. Al_2O_3 (Fig. 7.11) is widely used as a raw material in ceramic mixtures as well as in the pure state. Single crystals of pure alumina or of ruby and sapphire which are colored by minute quantities of chromium, iron, and titanium are encountered in nature. However, the material for optical uses, as in the ruby laser (Chap. 15), is produced by melting alumina powder in an oxyacetylene flame and building up a single crystal from the droplets.

Because of its high melting point, alumina is widely used as a refractory. Grains of relatively high-purity alumina are pressed into a shape and sintered at high temperatures to produce brick, tubes, and crucibles. Small amounts of flux are used. In more recent applications alumina tubes have been used for sodium vapor illuminating lamps because of their resistance to attack and optical transmission of 90 percent. Also, cutting tools for machining metal have been developed because of the high hardness of alumina.

Magnesia. MgO is also an important raw material and refractory. The unit cell is the sodium chloride structure (Fig. 7.4). In many cases the refractories are prepared by heating the mineral dolomite, a carbonate of magnesium and calcium, $(Mg,Ca)CO_3$, to eliminate CO_2. This class of brick is known as "dolomite refractory." The advantage of this refractory type is that in making steel it is often necessary to use a slag on the surface of the metal which is high in CaO. If the walls of the furnace or ladle are made of SiO_2, they are fluxed by the CaO to form a low-melting-point glass and therefore are gradually dissolved. However, there is little reaction with dolomite refractories, since no low-melting-point materials are formed.

Silicates. Silica reacts with Al_2O_3, MgO, CaO, and FeO, as well as with Na_2O, K_2O, to form many useful compounds including the glasses, which will

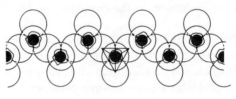

(a)

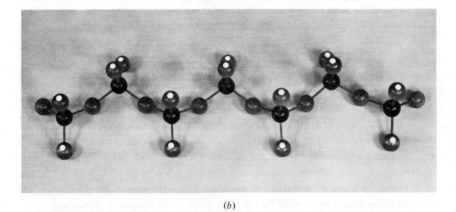

(b)

Fig. 7.23 *A single-chain silicate structure. (a) Sketch. (b) Photograph of the silicate model. The oxygen atoms with white dots still have unsatisfied bonds.*

[Part (a) from L. H. Van Vlack, "Elements of Materials Science," 2d ed., Addison-Wesley Publishing Company, Inc., Reading, Mass., 1964, Fig. 8-11, p. 212.]

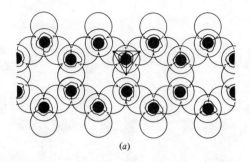

(a)

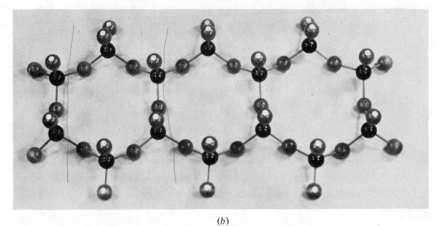

(b)

Fig. 7.24 *A double chain of* SiO_4 *tetrahedra. (a) Sketch. (b) Photograph of the silicate model. Again the unsatisfied oxygen atoms are shown with white dots.*

[Part (a) redrawn from L. H. Van Vlack, "Elements of Materials Science," 2d ed., Addison-Wesley Publishing Company, Inc., Reading, Mass., 1964, Fig. 8-12, p. 212.]

be discussed last. To illustrate the range of properties and how they depend on the structure, let us consider the familiar minerals, asbestos, mica, and clay.

ASBESTOS (COMPLEX HYDROUS MAGNESIUM SILICATES). The original building block for silica is the tetrahedron $SiO_4{}^{4-}$ (Fig. 7.13). Now consider the addition of several of these islands to make a single or double chain, as shown in Figs. 7.23 and 7.24. The unsatisfied oxygen atoms lie only at the edges of the chains and can be bonded with foreign atoms. In asbestos, Mg^{2+} ions can join these chains, and these chains can also be terminated by OH^- ions (from water). The natural occurrence, therefore, is for the structure to be fibrous, so that when stress is applied, the chains do not fracture but rather failure between chains can occur.

The fact that asbestos is hydrous and hence can be dehydrated at elevated temperatures should be remembered if the material is being considered for high-temperature vacuum systems.

Fig. 7.25 *Photograph of a sheet structure silicate model. Only the fourth (upper) oxygen atoms have unsatisfied bonds (shown by light-colored dots).*

MICA. This material is a complex silicate in which the bonding produces a sheet structure, as for example in muscovite, $[KAl_3Si_3O_{10}(OH)_2]_2$. The double chain shown in Fig. 7.24 is extended into a plane (Fig. 7.25), and thus there is strong silicon-oxygen bonding in two dimensions rather than in a line. The extra bonds for the oxygen atoms above the sheet plane are again satisfied by foreign ions. In this case we have platelike cleavage, enabling the mica to be split into very thin sheets.

CLAY. In this material we have a sheet structure similar to that of mica. In kaolinite, $[Al_2Si_2O_5(OH)_4]_2$ (Fig. 7.26), for instance, we have alternating layers of Si^{4+} ions and Al^{3+} ions, but the SiO_4^{-4} tetrahedra are still distinct. The complex unit cell leads to the sheetlike cleavage and greasy-feel characteristic of clay. This structure also accounts for the excellent moldability of this mineral when water is added. Water molecules, being polar, can attach themselves to the clay layers by van der Waals forces, hence providing the plasticity. Too much water allows for water-to-water bonds and thus a loss in plasticity or a soupy consistency.

Mullite. Important ceramics are encountered at various percentages of silica and alumina, as indicated in the phase diagram in Fig. 7.27. When a

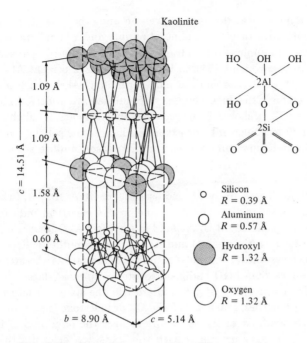

Fig. 7.26 *Sheet structure of kaolinite clay. (Note the planes of potential easy cleavage.)*

(After W. E. Hauth.)

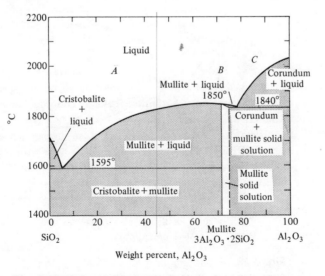

Fig. 7.27 SiO_2-Al_2O_3 *phase diagram. A = fire clay compositions, B = mullite refractory, C = high-alumina refractory.*

fire clay (kaolin) is heated, water is given off and other complex changes take place, so that above 1595°C the mineral mullite and a liquid are present. This limits the use of alumina-silica refractories with less than 72 percent Al_2O_3 to temperatures below about 1550°C. However, above 72 percent Al_2O_3 we have mullite ($3Al_2O_3 \cdot 2SiO_2$) or mullite plus corundum (Al_2O_3) with a solidus at 1840°C. Therefore, the high-alumina refractories are widely used for steel making (1600°C). It should be emphasized that a relatively slight change in Al_2O_3, from 70 to 80 percent, for example, changes the solidus by 255°C in a very important temperature range. For this reason mullite is an important refractory.

Spinels. There are two broad uses of spinels, $A^{2+}B_2^{3+}O_4$ (where A and B are metal ions): in the refractory and in the electrical industry. In the refractories the phase $MgAl_2O_4$ is found in the MgO-Al_2O_3 diagram (Fig. 7.28), at an equal molar ratio of MgO and Al_2O_3, just as mullite is encountered in the Al_2O_3-SiO_2 system at a 3:2 molar ratio. The material has a greater resistance to thermal shock than MgO and good resistance to most slags.

In the electrical industry the interest in the spinel structure is due to its magnetic properties. Magnetite, $FeFe_2O_4$, or lodestone is a naturally occurring spinel. In contrast to the refractory spinels, the only ones of interest in the electrical industry are those with the transition elements that lead to the formation of magnetic materials such as $CoFe_2O_4$, discussed in Chap. 14.

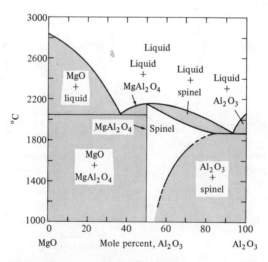

Fig. 7.28 *Phase diagram for MgO-Al_2O_3. The spinel type structure is important as a refractory and in ferrimagnetic materials.*

Barium Titanate. This material, $BaTiO_3$, has the perovskite structure discussed earlier (Fig. 7.12) with barium instead of calcium. Its electrical uses depend on the fact that the Ti^{4+} ion does not fit at the center of the unit cell formed by barium and oxygen ions, and this results in an off-center charge for the cell. We shall see how this leads to use in transducers in Chap. 13.

Other Ceramic Compounds. Although the total tonnage is not very large, an interesting group of hard refractory materials has been developed in the carbides, nitrides, and silicides and in graphite, as shown by the following examples.

Carbides of tungsten, titanium, tantalum, and chromium are made commercially by reacting the metal or oxide with carbon, while the others are usually produced by special high-temperature processes. Typical hardness values are given in Table 7.3.

In many cases it is necessary to heat a mixture of the carbides, etc., with a softer metal such as 10 percent cobalt to obtain the toughness necessary in a cutting tool. This is called a "cermet." A typical microstructure is shown in Fig. 7.29. A few other compounds should be mentioned at this point. Boron carbide, B_4C, has a rhombohedral structure close to diamond in microhardness, VHN 3700, and is extremely inert. Boron nitride, BN, on the other hand, has a soft hexagonal structure similar to graphite, but is an electrical insulator and more oxidation-resistant. Because of its structure, like graphite it is relatively easy to machine. Other carbides and nitrides have generated interest because their high melting points and high hardnesses have suggested possible uses ranging from nose cones of space vehicles to refractories and cutting tools.

Carbon. This can be produced in many forms important in ceramics. Particles of carbon can be bonded with pitch, and the mixture can be heated

Table 7.3 MICROHARDNESSES OF
SEVERAL CERAMIC MATERIALS

Material	Microhardness*	
Hardened tool steel	VHN	600
Quartz	VHN	1250
WC	VHN	2400
Al_2O_3	VHN	2800
TiC	VHN	3200
SiC	VHN	3500
TiB_2	VHN	3400
B_4C	VHN	3700
Diamond	VHN	$\sim$8000

*These values are somewhat different from those given in Fig. 7.19 because of the different load used. Also, micro- and macrohardness values are different (Chap. 3).

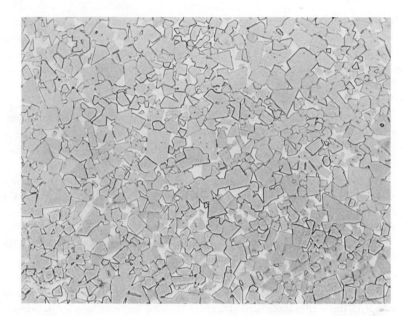

Fig. 7.29 *Cobalt-bonded tungsten carbide tool material. The gray crystals are tungsten carbide, the matrix cobalt. To make this structure, tungsten carbide crystals were placed in a ball mill with cobalt powder. After milling, 1 percent paraffin was added. Then the mixture was pressed in a die and sintered for 0.5 hr at 2650°F (1452°C). The cobalt melted and the entire mass was sintered. 1,500X, alkaline ferrocyanide etch.*

to decompose the pitch to "amorphous carbon." When this mixture is heated to a higher temperature, as by having heavy electric currents passed through it, graphite crystallizes in the mass. The layered structure has strong covalent bonds within the sheets (Fig. 7.30) but semimetallic bonds between the sheets. This arrangement leads to cleavage between the sheets, giving excellent properties as a lubricant as well as semimetallic properties such as high electrical and thermal conductivity parallel to the sheets. Between these sheets there are barriers to electron motion. In the diamond form the perfect covalent bonding throughout the mass leads to the highest hardness known. However, the cleavage on (111) planes is perfect so that a crystal may be split with a sharp edge. Nonetheless, diamond tools and wheels impregnated with diamond are invaluable in shaping other hard ceramics.

Glasses. In contrast to the crystalline ceramics we have just discussed, the glasses cover a wide range of compositions. Fused silica is an important material, but for ease in working, other oxides such as Na_2O and CaO are added

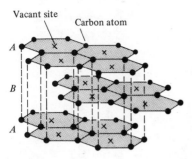

Vacant site Carbon atom

Fig. 7.30 *Sketch of the layer structure of graphite (carbon).*

to SiO_2. We can anticipate some of the effects of these additions to glass, just as we anticipated the effects of adding metals as alloys in other metals. Therefore, we analyze these additives as follows:

1. Glass formers. In general, these elements have valences of 3 or greater and are relatively small ions, thus giving lower coordination numbers. Typical oxide glasses are SiO_2, B_2O_3, GeO_2, P_2O_5, As_2O_5, and As_2O_3.
2. Modifiers. In contrast to the glass formers, the oxides of elements with low valences, such as sodium and calcium, do not form glasses themselves but can be added in limited amounts to give different characteristics to a glass. For example, pure silica glass has a high melting point and the addition of Na_2O modifies this to a convenient working temperature (Fig. 7.31).
3. Intermediates. In this group are elements which do not form glasses themselves under normal conditions but which can be added in very large amounts to silica. For example, lead oxide is added to silica in large amounts to produce the so-called "crystal" which has great brilliance resulting from its high index of refraction.

The compositions and characteristics of important commercial glasses are given in Table 7.4. Fused silica is difficult to form but is outstanding for laboratory glassware exposed to thermal shock and high temperatures. To overcome the difficulty of producing a pure silica material, ceramists developed Vycor. First a melt of 75 percent SiO_2, 20 percent B_2O_3, 5 percent Na_2O is made and poured. By heat treatment a two-phase structure composed of two glasses

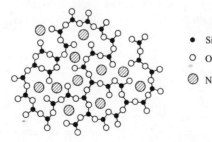

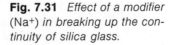

• Si^{4+}

○ O^{2-}

⊘ Na^{1+}

Fig. 7.31 *Effect of a modifier (Na^+) in breaking up the continuity of silica glass.*

Table 7.4 ANALYSES AND PROPERTIES OF REPRESENTATIVE GLASSES

| Type of Glass | Analysis, percent by weight | | | | | Softening Temperature, °C | Coefficient of Expansion, $(°C)^{-1} \times 10^{-7}$ | Characteristics | Use |
	SiO_2	Modifiers*	Al_2O_3	B_2O_3	PbO				
Fused silica	99.9					1667	5.5	Thermal shock resistant	Laboratory equipment
96 percent silica (Vycor)	96.0			4.0		1500	8.0	Thermal shock resistant	Laboratory equipment
Borosilicate (Pyrex)	80.5	4.2	2.2	12.9		820	32.0	Thermal shock resistant, easy to form	Cooking utensils
Aluminasilicate	57.7	9.5	25.3	7.4		915	42.0	Thermal shock resistant	Thermometers
Soda-lime silica	73.6	25.4	1.0			696	92.0	Easy to form	Plate, bulbs
Lead-alkali	54.0	11.0			35	630	89.0	High index of refraction	Cut glass
Lead-alkali	35.0	7.0			58	580	91.0	Dielectric	Capacitors

*Sum of Na_2O, K_2O, Ni_2O, CaO, MgO, and BaO.

is developed. One phase is a honeycomb of practically pure silica, and the other is high in sodium borate. Next the sodium borate phase is leached out with acid, leaving the relatively pure silica honeycomb. Finally, the remaining mass is fired at 900 to 1200°C to close the pores, and a reduction in volume of 30 percent is encountered. Despite the complex process, this glass is considerably cheaper than fused silica and has slightly less resistance to heat and thermal shock, as shown by the coefficient of expansion.

The borosilicate (Pyrex) and aluminosilicate glasses are used in less expensive laboratory glassware and in some glass cookware used in the home.

The soda-lime–silicate glasses are by far the most widely used because the low softening temperature and wide working range lead to excellent formability.

Glass Ceramics (*Pyroceram*). We have discussed the pronounced effects of small cracks in lowering the strength of glass. A crack is particularly disastrous in glass because there are no "crack stoppers" such as grain boundaries and the mass is homogeneous. Recently much stronger materials have been developed by producing a glass shape first and then by heat-treating to develop a very fine crystalline phase to strengthen the glass. The material can be made very resistant to thermal shock as well if the precipitate has a very low coefficient of expansion. This is a new field called "glass ceramics," and the Pyroceram materials are outstanding examples of products that have been developed. To obtain the fine crystalline precipitate, nucleants such as titanium dioxide may be used. The frontispiece for this chapter is a photomicrograph showing nucleation of a Pyroceram by rutile (TiO_2).

Miscellaneous. The composite structures such as concrete will be discussed in the next chapter because the method of processing is important in determining the usage.

Finally, a number of glasses such as the lead glasses are used for their electrical and optical properties. Specifics will be discussed in subsequent chapters; however, it should be pointed out here that chemistry and residual stress control are important to the performance of a glass. For example, residual stress in old window panes causes distortion. Also, lenses should be low in dispersion (i.e. have nearly the same index of refraction for different wavelengths). Since this is often not possible, several lenses of different chemistry may be used in combination to minimize the dispersion.

As a final comment, it is apparent that the dwindling of supplies of natural resources has promoted a greater interest in ceramics, since the earth's crust is rich in many of these materials. In the next chapter we will see how these ceramics can be formed into engineering shapes.

SUMMARY

The bonding in ceramics differs from that in metals in two important aspects. Ionic and covalent bonds, instead of metallic bonds, are predominant. The role of the coordination number is important because most ceramics are made up of different size atoms, as we see, for example, in the metal-oxygen combinations.

The density of a ceramic crystal can be calculated from its unit cell dimensions, as can the density of a metal.

The unit cells of many ceramic materials can be described by reference to a few key structures: sodium chloride, cesium chloride, calcium fluoride, alumina, perovskite, spinel, and silicate groups.

The glasses form another important family of ceramics. Although the structure of these is a well-bonded three-dimensional network in space, the side-by-side unit cell grouping of a crystal is absent. The small units such as SiO_4^{4-} tetrahedra are present and may be bonded to each other or to ions of other metals at the corners.

Solid solutions are encountered in crystals and in glasses. Lattice vacancies are important and can be related to the presence of ions of different charges.

Thermal properties are often important in conjunction with mechanical properties because of the possibility of spalling and the use of ceramics for insulation.

Solid-state transformations occur and two types are distinguished: the displacive which requires only minor shifts in ion positions, and the nucleation and growth of a new phase, such as quartz to cristobalite.

The structure and properties of simple ceramic materials are described.

DEFINITIONS

Refractory A material that resists exposure to high temperatures, as in a furnace lining or a gas turbine blade.

Cleavage The fracture path when a material is broken. For example, sodium chloride cleaves into cubes, whereas glass has no regular cleavage.

Coordination number, CN The number of equidistant nearest neighbors to a given atom in a unit cell. The sodium ion, for instance, has six equidistant chloride ions in sodium chloride; therefore, CN = 6. The CN is related to the radius ratio: ionic radius of the smaller atom/ionic radius of the larger atom.

Octahedral void or site The space within a group of *six* atoms which, when connected, form an octahedron. For example, the hole in the center of an FCC unit cell is an octahedral void. The *tetrahedral void* is the center of four atoms making a tetrahedron.

Glass A noncrystalline solid that softens on heating rather than showing a sharp melting point. The structure is composed of small units such as SiO_4^{4-} tetrahedra, linked to each other or by other ions in a mostly random three-dimensional network.

Viscosity coefficient The force required to slide one layer of a liquid past another under specified conditions. This will affect the flow rate of liquid glass in a mold or tube because the layer next to the tube is stationary and the inner layers must flow past it. The unit is the poise (1 poise = 0.1 Pa-sec).

Glass characteristics

Working point The temperature at which glass is easily formed, 10^4 poise.

Softening point The temperature at which glass will sag appreciably of its own weight.

Annealing point The temperature at which locked-up stresses can be relieved.

Strain point The temperature at which glass becomes rigid.

Transition temperature The temperature at which, on cooling, the rate of contraction changes to a lower value.

Spalling The cracking and breaking away of surface layers due to exposure to heat or cold, sometimes combined with a chemical action.

PROBLEMS

7.1 Calculate the density of CsCl (Fig. 7.5). Use the ionic radii given in Table 7.1.

7.2 Calculate the density of CaF_2 (Fig. 7.9). (*Note:* There are twice as many F^- ions per unit cell as Ca^{2+} ions). Would you expect your value to be higher or lower than the handbook value? Why?

7.3 Check the limiting value for CN's of 6 and 8. (Refer to Fig. 7.6.)

7.4 Check the CN for CsCl (Fig. 7.5). What is the atomic packing factor (i.e. what percentage of space is occupied by the ions)? What other ions might substitute for Cs^+ and still give the same CN?

7.5 What type of structure, that is, NaCl, etc., would you expect for the following materials: KCl, CsBr, $NiAl_2O_4$, Fe_3O_4, UO_2, $CaZrO_3$? Indicate which structure given in the text each material would resemble. (For example, KCl has either the NaCl or CsCl structure. Since the radius ratio fits the CN = 6 group, KCl has the NaCl structure.)

7.6 Calculate the hole size, i.e. atomic diameter, that would fit inside the barium titanate ($BaTiO_3$) unit cell if the barium and oxygen ions touched across the face diagonal. (*Hint:* Calculate a_0 and note that the diameter of the hole equals $a_0 - 2r_{O^{2-}}$.) Compare the hole size with the ionic diameter Ti^{4+}. How can the titanium position adjust to the misfit? (See

Chap. 13 on electrical properties for further explanation and utility of this phenomenon.)

7.7 Sketch the structure of lead-zirconium-titanate (PZT).

7.8 Make a sketch and show whether Fe_2SiO_4 (fayalite) and $MgSiO_3$ (enstatite) are island, chain, or sheet type silicate structures.

7.9 Zirconia (ZrO_2) is often stabilized with calcium to provide an important refractory. The basic cell is ZrO_2 with 1 Ca^{2+} ion present for every 10 Zr^{4+} ions. Will the vacant sites be anion or cation? What percentage of the total number of all sites will be vacant?

7.10 Draw a graph showing the softening temperature (temperature at which liquid begins to appear) in refractories made of different amounts of SiO_2 and Al_2O_3 as a function of composition. Indicate the significance of these values.

7.11 Kaolinite clay, $Al_2(Si_2O_5)(OH)_4$, is heated to drive off the hydrogen as water.

   a. What is the percentage loss in weight?

   b. What would be the liquidus and solidus temperatures of the resultant mixture of Al_2O_3 and SiO_2 (Fig. 7.27)?

7.12 Referring to Fig. 7.21, draw a graph showing the thermal expansion of a silica brick made up of approximately equal quantities of quartz, tridymite, cristobalite, and vitreous silica. Would this be better than an all-quartz brick? Why?

7.13 A spalling resistance index for refractories is given by the formula

$$\text{Spalling resistance} = \frac{kS}{\alpha E C_p \rho}$$

where k = thermal conductivity
   S = tensile strength of the material
   α = coefficient of thermal expansion
   E = Young's modulus (modulus of elasticity)
   C_p = specific heat
   ρ = density

Explain how each term contributes to a reduction or increase in the tendency to spall.

7.14 The following general rules apply to cleavage:

   1. With layer lattices such as those in mica and MoS_2, the cleavage is parallel to the layers.
   2. Cleavage planes are the most widely spaced.
   3. Cleavage does not cut through radicals or ionic complexes.
   4. Cleavage occurs so as to expose planes of anions if this does not violate rule 3.
   5. In AX crystals cleavage occurs on the (100) planes.

Explain the following observations using rules 1 to 5.

a. The cleavage of NaCl is cubic.
b. The cleavage of CaF_2 is octahedral.
c. The cleavage of graphite is parallel to the basal plane.
d. The cleavage of zinc is on the (0001) planes.

7.15 Many minerals and glasses with complex formulas such as $LiAlSiO_4$ are called "stuffed derivatives" of SiO_2. We note first that Li^+ and Al^{3+} substitute for one Si^{4+}. Second, the Al^{3+} ion takes the place of the Si^{4+} in the unit cell and the Li^+ ion is "stuffed" into a hole. In the allotropic form of silica called tridymite, important stuffed derivatives are $KNa_3Al_4Si_4O_{16}$ (nepheline) and $KAlSiO_4$. Which ions replace silicon in the tridymite lattice and which are stuffed?

One grade of Pyroceram has the formula of β spodumene, $Li_2O \cdot Al_2O_3 \cdot 4SiO_2$, plus quartz. Which are the stuffed ions?

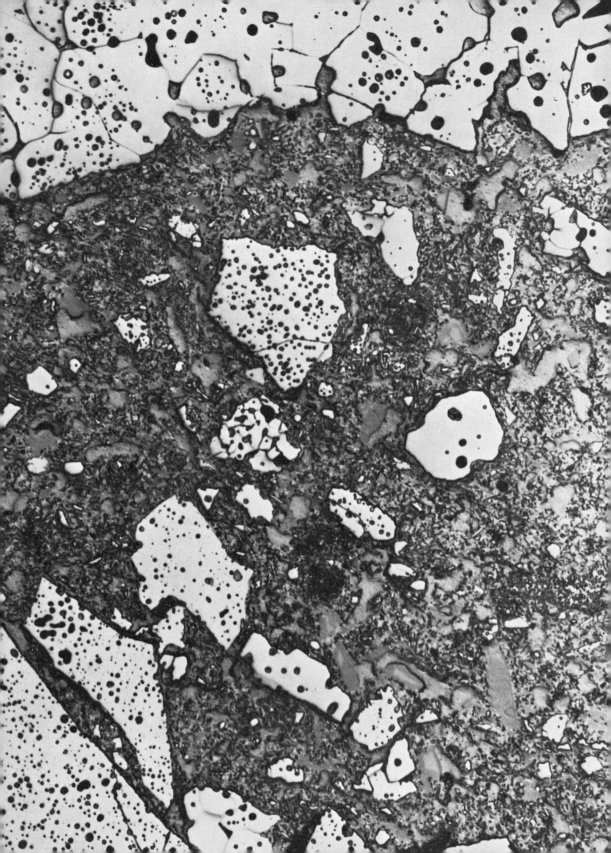

8

PROCESSING, SPECIFICATIONS, AND APPLICATIONS OF CERAMICS

THIS illustration shows a typical 90 percent alumina refractory. The brick consists of coarse synthetic alumina (white) bonded by a glassy aluminosilicate matrix. Magnification 100X.

In this chapter we will discuss common processing methods for different types of ceramic products and then the structures and uses of the different products.

8.1 General

Ceramics are used over a tremendous range of applications from common building brick to delicate porcelain to special optical glass. The task of discussing the processing and applications of these products in one chapter may seem difficult, but since we are now familiar with most of the important structures, it only remains for us to see how these structures or mixtures of them are assembled into usable shapes.

We need to study the processing of ceramics in more detail than that of metal products because, in general, ceramics are not machinable or forgeable from bar or plate stock into a finished part. In most cases we have to take the raw materials, press them into a shape such as a brick or plate or magnet, and then develop the desired structure in the shape by heating. If we reflect on this a little, we will see the reason is that most ceramics are hard—precluding machining—and brittle, eliminating cold working. (There are a few exceptions, such as the machining of pressed ceramic before firing in the "leather hard" condition and soft materials such as graphite and boron nitride. Also, the formability of glass while hot is extremely useful.)

Having noted these exceptions, we must conclude that in most cases the successful use of a ceramic calls for the production of both the finished part and the desired structure in a closely organized sequence. One more example: The beer can manufacturer buys steel or aluminum sheet from another company and at his leisure forms cans, while the bottle maker must start with raw materials, melt them, and then form the bottles to exact dimensions in one sequence.

In this chapter we will discuss first the manufacture and production of materials that are not glass—brick, refractories, art ware, cookware, dinnerware, tile, chemical ware, plumbing fixtures, ceramic bonded abrasives, magnetic ferrites, ferroelectrics—and then the products based on glass—containers, plate glass, chemical equipment, and optical glass. For the nonglasses there is a group of standard processes such as pressing and sintering, while for the glassy materials the processing is closely related to the fluidity in the vitreous state. Therefore, these groups will be discussed separately.

To begin with the first group, let us list the different processing methods and see how the various products are involved (Table 8.1). We note from this list that the same processes are used for a variety of products. Let us therefore discuss the processes listed, not in great mechanical detail, but to obtain a feel for how these can affect the final structure and shape.

8.2 Molding followed by firing

In this group a shape is formed by different methods and then fired to give it strength.

Table 8.1 PROCESSES USED FOR DIFFERENT CERAMIC PRODUCTS

Product	Molding plus Firing Processes				Viscous Fluid	Chemical Bonding	Single Crystal
	Slip Casting	Wet Plastic Forming	Dry Powder Pressing	Hot Pressing			
Cemented products		×				×	
Brick		×	×		×		
Refractory and insulation		×	×	×	×	×	
Whiteware	×	×					
Vitreous enamelware		×					
Abrasive wheels			×				
Molds for metal castings	×	×				×	
Special (magnets, laser crystals, etc.)			×	×			×

Slip casting is an interesting and rather unique method in which a suspension of clay in water is poured into a mold (Fig. 8.1). The mold is usually made of plaster of Paris *with controlled porosity* so that some of the water of the suspension enters the mold wall. As the water content of the suspension decreases, a soft solid forms. The remaining fluid is poured out and then the hollow form is removed from the mold. The bond at this point is clay-water. Next the part is fired. Although this method is usually used for low production rates, it is important because it is now being employed for obtaining shapes with fine dispersions of refractory *metal* powders on curved surfaces. A similar technique is used for forming a refractory shell on the outside of wax patterns. In this case the pattern is melted away and the mold is fired to leave a precise cavity in the refractory for subsequent casting of metal.

Wet plastic forming is done by several methods. In one case a wet or damp refractory is rammed in a mold and then forced out to give the required shape. In another example extrusion is used for simple shapes such as brick or pipe (Fig. 8.2). The plastic mass is forced through a die to give a long shape which is then sliced to the desired lengths. On the other hand, when circular shapes such as plates are to be formed, a mass of wet clay is placed on a rotating potter's wheel and shaped by a tool. This old process has now been highly mechanized.

Dry powder pressing is accomplished by loading a die with a charge of powder and pressing. The powder usually contains some lubricant such

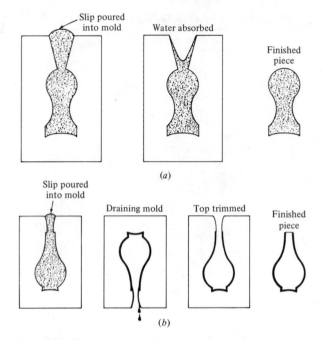

Fig. 8.1 *Slip casting of ceramic shapes.*

(W. D. Kingery, "Introduction to Ceramics," John Wiley & Sons, Inc., New York, 1960, Fig. 3.17, p. 52.)

as stearic acid or wax. A variety of shapes can be made by thoughtful design of multipart dies (Fig. 8.3).

After any of the preceding processes, the "green" part is placed in a kiln and fired. During firing the water and volatile binders are driven off, low-melting-point fluxes melt and bond the refractory, and sintering of the refractory grains takes place at 700 to 2000°C. Sintering was discussed briefly under powder metallurgy in Chap. 5, and we should recall that this is a diffusion process. It is especially important in ceramics because the high melting point of the material often makes it impossible to actually melt the material to bond the grains together. Sintering provides the solution since the diffusion of material results in true bonding between grains.

Hot pressing involves the operations of both pressing and sintering simultaneously. The advantages are greater density and finer grain size. The problem is to obtain adequate die life at elevated temperatures, and protective atmospheres are often used.

An example of cold powder pressing and sintering is shown in Fig. 8.4. It was necessary to produce a crucible with an internal thermocouple protection tube that could withstand superheated magnesium under an inert gas pressure to prevent its boiling. An $MgO\text{-}MgF_2$ mixture was selected. We

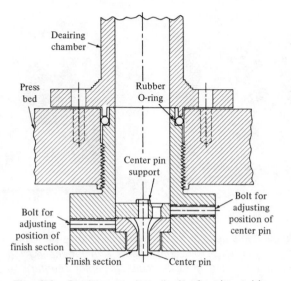

Fig. 8.2 *Piston-extrusion die for forming tubing.*
(W. D. Kingery, "Introduction to Ceramics," John Wiley & Sons, Inc., New York, 1960, Fig. 3.14, p. 48.)

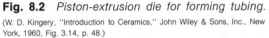

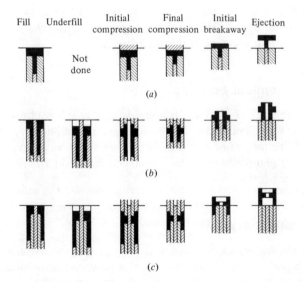

Fig. 8.3 *Die motions required for three different shapes.*
(W. D. Kingery, "Introduction to Ceramics," John Wiley & Sons, Inc., New York, 1960, Fig. 3.9, p. 42.)

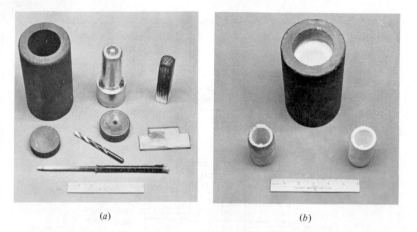

(a)   (b)

Fig. 8.4 *Components used in making ceramic shapes (MgO) by pressing and sintering. (a) Crucible die components and fabricating tools. (b) A powdered compact in the graphite mold-susceptor (center). A sintered crucible is shown on the left after removal from the mold. The one on the right is after resintering in an oxidizing atmosphere. Note the extreme shrinkage from the "green" in the mold-susceptor to the sintered shape.*
(Courtesy of Guichelaar, Flinn, and Trojan, University of Michigan.)

should note that the shrinkage was very severe and had to be accounted for in the original design. The use of a graphite external die allowed for induction heating to accomplish the sintering.

8.3   Chemical bonding

The advantage of chemical bonding is that it is a cold process and can produce precise dimensions. The most common and important bond is the setting of Portland cement, but other processes involve plaster of Paris, ethyl silicate, or phosphate cements. The structures produced are described in detail in Sec. 8.5.

8.4   Single crystals

To obtain desired optical, electrical, or magnetic properties, it is often essential to have a single crystal instead of a polycrystalline material. In all commercial cases the solid is crystallized from a melt under controlled conditions. It is most important to nucleate the crystal at a colder point and to avoid the formation of other nuclei. In one case the melt is contained in a tube that passes slowly from the furnace. In another the crystal is started from a nucleus on a rod touching the bath and then literally pulled from the bath. In powder methods for corundum crystals the powder is fused and crystallized on an existing nucleus.

8.5 Cements and cement products

In many ceramic shapes we will find the crystalline phases held together by the glass phase, so that we can consider glass a special high-temperature cement. There is another large class of cements, however, in which a mixture may be formed into a shape at low temperature and, because of interaction with water, a hydraulic bond develops. This is the field of common cements and plasters.

In general, the action of these cements is that when the water is added, the existing minerals either decompose or combine with water and a new phase grows throughout the mass. Examples are the growth of gypsum crystals in plaster of Paris and the precipitation of a silicate structure from Portland cement. It is usually quite important to control the amount of water to prevent an excess which would not be part of the structure and therefore would weaken it.

Portland Cement and High-alumina Cement

There are several grades of these cements, as shown in the ternary diagram in Fig. 8.5. A typical Portland cement will contain 19 to 25 percent SiO_2, 5 to 9 percent Al_2O_3, 60 to 64 percent CaO, and 2 to 4 percent FeO, while a high-alumina cement will be mostly CaO and Al_2O_3. Both types are prepared by grinding clays and limestone of the proper mixture, firing in a kiln, and regrinding.

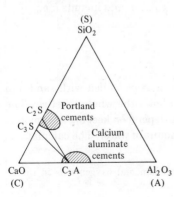

Fig. 8.5 *Cement compositions. [Ternary diagrams are explained in Chap. 4 (see in particular Fig. 4.13).]*

In Portland cement several minerals are present: tricalcium silicate (C_3S), $3\,CaO \cdot SiO_2$, dicalcium silicate (C_2S), and tricalcium aluminate (C_3A) (see Fig. 8.5). Tetracalcium aluminum ferrite, $4\,CaO \cdot Al_2O_3 \cdot FeO$, is also present.

The reaction is solution, recrystallization, and precipitation of a silicate structure.† To retard setting, 2 percent gypsum ($CaSO_4 \cdot 2H_2O$) is added which reacts with the tricalcium aluminate and reduces shrinkage. The heat of hydration (heat of reaction in the adsorption of water) in setting can be large and can damage massive structures. Low-heat cements are made by reducing the amount of tricalcium aluminate.

High-alumina cement has approximately 35 to 42 percent CaO, 38 to 48 percent Al_2O_3, 3 to 11 percent SiO_2, and 2 to 15 percent FeO. The setting is produced by forming hydrated alumina crystals from the tricalcium aluminate. It is quick-setting and attains the same strength in 24 hr that Portland cement attains in 30 days.

Mortar which is used in bricklaying combines the effect of Portland cement with a lime reaction. When lime (CaO) is mixed with water, it forms hydrated lime $Ca(OH)_2$. This reacts with air to form $CaCO_3$ and also with sand to form a calcium silicate. The formation of these new crystals causes bonding. A typical mortar is one part Portland cement, two parts lime hydrate, and eight parts sand by volume.

Silicate cements are made from sodium silicate and fine quartz. The quartz is bonded by the formation of a silica gel from the breakdown of the sodium silicate. The process is used in the manufacture of molds and cores in the foundry, and the reaction is accelerated by passing carbon dioxide gas through the sand to form sodium carbonate and silica rapidly.

Gypsum cement is used widely in wall construction. The mineral gypsum is heated to change the double hydrate to the hemihydrate: $CaSO_4 \cdot 2H_2O \rightarrow CaSO_4 \cdot \frac{1}{2}H_2O + \frac{3}{2}H_2O$ (gas). The hydrogen and oxygen atoms are an integral part of the crystal structure. When the hemihydrate is mixed with water, the gypsum phase reforms and hardens the shape.

Other cements involve the same type of reaction and include magnesium oxychloride and phosphate minerals.

8.6 Concrete shapes

In concrete we have a coarse aggregate of gravel filled with sand and a water-cement mixture that acts as a glue to bond the whole together. The economics as well as the strength of the mixture depend on keeping the amount of cement at a minimum and controlling the amount of water. The calculations

---

†This is a sheet structure of silicate groups with calcium and oxygen ions in the interstices. Water molecules separate the sheets.

for the proper mixture to fill a desired volume are related to a set of definitions involving packing factor, volume, and density. Some of these are merely large-scale analogs of the atomic scale factors already mentioned, but they should be discussed carefully because of their importance in manufacturing operations.

8.7 General packing in solids

Although we have mentioned solid packing before, in discussing powdered metals and in earlier sections in this chapter, we have deferred the analysis until now since very important applications arise in the manufacture of cement products. Basically we have to define precisely several terms that are illustrated in Fig. 8.6, in which we have some limestone rock in a box.

1. *Porosity*   Open pores: Spaces into which water can penetrate
Closed pores: Spaces into which water cannot penetrate
True porosity: Volume of (open + closed pores)/total volume of box
Apparent porosity: Volume of open pores/total volume of box
2. *Volume*   Bulk volume: True volume of rock + (open + closed pores)
Apparent volume: True volume + closed pores only
3. *Density*   Bulk density: Mass/bulk volume
True density: Mass/true volume
4. *Packing factor* True packing factor: True volume/bulk volume or bulk density/true density
Apparent packing factor: Apparent volume/bulk volume or bulk density/apparent density

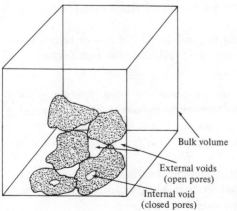

Bulk volume

External voids (open pores)

Internal void (closed pores)

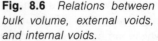

Fig. 8.6 *Relations between bulk volume, external voids, and internal voids.*

The normal engineering problem is to find methods of optimizing or controlling the packing factor or, more importantly, the porosity, which is 1 minus the packing factor.

EXAMPLE 8.1 In the packaging industry container shape is important. What is the apparent porosity if basketballs are shipped inflated so that they just fit into the box?

ANSWER With a radius r and a cubic box of side length a,

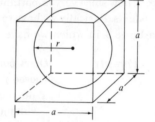

$$2r = a$$

$$\text{Packing factor} = \frac{\frac{4}{3}\pi r^3}{a^3} = \frac{\frac{4}{3}\pi r^3}{8r^3}$$

$$= \pi/6 = 0.524$$

$$\text{Porosity} = 47.6 \text{ percent}$$

The same answer would be obtained for any spherical shape placed in a cubic container into which it would just fit.

The next question is: "How can the amount of porosity be changed?" There are essentially three different ways.

1. Shape changes. Bricks can be packed with less porosity than spheres. A greater weight of pencils of hexagonal cross-section can be placed in a box than dowels. Even the honeybee has optimized the shape of its honeycomb.
2. Compression and impregnation. The exclusion of air pockets, as in stuffing a sleeping bag into a tote bag or compression with a garbage compactor, increases packing. The use of wood fibers and resins to make plywood is an example of impregnation.
3. Mixed sizes. Small spheres can be used to fill the spaces between large spheres. The manufacture of concrete is probably the best example.

EXAMPLE 8.2 Calculate the materials needed to make a retaining wall requiring 300 ft³ of concrete. The formula for the mix is 2.75 ft³ of gravel, 2 ft³ of sand, 1 sack (94 lb) of cement, and 6 gal of water (1 ft³ = 0.02832 m³, 1 gal = 3.785 × 10⁻³ m³).

ANSWER We *cannot* add the bulk volumes of the constituents given to obtain the volume of the mix. We need to find the true volumes of the components, which can be calculated from the following table:

Component	Specific gravity	Bulk density, lb/ft³
Gravel	2.60	110
Sand	2.65	105
Cement	3.25	94

We assume that by using this mix there will be no open or closed voids. (That is, sand fills voids in gravel, cement fills voids between sand, and water fills the remaining voids.) Therefore, we calculate the space occupied by each component:

$$\text{True volume of gravel} = \frac{2.75 \text{ ft}^3 \times 110 \text{ lb/ft}^3}{(62.4 \text{ lb/ft}^3)(2.60)} = 1.86 \text{ ft}^3 \quad (0.0526 \text{ m}^3)$$

In other words, the weight of gravel in the mix is 2.75 ft³ bulk $\times$ 110 lb/ft³ bulk = 302 lb. If the specific gravity were 1, then 62.4 lb would occupy 1 ft³. With a specific gravity of 2.6, a cubic foot would weigh 162 lb. Therefore, the true volume is $\frac{302}{162}$ = 1.86 ft³.

Similarly,

$$\text{True volume of sand} = \frac{(2.00 \text{ ft}^3)(105 \text{ lb/ft}^3)}{(62.4 \text{ lb/ft}^3)(2.65)} = 1.27 \text{ ft}^3 \quad (0.0359 \text{ m}^3)$$

$$\text{True volume of cement} = \frac{94 \text{ lb}}{(62.4 \text{ lb/ft}^3)(3.25)} = 0.46 \text{ ft}^3 \quad (0.0130 \text{ m}^3)$$

$$\text{True volume of water} = \frac{(6 \text{ gal})(8.33 \text{ lb/gal})}{62.4 \text{ lb/ft}^3} = 0.80 \text{ ft}^3 \quad (0.0226 \text{ m}^3)$$

$$\text{Total true volume} = 4.39 \text{ ft}^3 \quad (0.1241 \text{ m}^3)$$

$$\text{Desired true volume} = 300 \text{ ft}^3 \quad (8.5 \text{ m}^3)$$

Therefore, multiply the unit mix by 300/4.39 = 68.33 to obtain the total materials required for the retaining wall.

Ceramic Products

8.8 Brick and tile

With these processes in mind we can now discuss the special features of ceramic products, beginning with brick and tile. For construction brick, a low-cost easily fused clay is used as the base, and this contains high-silica, high-alkali, high-FeO, gritty materials found in natural deposits. The products are formed by either dry or wet pressing and are fired at relatively low temperatures.

The specifications are simple, involving only compressive strength ranging from 2,000 to 8,500 psi (13.8 to 58.6 MN/m²) (depending on grade) and dimensional tolerances.

Building tile (unglazed), clay pipe, and drain tile are similar in processing.

8.9 Refractory and insulating materials

For furnace and ladle linings either brick or monolithic (rammed in place) linings are employed. In dealing with liquid metals and slag it is essential to distinguish between the acidic, neutral, and basic refractories. In general, the refractories composed of MgO and CaO are called "basic" because the water solutions are basic. The refractories based on SiO_2 give very weak acid solutions. The really important effect is that acid slags will attack basic refractories, and the converse is also true.

The most important characteristics of these bricks are resistance to slag, resistance to temperature effects, and insulating ability. The relation of thermal conductivity to temperature is shown in Fig. 8.7 for various materials.

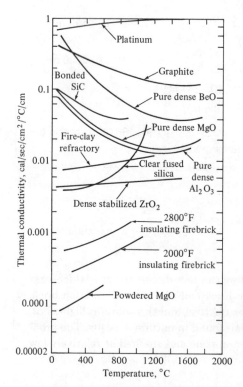

Fig. 8.7 *Thermal conductivity (logarithmic scale) of ceramic materials over a wide temperature range.*

(W. D. Kingery, "Introduction to Ceramics," John Wiley & Sons, Inc., New York, 1960, Fig. 14.37, p. 507.)

The acidic bricks are less expensive, but in many furnaces slags that are high in CaO and MgO are used to refine the metal (remove phosphorus and sulfur). These slags react with SiO_2 to form low-melting-point materials which erode the bricks. Therefore, when basic slags are used, the refractories must be basic. In intermediate cases alumina and chromite brick are used. The furnace or ladle linings can be formed from brick or by ramming a stiff refractory mud in place and firing it with a gas flame or coke fire.

Insulating brick contains a great deal of pore space and therefore is not as resistant to slag as the inner lining of a vessel. It is usually used behind the working lining. Insulating brick for lower temperatures is made of asbestos and plaster mixtures.

8.10 Earthenware, stoneware, china, ovenware, and porcelain

A wide variety of products, from the simplest earthenware shapes to fine china and electrical ceramics, are included in this group. In each case we will review first the raw materials used to make the products and then the reactions that occur during processing.

Earthenware is made of clay (kaolin, for example), but in some cases silica (SiO_2) and feldspar, such as $K(AlSi_3)O_8$, are present. The important feature is that it is fired at a low temperature compared with the other products in this group. This gives a relatively porous earthy fracture. For example, the interesting cup shown in Fig. 8.8*a* is made today by Indians in Chile simply by mining clay from the nearby hillside and using a wood-fired kiln. While the cup is quite serviceable, it is not high in strength because very little fusion or sintering has occurred. A good deal of widely used clay "soil pipe" for drains has similar characteristics.

Finer grades called *semivitreous earthenware* are made using clay-silica-feldspar mixtures, and because of the three constituents these are called "triaxial." The firing temperature is higher, resulting in the formation of some glass, lower porosity, and higher strength. All grades may be unglazed or coated with a separate glassy-surface-forming material which leads to a glazed surface.

Stoneware differs from earthenware in that a higher firing temperature is used, giving a porosity of less than 5 percent compared to 5 to 20 percent for earthenware (Fig. 8.8*b*). The composition is usually controlled more carefully than that of earthenware, and the unglazed product has the matte finish of fine stone. In some variations, such as the well-known Wedgewood jasper stoneware, barium compounds are added. This is an excellent material for ovenware or chemical tanks and coils. It is essentially unattacked by most acids but is corroded by alkalis.

China is obtained by firing the triaxial mixture mentioned above or other mixtures to a high temperature to obtain a translucent object. This

(a)

(b)   (c)

Fig. 8.8 (a) Earthenware cup, Chile. The clay is obtained from nearby hills and shaped. Then the mass is baked in a charcoal-fired kiln. The structure is relatively porous beneath the glaze. (b) Stoneware mug, Annapolis, Md. The clay-silica mix is fired in an electrically heated kiln at 2100°F. The product is partly fused and is stronger than earthenware. (c) English bone china, Minton. The mixture of flint, clay, and bone ash is fired at a high enough temperature to produce a glassy phase, giving the translucent product.

is why an expert will hold a plate toward the light and see how clearly the shadow of a hand comes through. The reason for the translucency is that a large share of the mixture of quartz, clay, and feldspar crystals has been converted to a clear glass. The expression "soft paste" porcelain is sometimes used. The firing temperature is lower than for "hard" porcelain because a small amount of CaO is present as a flux. English bone china (Fig. 8.8c) is different in composition, containing about 45 percent bone ash (from cattle bones), 25 percent clay, with the balance feldspar and quartz. The calcium phosphate from the bones gives a lower-melting-point material, and the phosphate group substitutes for part of the silica as a glass former. Another well-known china, "beleek," is made by adding glass as a flux to the original mixture, giving high translucency to the final product.

The specifications for *ovenware* and flameware are interesting. It has been found that the ceramics with a coefficient of expansion of about $4 \times 10^{-6} (°C)^{-1}$ will withstand heating in an oven and cooling in air, but that a coefficient below $2 \times 10^{-6} (°C)^{-1}$ is needed for parts in direct contact with a flame, such as a frying pan. As a result, some triaxial compositions with coefficients of about $4 \times 10^{-6} (°C)^{-1}$ are used in the oven. To obtain the lower coefficients, it is necessary to add Li_2O as contained in the mineral spodumene or cordierite.

Porcelain is fired at the highest temperatures of the group and is closely related to the china just discussed. In general, the avoidance of fluxes and the higher temperatures give a very hard, dense product.

Since the clay-silica-feldspar system is of such great importance to all these products, the ternary phase diagram should be reviewed briefly. Figure 8.9 is essentially a contour map showing the decrease in liquidus temperature as we go from hard porcelain which is high in silica and alumina to dental porcelain which is high in alkali (K_2O). With the low melting point of the *dental porcelain* it is possible to fuse most of the tooth and convert it to a glassy translucent material similar to a real tooth.

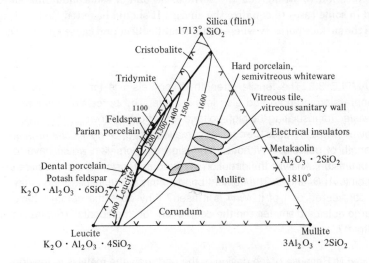

Fig. 8.9 *Areas of triaxial whiteware compositions shown on the silica-leucite-mullite phase equilibrium diagram.*

(W. D. Kingery, "Introduction to Ceramics," John Wiley & Sons, Inc., New York, 1960, Fig. 13.10, p. 419.)

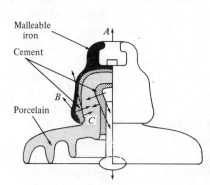

Fig. 8.10 *Cross-section of a disk insulator unit. The mechanical strength of porcelain, like most ceramic materials, is greater in compression than in tension. For this reason, disk insulator units, which are in tension during service, must be designed to ensure that the main stresses on the porcelain component are of a compressive nature.*

(Courtesy of Doulton Insulators, Ltd.)

Ceramics for Electrical Use. Electrical ceramic products may be divided into the large insulators such as those used for power lines and the small electronic components such as capacitors and magnets.

The large insulators (Fig. 8.10) are produced from a triaxial porcelain such as 60 percent kaolin, 20 percent feldspar, and 20 percent silica, as shown on the ternary diagram in Fig. 8.9. Either specialized slip casting or plastic forming is used. Special glazes are used for two reasons. First, the coefficient of expansion of the glaze is lower than that of the porcelain to put the glaze in compression to strengthen the surface. Second, a semiconducting glaze is used in some cases to equalize the charge. It should be noted that a glaze lowers the surface porosity where water might collect and cause an electrical short.

EXAMPLE 8.3 The tensile strength of porcelain is far less than the compression strength and also less reliable. The porcelain is needed, however, for insulating the high-voltage wires from the towers. The design shown in Fig. 8.10 uses the ductility of malleable iron and the compressive strength of porcelain. Without this design the porcelain would have to be in tension. Analyze the direction of stresses (tension or compression) at points *A*, *B*, and *C*. The cement bonds firmly to the insulator and may be considered part of it. Why is it used? Why does the ceramic have a large outside diameter and the deep sheds or "petticoats" (ripples on bottom)?

ANSWER Because of the design of the part, only the metal is in tension and the ceramic is in compression, as shown in the figure. The use of the cement permits the insertion of the metal part first into the insulator. The large outside diameter gives a long path for an electrical discharge to travel on the surface and therefore prevents the discharge. The deep sheds are to avoid build-up of a continuous sheet of water or ice which could cause a short circuit.

In many ways the production of *special electronic ceramics,* the ferroelectrics such as barium titanate and the ferrimagnetics such as cobalt-iron spinel, is similar to that of conventional ceramics employing dry pressing and firing. In other ways the processing is different, because of the need for high purity in the materials and usually for the avoidance of a glassy phase in sintering. Also, dimensional tolerances are extremely limited in many parts, to the order of 0.005 in. (0.0127 cm).

The most common material, $BaTiO_3$, is produced by reacting $BaCO_3$ and TiO_2, eliminating CO_2. A small amount of SiO_2 or Al_2O_3 (2 percent) which might be tolerated in a normal refractory lowers the dielectric constant appreciably. In the production of capacitors for miniaturized circuits, wafers are pressed or slip-cast only 0.008 in. (0.0203 cm) thick. After firing at 1300°C a conducting circuit is printed on the surface consisting of a powdered silver-glass mixture. Resistors are added by printing on a carbon mix. The whole part is then fired at low temperature, leads are soldered on, and the part is encased in water-resistant plastic or wax. In a package smaller than a postage stamp a complete circuit may be constructed.

In producing magnetic ferrites with a spinel structure, the pure single oxides such as MnO, CoO, and Fe_2O_3 are mixed in a blender and then fired in a tunnel kiln at 650 to 1048°C, depending on the composition for solid-state reaction. After the desired compound is obtained, the product is reground, pressed, and sintered under carefully controlled conditions (± 1°C) to control grain size, which determines magnetic characteristics.

Further consideration of these materials will be given in Chaps. 13 and 14.

8.11 Abrasives

In abrasive wheels or papers the idea is to grip the hard particles firmly enough so that they do not leave the wheel until they have become rounded. This is accomplished by two methods: using a softer matrix which is worn away, and providing porosity which weakens the support. The porosity is also important in conveying coolant to prevent "burning" of the part. It is easily shown in the case of steel, for example, that heavy grinding heats the surface to the austenitic range, and the resulting structure depends on the cooling rate and the composition. Many a carefully heat-treated part has been ruined by lack of coolant during finish grinding, because a brittle fresh martensite layer has been formed.

About 85 percent of the abrasive wheels are synthetic alumina and 15 percent are silicon carbide. The carbide is harder but more fragile, so that a great deal of competition exists, depending on the application. The bond may be either a ceramic glass, resin, or rubber. The wear is faster with the softer bonds but the wheels can be operated at higher speeds and produced in thinner sections, such as cutoff wheels.

8.12 Molds for metal castings

Although not formally recognized as a branch of ceramics, the preparation of ceramic molds for 20 million tons of metal castings deserves attention because of both the volume of ceramics used and the intricacy of the problems involving metal-mold reactions. The problems are really at the interface between ceramics and metallurgy. The objective of the ceramist is to produce an expendable mold that will provide adequate surface finish and dimensional accuracy. From the metallurgical point of view the action of the metal on the mold should be as neutral as possible. For example, the reaction of the metal with a mold containing an excess of water will result in the solution of hydrogen in the metal and gas holes in the casting. Another reaction can be the combination of aluminum dissolved in liquid steel with SiO_2 in the mold wall. This can affect the casting surface, as shown in Fig. 8.11. A wide variety of ceramic materials and molding methods have been

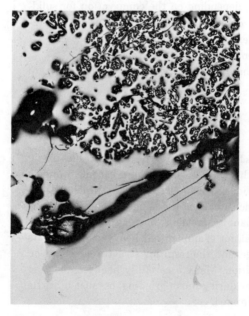

Fig. 8.11 *Photomicrograph of a cross-section taken through an inclusion attached to a steel casting. The aluminum dissolved in the liquid steel reacted with SiO_2 from the silica sand mold, producing Al_2O_3 (corundum) crystals as well as a glassy iron silicate material. 100X, unetched.*

(After G. A. Colligan.)

developed, including clay-bonded green sand (i.e. undried), dry sand, oil-bonded sand, sodium silicate-bonded sand, as well as mixtures using alumina, zirconia, and olivine, $(Mg,Fe)_2SiO_4$. The different "mixes" provide variation in casting surface finish and dimensional tolerances.

8.13   Glass processing and glass products

Because of the unique properties of glass, it can be cast, rolled, drawn, and pressed like a metal and, in addition, it can be blown.

Casting is accomplished by pouring the liquid into a mold. A famous case is the pouring of the 200-in.-diameter telescope disk by Corning Glass Works. Under normal circumstances if the disk were to reach the elastic range with temperature gradients present, the material would crack as these equalized later on. Also, as material was removed in grinding and polishing, the surface would distort. Therefore, the disk was cooled over a very long period of time to avoid temperature differences in the mass.

Rolling is widely used for window-glass and plate-glass production. The raw materials are melted at one end of a large furnace called a "tank furnace," and the liquid flows to the other end over a period of time to allow bubbles to float out. The temperature at the end where the rolls are located is controlled so that the glass is of the right viscosity to be rolled into a sheet. The sheet then passes through a long annealing furnace called a "lehr," where residual stresses are removed. For ordinary window glass the rolled material is usable in its original form, but for plate glass extensive grinding and polishing are needed.

A novel method for forming plate glass has become quite important (Fig. 8.12). The glass flows from the melting furnace onto a float bath of liquid tin. The float bath is covered by a refractory roof and a controlled, heated atmosphere is maintained to prevent oxidation. In the float chamber

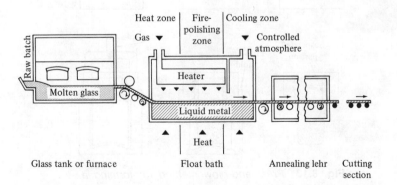

Fig. 8.12 *Method of making "float" glass. (Pilkington process.)*

both surfaces of the glass become mirror smooth, and then the sheet passes into the annealing furnace which has smooth rollers to avoid harming the finish. This has replaced the costly grinding and polishing operations of the old plate-glass roller method.

Centrifugal casting is used to make the funnels at the back sides of television tubes. A gob of glass is dropped into a metal mold which is rotated so that the glass rises by centrifugal force. The upper edge is trimmed, and then the glass faceplate is sealed to the funnel using a special low-melting-point solder glass.

Drawing of glass tubing is accomplished in a way similar to rolling. The glass of the proper viscosity flows directly from the melting furnace around a ceramic tube or mandrel pulled by asbestos-covered rollers. Air blowing through the mandrel keeps the tube from collapsing after passing the mandrel. Annealing is necessary, as for plate.

Pressing is accomplished by metering a gob of glass into a metal mold, compressing, and removing for annealing.

The *press-and-blow* method is widely used to make containers, as shown in Fig. 8.13. A gob is fed into a mold and pressed. Then the bottom half of the mold is removed, substituting a mold of the final shape. The blow

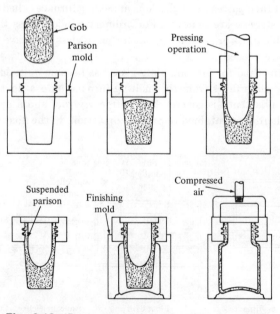

Fig. 8.13 *Press-and-blow method for forming a wide-mouth container.*

(W. D. Kingery, "Introduction to Ceramics," John Wiley & Sons, Inc., New York, 1960, Fig. 3.25, p. 67.)

operation gives the desired contour. The glass in the partly formed condition is called a "parison."

The fiber for fiberglass is now a very important product. This is made, in one method, by remelting glass marbles which flow through a heated platinum plate with orifices, giving filaments. Traction is provided by rotating the winding tube at surface speeds up to 12,000 ft/min (61 m/sec). Sizing material to separate and lubricate the fibers is applied as they are wound. The use of fiberglass with plastic is discussed in Chaps. 10 and 11.

The specifications for glass products vary widely, depending on the end use. For window and plate glass the chief requirements are flatness, transparency, and freedom from bubbles and harmful stresses which may not only cause breakage but also effect distortion. For containers accuracy of volume is usually important. In chemical ware it is necessary to maintain compositions that will not corrode. In cases where thermal shock is a consideration, the coefficient of expansion is important and silica or high-silica glass is specified. In the optical glasses the index of refraction is most important (Chap. 15), while in the electrical industry the dielectric constant is of high importance (Chap. 13).

Finally, the current interest in recycling materials has been readily applied to glass products. Because of the fabrication methods of such things as glass containers, it is advantageous to begin with a "prealloyed" glass, that is, one with a controlled chemistry and viscosity. Rather than starting from raw silica and adding other materials to reduce the softening point, manufacturers need only remelt the recycled glass and reform it to the necessary shape at a considerable saving.

8.14 Residual stresses and contraction

The development and control of residual stresses are more important in ceramics than in metals because no ductility is present to allow plastic deformation. However, residual stresses, if properly controlled, can be used to improve performance, as pointed out in previous sections.

In glasses the stress can be controlled by both mechanical and chemical methods. In tempered glass we take advantage of the fact that the compressive strength of glass is many times that in tension. Therefore, in a sheet of glass that is subjected to bending we would like to have residual compression in the surface layers. This is accomplished by quenching (normally by an air jet) the surfaces while the glass is in the plastic state. The surfaces are at a lower temperature as a result of the quench, but there is no residual stress immediately after the quench because the core is plastic. However, on cooling thereafter the core will attempt to contract a greater amount than the surface because it falls through a greater temperature interval. On reaching room temperature there is tension in the core and compression in the surface (Fig. 8.14).

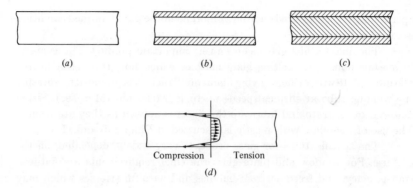

Compression | Tension

(d)

Fig. 8.14 *Dimensional changes in tempered glass. (a) Hot glass, no stresses. (b) Surface cooled fast, surface contracts, center adjusts, only minor stresses. (c) Center cools, center contracts, surface is compressed, center in tension. (d) Surface compression of tempered glass. These compressive stresses must be overcome before the surface can be broken in tension.*

(L. H. Van Vlack, "Elements of Materials Science," 2d ed., Addison-Wesley Publishing Company, Inc., Reading, Mass., 1964, Fig. 8-36, p. 228.)

The chemical method is to expose the surface of a glass containing sodium ions to a solution of (larger) potassium ions. Chemical exchange takes place and the "wedging in" of the larger ions causes surface compression.

Present government safety regulations require tempered glass in many applications. Windows other than windshields in automobiles are tempered glass. Even when such a window fails, it breaks up into many small, almost dustlike pieces rather than large splinters which could become projectiles in an accident. On the other hand, windshields are laminated with a polymer to contain any glass splinters since the use of highly tempered glass here might result in a total loss of transparency during an accident.

In reinforced concrete, prestressed beams have been developed to take advantage of the high compressive strength of the concrete. Before the concrete is poured, steel rods are positioned in the mold in the side of the beam under tension. The steel rods are stressed in tension by jacks, and then the concrete is poured. After the concrete is set, the jacks are removed. The steel contracts, producing residual compression in the concrete. When the beam is loaded, the first portion of the load merely reduces the stress in the concrete to zero on the tension side of the beam. Therefore, the total load can be greater before the capacity of the beam is exceeded.

Stresses due to thermal expansion and spalling are related problems. When the surface layers of a refractory are heated, they expand and tend to spall off the surface. Two techniques to combat this problem have been developed.

Glazes that have a lower coefficient of expansion than the base material are applied and fired. Since the base material tends to contract more than the glaze, the glaze is in compression and heating merely lowers the stress level. The other method is the use of a glass with a very low or negative coefficient of expansion, such as Pyroceram.

Therefore, even though ceramics inherently have low tensile strengths, they can still be used in tensile applications by the imposition of residual surface compressive stresses.

SUMMARY

In order to produce a part of a ceramic material, it is usually necessary to press or form the part from powder or paste and then fire the part to sinter the grains together and eliminate moisture. In many cases, such as brick, pipe and whiteware, it is uneconomical to attempt to change dimensions after firing, so careful processing is needed.

In the production of shapes from concrete no firing is required and the strength is due to a hydraulic bond; that is, a new rigid structure is developed by crystallization of the cement. In the calculation of amounts of material for production of masses of concrete, the concepts of packing and pore volume are important.

Glass processing takes advantage of the plastic nature of the material between the softening point and the working point temperatures.

Residual stresses may be eliminated by annealing to allow conversion of elastic strain to plastic strain, or they may be controlled at desired values by prestressing steel bars in concrete or by controlling cooling rates.

DEFINITIONS

Slip casting A shape-forming method in which a suspension of a solid such as clay in water is poured into a porous mold. The water diffuses from the layers next to the mold surface, leaving a solid shape. The liquid is poured out of the interior of the mold.

Wet plastic forming A method in which a wet plastic refractory mix is shaped by extrusion or other forming.

Dry powder pressing A process in which ceramic powder plus lubricant and binder are pressed into a die.

Hot pressing A method similar to dry pressing except that the heated die causes sintering of the part.

Chemical bonding A process in which a ceramic shape is bonded by the formation of a new structure between the grains (usually a hydraulic bond).

Portland and high-alumina cement Fine powders produced by sintering mixtures of clays and limestone followed by grinding. The addition of water causes a strong sheet structure to be formed.

Concrete An aggregate of sand and gravel bonded by cement to which water has been added.

Open pores The spaces in a container of rock or other material that are filled when a liquid is added.

Closed pores The internal cavities in a material that are not penetrated when the material is immersed in a liquid.

Bulk density The mass of a loosely packed container of material per unit volume.

True density The mass per volume of material without any voids.

True packing factor The true volume (no voids)/bulk volume or bulk density/ true density.

Residual stress A part at rest which contains elastic strain. If the elastic strain is known, the residual stress can be calculated from the modulus of elasticity.

PROBLEMS

8.1 Calculate the amounts of materials needed to make a concrete driveway 100 ft (30.48 m) long, 12 ft (3.66 m) wide, 6 in. (0.15 m) thick using the mix given in the text. Assume that 5 percent air will be entrapped.

8.2 What process and materials would you use to make the following?

    a. A coke bottle
    b. Window glass for a bank window
    c. Concrete pipe sections 4 ft (1.22 m) in diameter, 6 ft (1.83 m) long, 2 in. (5.08 cm) thick
    d. A ceramic hot plate 0.5 by 6 by 6 in. (1.27 by 15.24 by 15.24 cm)
    e. A crucible for melting 100 lb (45.4 kg) of steel at 3000°F (1649°C) with basic slags
    f. A resin-bonded grinding wheel
    g. A vitreous (glass)-bonded grinding wheel
    h. A three-floor parking structure for 600 cars
    i. A cut-glass decanter
    j. A mold for pouring a 50-ton (45,400-kg) steel roll for a steel mill
    k. Insulating brick for an oven to operate at 1000°F (540°C)
    l. A high-voltage power line insulator with a glazed surface

8.3 In comparing some small pure Al_2O_3 crucibles it is noted that there is a significant difference in physical properties between manufacturers.

Suggest why this might occur. Also, why might one manufacturer have size tolerances of ±2 percent while another's might be twice that high?

8.4 Carbonated beverage bottles have been known to explode, particularly if the gas pressure increases as the result, for example, of leaving a full bottle in the sun. As a manufacturer, how might you minimize this possibility?

8.5 Show schematically (by drawing curves) how the compressive strength of concrete might vary as a function of

 a. The amount of water used
 b. The curing time

8.6 In the use of steel rods to produce prestressed concrete, what controls the maximum amount of prestress that can be obtained?

8.7 As noted in the text, barium titanate is formed from $BaCO_3$ and TiO_2. What is the percentage loss in weight in this process?

8.8 You are going to put a "ceramic enamel" on a cast-iron bathtub. How could you do it and what precautions are necessary?

8.9 An insulating brick measures 9 by 4 by 2.5 in. (22.86 by 10.16 by 6.35 cm). When placed in a container of water, it is found to displace 2.4 lb (1.09 kg) of water. What would be the apparent porosity?

8.10 A ceramic tube is made by compaction of a powder. In the "green" state it is 5-cm OD by 3-cm ID by 5-cm high and is allowed to absorb water and become saturated. It absorbs 35 g of water. If the final porosity after sintering is 2 percent, what would be the final dimensions if the shrinkage is uniform in all directions?

8.11 An insulating brick (true specific gravity $= 2.58$) weighs 3.90 lb (1.77 kg) dry, 4.77 lb (2.17 kg) when the open pores are saturated with kerosene, and 2.59 lb (1.18 kg) when suspended in kerosene (specific gravity $= 0.82$). What are

 a. The true volume, apparent pore volume, and bulk volume?
 b. The closed porosity?

9

PLASTICS (HIGH POLYMERS)—STRUCTURES, POLYMERIZATION TYPES, PROCESSING METHODS

THE intertwined strings of beads in the illustration portray the salient difference between the high polymers or plastics we will discuss in this chapter and the predominantly crystalline materials we have studied thus far.

Although we will find important occurrences of crystallinity in the polymers, the basic building block is the molecule, and the nature and interaction of the molecules have a predominating influence on properties. We will encounter interesting analogies between the structures of glass and high polymers.

9.1 Introduction

The high polymers are a fascinating group of materials not only because of their increasing engineering importance, but also because the structures can be altered and tailor-made to provide a wide spectrum of color, transparency, and formability into different shapes.

We should mention for completeness that the study of high polymers includes two major fields: the biological large molecules that are the basis of life and food, and the engineering polymers such as the synthetic materials—plastics, fibers, and elastomers—and the natural materials—rubber, wool, and cellulose. We will be concerned here with only the engineering materials, principally the plastics and elastomers (rubbers).

Since the polymers are different in some ways from metals and ceramics, let us take a general look at these materials before going into detail.

The backbone of every organic material is the chain of carbon atoms, so let us first review carbon alone. We see that there are four electrons in the outer shell (Fig. 9.1a), giving a valence of four. With each electron a covalent bond can be formed to another carbon atom or to a foreign atom (Fig. 9.1b). The elements encountered most frequently and their valences are

Valence 1: H, F, Cl, Br, I
Valence 2: O, S
Valence 3: N
Valence 4: C, Si

At the heart of the polymer structure is the fact that two carbon atoms can have one, two, or three common bonds (i.e. share one, two, or three pairs of electrons) as well as bonding with other atoms. This is often shown in two dimensions (Fig. 9.2), but it is important to keep in mind that we have a three-dimensional structure similar to silicon in the center of the SiO_4^{4-} tetrahedron (Fig. 9.3), which we studied in ceramics. Just as we needed the three-dimensional picture to understand the sheet and fiber structures of the ceramics, we need a three-dimensional model to understand the properties of the plastics. However, many important features can be shown with the simpler two-dimensional view.

What are the results of bonding carbon atoms and forming chains of increasing length? If we start with ethane, C_2H_6, with two carbon atoms per molecule, we have a gas. Then as we increase the number of carbon atoms in the chain to several hundred, we pass through liquids of increasing boiling points to greases and waxy solids. Finally, when we reach over 1,000 carbon atoms, we obtain the materials with the characteristics of plastics, a combination of strength, flexibility, and toughness. These phenomena are due to the basic point that as the length of the molecules increases, the total binding force between

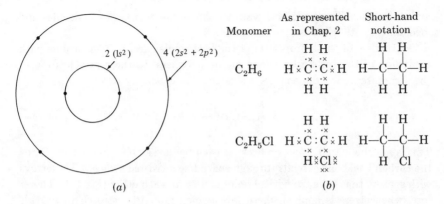

	Monomer	As represented in Chap. 2	Short-hand notation

Fig. 9.1 (a) Simplified sketch of the electrons of carbon. Note that four valence electrons are available for covalent bonds. (b) Carbon bonds (sharing of electron pairs) in the simple monomers C_2H_6 and C_2H_5Cl. Note that carbon may bond to another carbon atom, to an electropositive element such as hydrogen, or to an electronegative element such as chlorine.

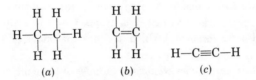

Fig. 9.2 Single, double, and triple carbon bonds. (a) Ethane. (b) Ethene (also called ethylene). (c) Acetylene.

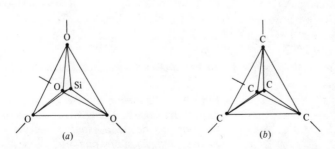

Fig. 9.3 Comparison of tetrahedra in (a) silicate and (b) carbon structures.

molecules rises. There are both van der Waals forces between molecules and mechanical entanglements between chains.

In view of these properties of the large molecules, let us continue with our study of the carbon backbone of the polymers. In the representation of the structure in two dimensions a straight line is usually used:

This is not the actual situation because each bond is at 109° to the next. Thus the carbon backbone extends through space like a twisted string of Tinkertoys with sockets that have cylindrical holes at 109° to each other (Fig. 9.4). Therefore, when stress is applied, these intertwined molecules stretch to provide elongation which can be thousands of times greater than it would be in a typical crystal of a metal or ceramic.

A vital point in making the long carbon chains is that we can take two molecules of a material such as ethylene in which there are two carbon-to-carbon bonds, open up one of the bonds in each, and join the molecules (Fig. 9.5). This is the source of the word "polymer" and the key to the science of polymer chemistry. The two original units are called "monomers" and, therefore, the large molecule is a polymer. Note that after the two mers are joined, there are still two free bonds for joining to other mers, so that the process can go on, linking mers together until it is stopped by the addition of another chemical called a "terminator," which satisfies the bonds at the ends of the molecules.

Thus in this simple case we form a number of large molecules of a linear polymer, polyethylene. We have shown the molecule as a straight line and it is called a "linear polymer," but we should emphasize that actually the molecules are not straight but can be thought of as a mass of worms randomly

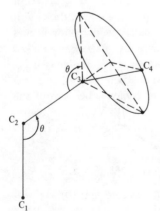

Fig. 9.4 *Formation of the carbon backbone. Atoms C_1, C_2, and C_3 define a plane; atom C_4 may lie at any one of several preferred positions on the circle shown.*

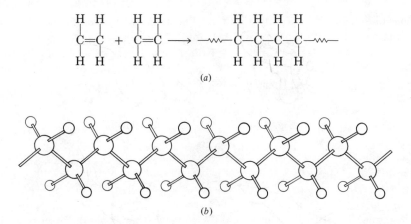

Fig. 9.5 *(a) Polymerization of ethylene by the opening of double bonds. (b) The geometry of a polyethylene chain.*

thrown into a pail. There would be considerably more intertwining than with worms, because if we scaled up the typical polymer molecule to a diameter of 0.25 in. (0.635 cm), it would be 20-ft (6.1-m) long!

We can continue our analogy a little further. We know that segments of the worms continually coil and uncoil, and a similar motion is noted with the polymer molecules. If we pulled slowly on the mass of worms, we would find a higher percentage with the long axis parallel to the direction of tension. The same effect occurs with nylon molecules, and after the alignment the material is stronger but less ductile.

The effects of temperature on a linear polymer are striking and are related to the behavior of a typical glass discussed in the preceding chapter. We can visualize a soda-lime glass made up of molecules of chains of SiO_4^{4-} tetrahedra terminated by sodium atoms. On cooling the thermal agitation of the molecules decreases and the material becomes viscous. Finally, the chains become locked in place and the glass is brittle. The same phenomenon occurs in polymers, and the temperature at which chain movement decreases to a low value is quite aptly called the "glass transition temperature," designated T_g. We should add that in addition to the plastic and brittle ranges, in some polymers we will find another temperature range of behavior because of the ability of the molecules to uncoil elastically; this is called the "rubbery" range. The whole basis for plastic, rubbery, or brittle behavior, therefore, is related to the amount of molecular movement in a given material.

So far we have introduced only one group of plastics, called the "thermo-plastics" because they become plastic when heated. There is another group (Fig. 9.6) in which a single large network, instead of many molecules, is formed during polymerization. Since this is usually accomplished by heating the basic

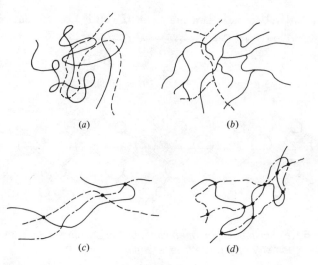

Fig. 9.6 *Different types of carbon backbone structures. (a) Linear. (b) Branched. (c) Loose network. (d) Tight network.*

materials together, this group is called the "thermosetting plastics." An example is a billiard ball made up of a thermosetting plastic; the molecular weight would be about 10^{28}! Since there is essentially one giant molecule, there is no movement between molecules once the mass has set. Furthermore, the material will not become plastic on heating, since there is no motion between molecules as there is in the thermoplastics. We shall see that in order for the network structure to form, the mers must have more than two places for bonding; otherwise only a linear polymer can be obtained. Both structures are useful. The thermoplastics are more easily formed and deformed elastically and can be remelted, while the thermosetting polymers are more rigid and generally have higher strength.

One final point: We have spoken of the carbon atom and the grouping of carbon and other atoms into many jointed chains (and even rings) giving the *molecules.* Now it may come as a surprise when we say that these chains, often containing 10,000 atoms, may be folded back and forth to form *crystals.* This is only possible in some cases, but where it occurs, the properties of the crystalline polymer are quite different from the disorganized structure of the amorphous material. The crystallization is usually only partially complete, but it is still possible to obtain x-ray diffraction measurements of the repetitive atom spacings.

Now that we have discussed the structures of polymers in general, let us plan our discussion in detail.

First we will consider how the molecular structures are built up. In the metals we found that the nature, amount, size, shape, distribution, and orientation of phases determined the properties of a given material. In the polymers we will find that in a related way the properties are functions of the mers of which the polymer is made up, of the arrangement of different mers in the polymer, of the size and shape of the molecules, and of their distribution (alignment, crystallinity).

Next we will discuss the processing of the polymers which makes possible rapid low-cost production of components with excellent surfaces and closely controlled dimensions.

Finally, in Chap. 10 we will summarize the mechanical properties of polymers and will take up some typical applications of polymeric materials.

Polymerization Methods

9.2 Review of the building blocks—the mers

Before we study the methods of polymerization, we should take a look at the mers, the building blocks we are going to use. If we were to set up a shelf of the most frequently used chemicals, it would include those shown in Table 9.1.

The nomenclature is based on the common saturated hydrocarbon in the majority of cases. Thus with one carbon atom we have the methyl group, with two carbon atoms the ethyl group, and so forth. The bonds are all considered covalent, so there is no question of sign, only of the quantity of shared electrons. Again these valences are 1: H, F, Cl, Br, I; 2: S, O; 3: N; 4: C, Si.

The most important group in the table, the *olefins,* is basic to the polymer industry. These differ from the saturated hydrocarbons by the double bond. This

$$\overset{\displaystyle |}{\underset{\displaystyle |}{C}}=\overset{\displaystyle |}{\underset{\displaystyle |}{C}}$$

bond can open to become

$$\times-\overset{\displaystyle |}{\underset{\displaystyle |}{C}}-\overset{\displaystyle |}{\underset{\displaystyle |}{C}}-\times$$

and other mers can join at the points marked $\times$ to form the polymer. We shall take up numerous examples shortly. The structural differences among the principal categories should be noted; that is, "ene" signifies an unsaturated hydrocarbon, "ol" an alcohol (with the OH group), "aldehyde" a —HC=O group, "acid" an H—O—C=O group, and "ester" a combination of an alcohol and an acid to yield the ester plus water. For example, ethyl alcohol plus acetic

Table 9.1 SOME COMMON CHEMICALS (MONOMERS) USED IN PLASTICS

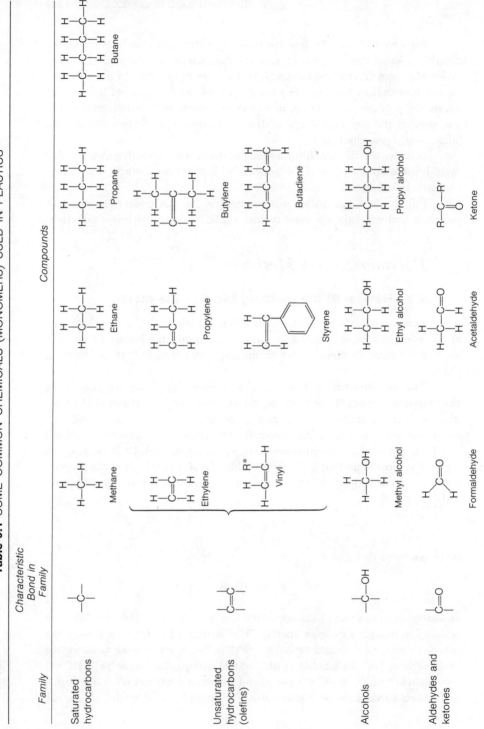

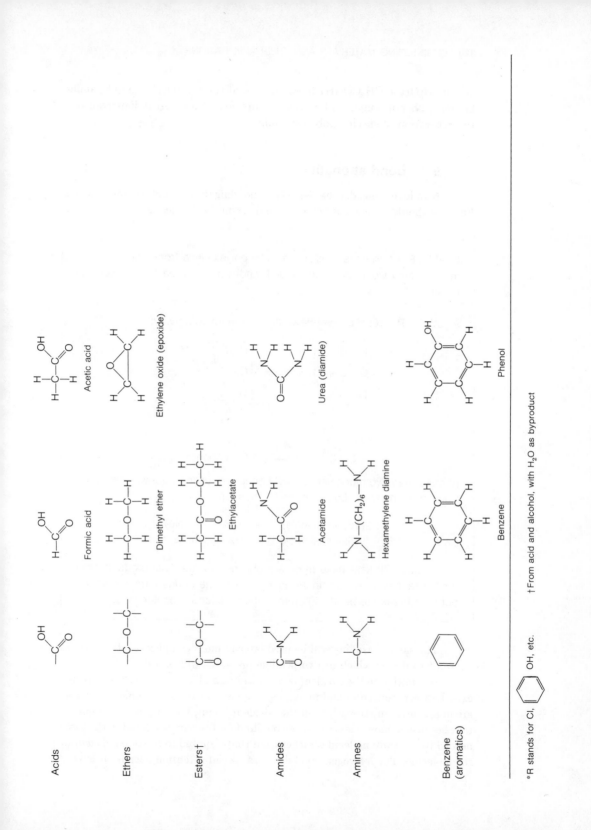

Acids

Ethers

Esters†

Amides

Amines

Benzene
(aromatics)

Acetic acid

Formic acid

OH

Ethylene oxide (epoxide)

Dimethyl ether

Ethylacetate

Urea (diamide)

Acetamide

Hexamethylene diamine

$N-(CH_2)_6-N$

Phenol

Benzene

*R stands for Cl,   OH, etc.   †From acid and alcohol, with H_2O as byproduct

acid (C_2H_5OH + CH_3COOH) gives ethylacetate and water. The amide, amine, benzene, phenol, vinyl, and styrene groups are also of great importance to the generation of specific polymer families.

9.3 Bond strengths

Now let us consider the elements of bonding these structures into plastics. First we should review the table of bond strengths, Table 9.2.

EXAMPLE 9.1 Suppose we are to make polyethylene from ethylene. How much energy would be interchanged, and would the reaction be spontaneous?

ANSWER The reaction can be written in the following way:

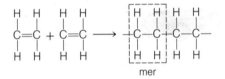

mer

or

$$2\ C{=}C\ \text{bonds} \longrightarrow 4\ C{-}C\ \text{bonds}$$

[If more than two monomers (C=C bonds) are used, there will always be twice as many C—C bonds formed.] Therefore,

$$2(162\ \text{kcal/mole}) \longrightarrow 4(88\ \text{kcal/mole})$$
or
$$324\ \text{kcal/mole} \longrightarrow 352\ \text{kcal/mole}$$

Since 28 kcal/mole more are obtained for the final products than for the reactants, we get more energy out of the bonds formed than we put in to break the bonds. Therefore, the reaction is spontaneous.

The same reasoning will be used later in more complex reactions where we want to decide which of two possible processes will prevail.

In addition to the covalent bonds within a molecule, van der Waals bonds exist between molecules and in overlapping parts of the same molecule. These are much lower in strength than the covalent bonds, but they are very important because in most cases the stress required for fracture is related to the force needed to *separate* molecules rather than that required to break bonds *within* the molecule. The hydrogen bond deserves special attention because it is very

Table 9.2 BONDING ENERGIES IN ORGANIC MOLECULES

Bond	Bond Energy,* kcal/g-mole (approx.)	Bond Length, Å	Bond	Bond Energy,* kcal/g-mole (approx.)	Bond Length, Å
C—C	88	1.5	O—H	119	1.0
C=C	162	1.3	O—O	52	1.5
C≡C	213	1.2	O—Si	90	1.8
C—H	104	1.1			
C—N	73	1.5	N—H	103	1.0
C—O	86	1.4	N—O	60	1.2
C=O	128	1.2			
C—F	108	1.4			
C—Cl	81	1.8	H—H	104	0.74

*The values vary somewhat with the type of neighboring bonds; e.g., the C—H is 104 kcal/mole in

methane, H—C—H, but only 98 kcal/mole in ethane, H—C—CH₃, and 90 kcal/mole in trichloromethane,

H—C—Cl. To obtain J/g-mole, multiply kcal/g-mole by 4.186×10^3.

Source: L. H. Van Vlack, "Materials Science for Engineers," Addison-Wesley Publishing Company, Inc., Reading, Mass., 1970.

strong in a number of cases, especially in cellulose (cotton) and polyamides (nylon and protein). Examples of this bond strength are:

Bond	—H···x— bond length, Å	Energy, kcal/g-mole†
O—H···O—	2.7	3 to 6
N—H···O—	2.9	4
N—H···N—	3.1	3 to 5

This bond does not have the strength of a covalent bond because there is no sharing of electrons. It arises from the fact that at the H side of an OH radical, for example, there is a positive polarity because the electron from the hydrogen is attracted closely to the oxygen. Similarly, an oxygen that has accumulated an electron from another source is of negative polarity.

9.4 Bonding positions on a mer, functionality

There must be a sufficient number of bonds in the mer that can be opened up for attachment to other mers of the same or different formula. This number

---

†To obtain J/g-mole, multiply kcal/g-mole by 4.186×10^3.

is called the "functionality."† The ethylene mer, therefore, is bifunctional, because the C=C bond can react with two neighboring monomers. For polymerization the mers must be at least bifunctional. Not only is a double carbon bond bifunctional, but the N—H bond in an amine, the O—H bond in an alcohol, and the C—OH bond in an acid may be split to form another bond. There are two or more of these bonds in amines and dialcohols. For example, in ethylene glycol (a dialcohol of ethylene), H—O—CH$_2$—CH$_2$—O—H, we find bifunctionality for polymerization by splitting both OH groups.‡ Therefore, we may form polymers without a carbon double bond in the mer. When the monomers are trifunctional or greater, it is possible to form network or thermosetting polymers.

The *degree of polymerization* (DP) is the molecular weight of the polymer divided by the molecular weight of the mer. It therefore tells the number of mers in the molecule or its average length. It is an important measurement because, as we have already seen, larger molecules result in higher bond strengths and therefore higher melting or softening points. There is always a variation in the DP in a given batch of polymers because not all chains start growing at the same time and the time of growth of individual chains is also variable. This variation can be measured and expressed statistically.

9.5 Polymerization mechanisms

Now let us take up the important polymerization mechanisms, since the type of structure we form is intimately related to the mechanism involved.

Addition Polymerization. We said earlier that in ethylene, for example, we break the C=C bond to form the polymer. We need to explain how this is done and also how the growth of a polymer molecule stops. The sequence is as follows.

An *initiator* is added to the polymer. The initiator, such as a radical R with either a free electron or an ionized group, will attract one of the electrons of the carbon double bond (Fig. 9.7). The other electron of the broken bond is unsatisfied and in turn attracts an electron from another mer, and so the molecule grows. Finally the chain is stopped when two growing segments meet or when a growing segment meets another R, the *terminator*. The rate of addition polymerization can be very rapid; for example, the polymerization of isobutylene using H$_2$O$_2$ can take place in a few seconds, yielding polymers made up of thousands of mers. For this reason the process is also called "chain-

---

†This discussion is so greatly simplified that it is not possible to predict many cases of functionality from these simple premises. See a text on organic chemistry for a fuller explanation.
‡This happens provided that other bifunctional molecules *having groups which can react with dialcohols in this manner* are present.

Step 1 $R_1 \times$ + $C \overset{\circ}{\underset{\circ}{\circ}} \overset{\bullet}{\underset{\bullet}{\bullet}} C$ ⎫

Step 1' $R_1 \overset{\circ}{\underset{\times}{}} C \overset{\circ}{\underset{\circ}{\circ}} C^{\bullet}$ ⎬ Initiation

Step 2 $R_1 \overset{\circ}{\underset{\times}{}} C \overset{\bullet}{\underset{\circ}{\circ}} C \overset{\circ}{\underset{\circ}{\circ}} C \overset{\bullet}{\underset{\bullet}{\bullet}} C^{\bullet} \cdots$ Growth (more monomers opening double bonds)

Step 3 $R_1 - C - C - C \cdots C^{\bullet} + \times R_2$ ⎫

Step 3' $R_1 - C - C - C \cdots C - R_2$ ⎬ Termination

Fig. 9.7 *Atom bonding in addition polymerization. For simplicity, the hydrogen or other atoms bonded to the sides of the carbon chain are not shown. R_1 and R_2 are respectively the initiator and terminator groups.*

reaction polymerization." A variety of chemicals including organic substances can be used as initiators and retarders to modify the reaction rate. Addition polymerization can also take place between two different mers, as in the formation of a copolymer of ethylene and vinylchloride. In this case it is called "copolymerization."

EXAMPLE 9.2 Sketch how the polymerization of styrene takes place and indicate whether it is thermoplastic or thermosetting.

ANSWER

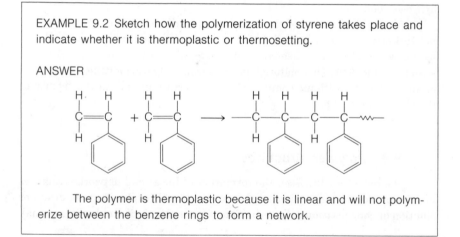

The polymer is thermoplastic because it is linear and will not polymerize between the benzene rings to form a network.

Condensation Polymerization. This process is also called "step-reaction polymerization" because two or more different molecules have to get together in each step for the growth of the molecule. For this reason and the difference in kinetics, it is much slower than the zipperlike addition polymerization. Furthermore, there is usually a byproduct which must condense; hence its name. This reaction can produce *either* a chainlike molecule, 6/6 nylon, *or* a network structure, phenol-formaldehyde (Fig. 9.8). In general, the linear polymers formed by condensation have a lower degree of polymerization. The chains are generally stiffer because of the backbone elements bonded between

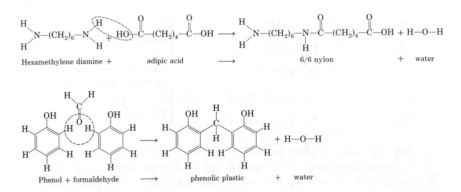

Hexamethylene diamine +   adipic acid   ⟶   6/6 nylon   + water

Phenol + formaldehyde   ⟶   phenolic plastic   + water

Fig. 9.8 *Formation of nylon and phenol-formaldehyde polymers. In the latter case further polymerization gives a network. (Note: The condensation reaction for phenol-formaldehyde is more complex than shown and involves several intermediate steps.)*

the carbon atoms, and the attractions between chains are greater due to hydrogen bonding.

We recall that one of the mers must be trifunctional or greater for the network structure to be formed. This can involve adjacent ring groups, as shown by a new development in heat-resistant materials in the formation of a ladder structure (Fig. 9.9). This material, derived from polyacrylonitrile, can be held in an open flame without change. These properties are due to the rigidity of the C—N ring backbone.

9.6   Polymer structures

We have just discussed the formation of linear and network structures in relation to polymerization processes. We should emphasize that in either addition or condensation polymerization, if the molecules are only bifunctional,

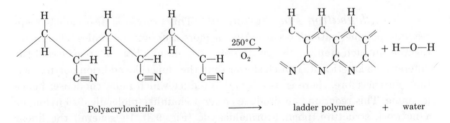

Polyacrylonitrile   ⟶   ladder polymer   + water

Fig. 9.9 *Formation of a ladder polymer from polyacrylonitrile which still has unsaturated positions. The original polyacrylonitrile was formed by an addition reaction (no byproduct).*

we form a linear or chainlike polymer. This is the thermoplastic polymer. On the other hand, if one of the molecules is trifunctional or greater, a three-dimensional network may be formed by either addition or condensation polymerization. This is a thermosetting polymer.

As an example of the first case, we form long chainlike polyethylene molecules. Upon heating, the van der Waals forces between these weaken and the material melts (thermoplastic). In the second case phenol-formaldehyde, for example, can continue to bond at other locations and form a network through space. Heat (and pressure) merely help to continue the bonding, and the material is therefore thermosetting. If the temperature is raised further, the primary bonds of the chains, rather than the bonds between molecules, are broken and the process called "degradation" takes place. The polymer is no longer useful. The material used in an electrical wall socket is a good example of a thermosetting plastic that may char but certainly will not melt in service (phenol-formaldehyde type).

EXAMPLE 9.3 The reaction between urea and formaldehyde is somewhat like the phenol-formaldehyde reaction (Fig. 9.8) but more complex. It takes place in two steps: first the formation of a methylol compound and then the condensation step.

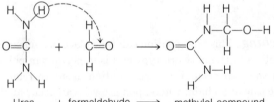

Urea　　+ formaldehyde ⟶　　methylol compound

How does the condensation step take place, and is the product thermoplastic or thermosetting?

ANSWER

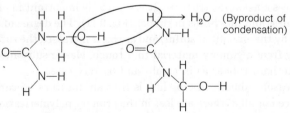

(The reaction is repeated at other N—H bonds.)

A complete network is formed, which makes the polymer thermosetting. The same network would inhibit crystallinity (see Sec. 9.7).

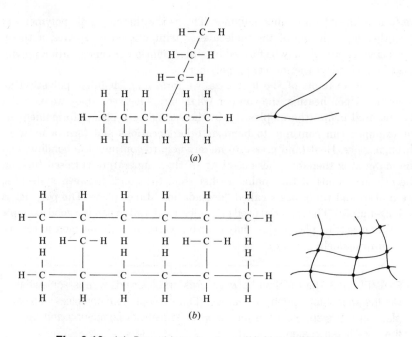

(a)

(b)

Fig. 9.10 *(a) Branching structure. (b) Network structure.*

Branching. This modification can take place in a linear, thermoplastic polymer as well as in the network type and is important in polyethylene. One way in which this can occur is for an addition agent to remove a hydrogen atom from the side of a chain, whereupon growth can occur at this point. The difference between branching and a network structure is shown in Fig. 9.6 and in a slightly different way in Fig. 9.10.

Crosslinking. Another modification of polymerization is crosslinking, which produces a network structure. It is especially important in obtaining the desired hardness and toughness in rubber (Fig. 9.11). The degree of crosslinking is controlled by the amount of sulfur, and as this is increased the rubber changes progressively from a gummy material to a tough, elastic substance as in tires and finally to hard rubber as in combs and battery cases.

The reason rubber can crosslink is that an unsaturated carbon bond is available since not all of these are lost in the primary polymerization, as shown in Fig. 9.11 for polyisoprene.

Ring Scission. This is related to addition polymerization and also to crosslinking in that two molecules are joined by a third. However, in this case

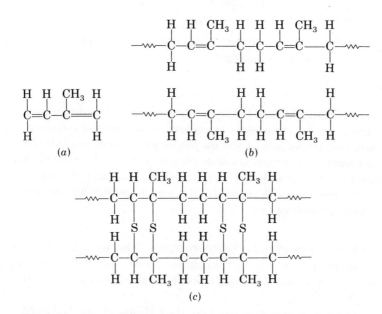

Fig. 9.11 *Crosslinking of rubber by sulfur (vulcanization). (a) Iso-prene monomer. (b) Isoprene polymer. (c) Item in (b) crosslinked with sulfur.*

a ring structure, such as an epoxide group,

is broken by combination with the linking reagent.

EXAMPLE 9.4 As many people know, epoxy cement comes in two tubes. In one tube is the epoxy itself, and in the other a material that will crosslink the epoxy molecules. Show how the polymerization could take place by the breaking of one of the rings as shown in Table 9.4. Also explain how the reaction leads to a tough crosslinked network.

ANSWER

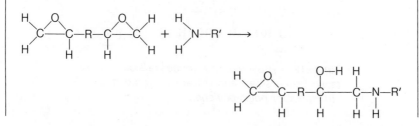

Crosslinking can occur at each $\overset{O}{\overset{\triangle}{C-C}}$ ring, and since there are two per molecule, a network is built up.

Location of Atom Groups. It is important to control the symmetry of distribution of a given group or element on the sides of the chain. There are three possibilities, and we will use polypropylene as an example (Fig. 9.12). In polypropylene the *atactic* (random side group) material is a waxlike product of little use, softening at 74°C. However, both the *isotactic* (same side of chain) and *syndiotactic* (opposite side of chain) types are useful, tough plastics, partly crystalline and melting near 175°C.

Other Important Configurations. The most important other configurations are *trans* and *cis* structures. These can be illustrated by the different configurations of the same chemical formula, gutta-percha and natural rubber (Fig. 9.13).

Note that in the trans structure the CH_3 and H are on opposite sides of the double bond, giving a relatively straight carbon chain. In rubber the chain curves (there cannot be later rotation of the *double* bond), leading to a helical springlike molecule in comparison to brittle gutta-percha.

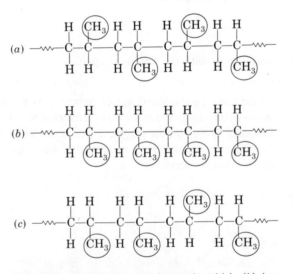

Fig. 9.12 (a) Syndiotactic (opposite side), (b) isotactic (same side), and (c) atactic (random sides) structures for polypropylene.

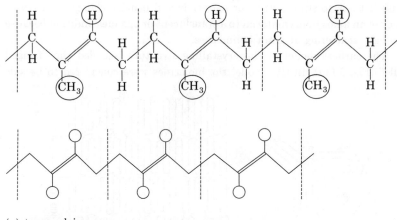

(*a*) trans-polyisoprene
[Possible rotations about indicated C—C bonds (-----------)
retain the linear structure of the molecule.]

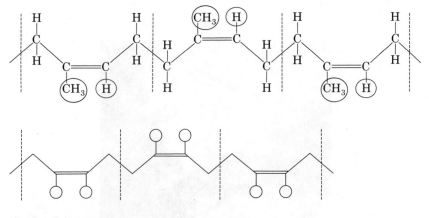

(*b*) cis-polyisoprene
[Possible rotations about indicated C—C bonds (----------)
allow the molecule to coil upon itself.]

Fig. 9.13 *Trans and cis configurations. (a) Trans: Gutta-percha. (b) Cis: Natural rubber. (Note: The structure cannot pivot around a C*=*C bond.)*

9.7 Crystallinity

Up to this point we have considered only two types of polymer structures: (1) the random arrangement of linear molecules, and (2) the network structure consisting of one large molecule with an amorphous (noncrystalline) structure. A third type of structure, the crystalline, is important in the linear polymers because it leads to stiffer, stronger materials and usually changes the transparency of an amorphous material to a translucent or opaque condition because of the light scattering at grain boundaries.

The earliest model of the crystalline structure is called the "fringed micelle" (Fig. 9.14a). In this model the molecules were considered to lie side

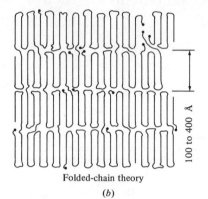

Fringe-micelle theory

(a)

Folded-chain theory

(b)

100 to 400 Å

(c)

Fig. 9.14 (a) *Fringed micelle model of crystallinity.* (b) *Crystallinity created by the folding back of chains. The growth mechanism involves the folding over of the planar zigzag chain on itself at intervals of about every 100 chain atoms. A single crystal may contain many individual molecules.* (c) *Spherulites in polycarbonate polymer. Transmission electron micrograph, 66,000X.*

[Parts (a) and (b) from L. E. Nielsen, "Mechanical Properties of Polymers," Litton Educational Publishing, Inc., New York, 1962. Reprinted by permission of Van Nostrand Reinhold Company. Part (c) courtesy of Jim de Rudder, University of Michigan.]

by side in some crystalline regions (the micelles) and to be at random in others. While this model is rather crude, it does help explain the strengthening which occurs when a fiber such as nylon is drawn: the molecules in the amorphous regions become stretched out and aligned in the direction of the applied stress, thus raising the yield strength. Recent research using electron microscopy has led to a more detailed knowledge of the crystalline regions. Single crystals of a number of polymers have been grown, and it has been found that the chain structures of the molecules fold back and forth in regular fashion to build up the crystal (Fig. 9.14b). It has also been found that the size can be increased either by annealing at elevated temperatures or by increasing the temperature of crystallization, as it can in the crystals of metals. For example, a polyethylene crystal at 100°C might have a typical thickness of 100×10^{-8} cm, and heating at 130°C for several hours will increase it to 400×10^{-8} cm. Formation of spherulitic crystals is shown in Fig. 9.14c.

The tendency to crystallize is very important and is closely related to the structure and polarity of the molecule. Regular molecules without bulky side groups or branches show strong tendencies to crystallize. It should be emphasized that partial rather than complete crystallization is obtained at best.

EXAMPLE 9.5 Explain the relative percent crystallinity in the following polymers for each numbered example.

Polymer	Percent crystallinity
1. Linear polyethylene	90
Branched polyethylene	40
2. Isotactic polypropylene	90
Atactic polypropylene	0
3. Random copolymers of linear polyethylene and isotactic polypropylene	0
4. *Trans*-1,4-polybutadiene	80
Cis-1,4-polybutadiene	80
Random cis and trans forms	0

ANSWER

Case 1: The branches interfere with regular arrangement.

Case 2: The random arrangement of the CH_3 side groups leads to non-crystalline material.

Case 3: Introducing a nonregular assortment of ethylene and polypropylene mers leads to a random spacing of the CH_3 side groups of the polypropylene.

Case 4: The situation is similar to part (2) in the random case.

9.8 Copolymers, blending, plasticizers

It is possible to synthesize useful polymers by joining different mers. For example, if we alternate mers of ethylene and vinylchloride, we form the ethylene-vinylchloride copolymer:

$$-(CH_2)_2-C_2H_3Cl-(CH_2)_2-C_2H_3Cl-$$

This is an alternating copolymer. Other types of copolymers are random, block (meaning an insert of a number of mers of one kind), and graft (in which the copolymer is added as a branch) (Fig. 9.15). Crystallization is only partial at best in copolymers. Blending or alloying is the blending of two or more distinct polymer molecules such as polyethylene and nylon to form a new product, in this case with unusual permeability.

Blending with a low-molecular-weight (approximately 300) material is called using a "plasticizer" because it is a common way to soften a polymer or make it flexible. Just as the addition of water to clay will plasticize, the addition of a small polar molecule can affect the van der Waals forces between the polymer chains. If too much plasticizer is added, the bond is chiefly between the small molecules and we have a liquid. An example is a common paint in which excess plasticizer allows brushing. Then the paint dries with evaporation of the plasticizer. This is usually accompanied by some polymerization and

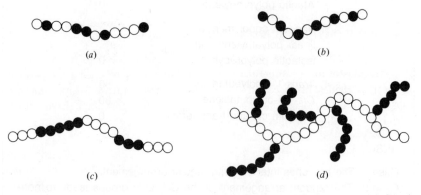

Fig. 9.15 *Different types of copolymers. (a) Random. (b) Alternate. (c) Block. (d) Graft.*

crosslinking with oxygen. The residual plasticizer makes the film tough and flexible. With time the combination of further oxidation and loss of plasticizer can result in brittleness and flaking of the paint.

EXAMPLE 9.6 Draw a possible structure of ABS (acrylonitrile-butadiene-styrene) if it is described as "graft of styrene and acrylonitrile on a butadiene backbone" (Table 9.4). Indicate whether the material is thermosetting or crystalline and whether it could form a network.

ANSWER

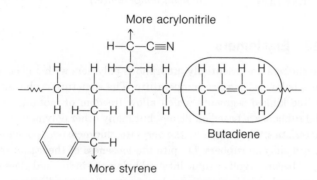

The material is thermosetting or thermoplastic, depending on whether it crosslinks to form a network. Since there can still be unsaturated carbon positions in the butadiene (as outlined in the second mer), crosslinking is possible. The material is too irregular to be crystalline (copolymers show little crystallinity).

9.9 The glass transition temperature T_g and the melting temperature T_m

We recall that when a piece of glass is cooled below the melting range, it is still plastic for a temperature interval but finally reaches a temperature where it is rigid. This is called the "transition temperature." The same phenomenon is encountered during the cooling of an amorphous thermoplastic material. Curiously, however, when the *plastic* reaches the transition temperature, it is called the "*glass* transition temperature" T_g. These characteristics are shown in Fig. 9.16. The specific volume (volume of 1 g) is plotted with change in temperature. Note that there is a sharp contraction at the melting point if the material crystallizes. In the case of the amorphous material the volume decreases more rapidly for a while below T_m, but then below T_g the amount of contraction is small.

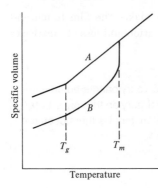

Fig. 9.16 *The glass transition temperature for a polymer. Specific volume vs. temperature for A, an amorphous polymer, and B, a partly crystalline polymer. (T_g = glass transition temperature and T_m = melting temperature.)*

9.10 Elastomers

The rubbers are essentially amorphous polymers with a glass transition temperature below the service temperature. The structure is characterized by highly flexible kinked segments which allow freedom of motion. The bulk of commercial rubbers are based on dienes, but many other polymers have rubber-like characteristics. These include the acrylate, fluorocarbon, polysulfide, polyurethane, and silicone rubbers. Despite the looseness of the structure of these materials, all become rigid at some interval below the individual glass transition temperature, and this is serious in low-temperature applications. A simple experiment is to immerse a sample of rubber in liquid air and then remove it and shatter it with a hammer blow.

We have already discussed the cis nature of the carbon chain and the use of sulfur as a crosslinking agent. Many other linkages are used. In fact, the crosslinking of rubber with oxygen and the resultant loss in elasticity is a common aging process in most rubber products.

9.11 Fillers

A high percentage of plastics are made with fillers. Phenolic and amino resins are almost always compounded or filled with substances like wood flour, short-fiber cellulose, powdered mica, and asbestos. These materials improve strength and dimensional stability and also decrease the cost of the polymers. The use of glass fibers in polyester resins is of great importance. Finally, carbon black is an important filler and strengthener of rubber. In a common tire mix the actual amount of polymer may be only 60 percent, with the remainder essentially made up of fillers. It should be apparent that a discussion of the properties of polymers should carefully distinguish whether plasticizers or fillers have been used.

Processing of Polymers

9.12 Processing of plastics

The processing of plastics has some of the features of the processing of both metals and ceramics. It is possible to buy sheets, rods, and fibers of many polymers, while in contrast it is usually necessary to buy the hard, brittle ceramic shapes in their finished form. On the other hand, in many cases in the fabrication of a plastic part a chemical mixture is compounded in the same equipment that forms it. For this reason we will take up the fabrication processes in this chapter while the polymerization methods are still before us.

Plastics fabricators use practically all the methods we have discussed for metals and ceramics plus a few tricks of their own. One of the key advantages of the plastics industry is the ability to produce accurate components with excellent surface at low cost and high speed. Part of this success is due to the lower temperatures at which plastics are liquid compared to metals and ceramics. (However, we shall see in the next chapter that this same property limits usage at elevated temperatures.)

We will discuss here only the processes for making typical components, leaving specialized processes for fibers, adhesives, etc., for the applications section of the next chapter.

The process to be used is largely dependent on whether we have a thermoplastic or thermosetting material. In Table 9.3 we have listed representative groups of plastics, and we see that injection and extrusion molding are predominant for the thermoplastics, and compression and transfer molding are most important for the thermosetting resins.

Compression molding is illustrated in Fig. 9.17. A slight excess of the material in a preformed blank is placed in the die, the mold is closed, and under heat and pressure the material becomes plastic and the excess is squeezed out as "flash." If this is used for a thermoplastic material, the mold must be cooled and, therefore, most of these plastics are formed by injection instead. However, for a thermosetting material the part may be ejected hot.

In transfer molding (Fig. 9.18), used for thermosetting compounds only, a partly polymerized material is heated to a high enough temperature to flow but not to crosslink in a premolding chamber. The material is then squeezed into the mold itself in the same equipment, where crosslinking takes place at a higher temperature and pressure.

Injection molding (Fig. 9.19) is done by feeding the polymer powder into a barrel, melting, and then injecting into a mold. It resembles the die casting of metals.

Table 9.3 SUMMARY OF PROCESSING METHODS FOR VARIOUS POLYMERS

Polymer	Compression Pressure,* psi × 10³	Compression °C	Injection Molding Pressure,* psi × 10³	Injection Molding °C	Extrusion °C	Transfer Molding Pressure,* psi × 10³	Transfer Molding °C
Thermoplastics							
Polyethylene			5 to 22	135 to 143	80 to 94		
Polypropylene			10 to 22	204 to 288	193 to 221		
Polystyrene			10 to 24	162 to 243	90 to 107		
Polyvinylchloride			7 to 15	160 to 175	162 to 204		
Tetrafluoroethylene†	0.5 to 2	140 to 175					
ABS‡		162 to 190	6 to 30	218 to 260	277 to 299		
Polyamides			10 to 20	271 to 343	177 to 232		
Acrylics			10 to 20	160 to 260	188 to 204		
Acetals		200 to 245	15 to 25	193 to 215			
Cellulosics	0.5 to 5		8 to 32	215 to 254	215 to 232		
Polycarbonates			15 to 20	274 to 330	247 to 304		
Thermosetting Polymers							
Phenolics	1.5 to 5	143 to 193	4 to 8	160 to 171		2 to 10	135 to 171
Urea-melamine	2 to 5	149 to 171				6 to 20	149 to 165
Polyesters						1 to 5	121 to 177
Epoxies						0.1 to 2	143 to 177
Silicones§	1 to 3	177				0.5 to 10	177

* Multiply psi by 6.9 × 10⁻³ to obtain MN/m² or by 7.03 × 10⁻⁴ to obtain kg/mm².
† Requires preforming of particles and fusion at 370°C.
‡ Acrylonitrile-butadiene-styrene.
§ Glass-reinforced.

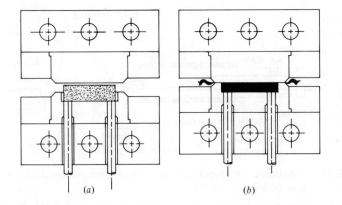

Fig. 9.17 *Compression mold. (a) Open, with preform in place. (b) Closed, with flash forced out between mold halves.*
("Modern Plastics Encyclopedia," McGraw-Hill, Inc., New York, 1966.)

Extrusion of tubing, rod, and other shapes of constant cross-section is similar to the metal process (Fig. 9.20). However, it is necessary to cool the part to near T_g to gain dimensional stability. This is done by running the part into cooling water or by an air blast. In the case of rubber extrusion the rubber is carefully compounded to permit forming to the desired shape before it is finally cured to stiffen it.

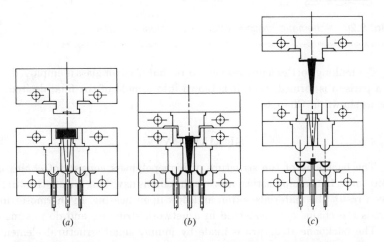

Fig. 9.18 *Typical transfer molding operation. Material is fed into the pot of the transfer mold and then forced under pressure when hot through an orifice and into a closed mold. After the polymer has formed, the part is lifted by ejector pins. The sprue remains with the cull in the pot.*
("Modern Plastics Encyclopedia," McGraw-Hill, Inc., New York, 1966.)

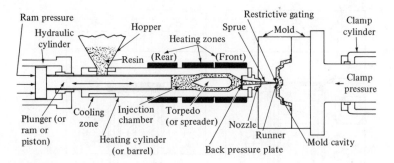

Fig. 9.19 *Schematic cross-section of typical plunger (or ram or piston) injection molding machine.*

("Petrothene Polyolefins—A Processing Guide", 3d ed., U.S. Industrial Chemicals Co., New York, 1965.)

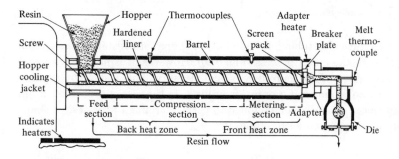

Fig. 9.20 *Schematic cross-section of a typical extruder.*

("Petrothene Polyolefins—A Processing Guide", 3d ed., U.S. Industrial Chemicals Co., New York, 1965.)

For making bottles a process similar to that used for glass is employed. First a parison is formed; then it is blown into a mold of the final desired shape, as shown in Chap. 8.

SUMMARY

The backbone of the structure of high polymers consists of chains of carbon atoms. In thermoplastic materials the chains have long linear structures, and as a result the materials soften and flow upon heating. In thermosetting materials the chains are connected in a network structure and do not melt.

The backbone structure is made by joining small structural elements called "mers." When only two bonding positions are present in the mer, the linear thermoplastics are formed, while if at least one variety of mer in the mix is more than bifunctional, a network will form.

The polymer can crystallize if the mers are linear and possess no side groups on the chain or small regularly oriented side groups (isotactic or syndiotactic).

Amorphous polymers exhibit behavior similar to glasses, passing from a viscous to a rigid structure at the glass transition temperature T_g. A rubbery range is also encountered above T_g. When a linear polymer is made of several different mers, it is called a "copolymer" and generally will not crystallize.

Most processing is accomplished by compression, transfer, injection, extrusion, and blow molding.

DEFINITIONS

Mer A unit consisting of relatively few atoms which are joined to other units to form a polymer.

Monomer The same unit as a mer standing alone, i.e. not part of a polymer.

Polymer A molecule made up of repeating structural groups or mers. For example, polyethylene is made up of $-CH_2-CH_2-$ groups.

Initiator A material that when added to a monomer acts to initiate polymerization.

Terminator A material that reacts with the end of a growing polymer chain to terminate growth.

Glass transition temperature, T_g The temperature at which a high polymer becomes rigid. T_g is similar to the transition temperature for glasses.

Thermoplastic A high polymer that flows and melts when heated. Scrap may be recovered by remelting and reusing.

Thermosetting plastic A high polymer that sets into a rigid network. Such a polymer does not melt when heated but chars and decomposes, a process called "degradation." These polymers are not reusable.

Linear polymer A polymer in which the mers are joined in a line rather than a network.

Branched polymer A linear polymer with forked branches.

Bond strength The energy required to break the bond under discussion.

Functionality The number of positions on a monomer at which bonding to another monomer can take place.

Bifunctional Having two bonding positions. ("Tri" signifies three positions and "tetra" signifies four.)

Degree of polymerization The molecular weight of a polymer divided by the molecular weight of the mer.

Addition polymerization Bonding between similar and/or different monomers by linking at functional positions without the formation of a condensation product. Also called "chain-reaction polymerization." When monomers are different, this is called "copolymerization."

Condensation polymerization Bonding between mers that is effected by a reaction in which the bonding takes place, with the emission of a byproduct such as H_2O, NH_3 gas, etc. For example, phenol + formaldehyde gives phenol-formaldehyde resin + water.

Network structure The structure formed when one or more of the monomers is tri- or polyfunctional.

Crosslinking Connections made between polymer molecules by a crosslinking agent. For example, butadiene rubber can be crosslinked with sulfur (vulcanization) or oxygen (oxidation).

Ring scission The break up of a ring structure to provide bonds for polymerization.

Atactic structure A structure in which the side groups such as CH_3 of a molecule are arranged nonpreferentially on the sides of the chain.

Isotactic structure A structure in which the side groups are all arranged on one side of the chain.

Syndiotactic structure A structure in which the side groups are regularly arranged on alternate sides of the chain.

Trans structure A relatively straight carbon backbone produced, for example, by positioning the H and CH_3 on alternate sides of the chain.

Cis structure A curved carbon backbone produced, for example, by positioning the H and CH_3 groups on the same side of the chain.

Crystallinity The alignment of a molecule or molecules in a regular array for which a unit cell can be described. Crystallinity is generally not complete.

Plasticizer A chemical of lower molecular weight added to a polymer to soften or liquefy it.

Solvent A liquid that dissolves the polymer, for example, in a paint.

Filler Foreign material used to strengthen or modify properties of polymers.

Blending An intimate mechanical mixture of several polymers.

Compression molding A process in which a metered amount of plastic is placed in a heated die and compressed to the desired dimensions. It is an important process for thermosetting resins.

Injection molding The injection of liquid resin into a die, followed by solidification and ejection.

Extrusion The forcing of liquid or plastic resin through a die to obtain the required shape, such as a rod.

Blowing A process in which a gob of plastic is formed and then blown into a mold of the desired shape.

Transfer molding A process in which material is fed into the pot of a transfer mold, heated, then forced into an adjoining cavity of the same mold under pressure.

PROBLEMS

Let us devote our efforts to synthesizing the important polymers not previously treated as examples. Many materials and chemicals needed have been included in Tables 9.1 and 9.4. Most are relatively simple, but there are a few complex cases.

9.1 Polyethylene is sold in two forms, low density (8 percent of the market) and high density (20 percent of the market). The low-density form shows a large amount of branching, while the high-density type is relatively free of branching and is ordered.

a. Sketch both types.
b. Which type will show greater crystallization?
c. Is it thermoplastic or thermosetting?

9.2 Consider polyvinylchloride (PVC).

a. Sketch the polymer and the copolymer with ethylene.
b. Could either crystallize? (*Hint:* Polyvinylchloride is syndiotactic.)
c. Is it thermoplastic or thermosetting?

9.3 The polymer made from phenol and formaldehyde is an example of the phenolics. Would it be

a. Crystalline?
b. Thermoplastic or thermosetting?

Sketch the structure.

9.4 Sketch the polypropylene molecule with

a. Atactic structure
b. Isotactic structure
c. Syndiotactic structure

Which would be amorphous and which could crystallize?

9.5 The term *ester* is applied to the product resulting from the reaction of an alcohol and an acid. If we let R and R′ stand for the remainder of the molecule, such as C_2H_5, etc., the reaction is

$$R \cdot OH + R'COOH \longrightarrow R \cdot \underset{\substack{\| \\ O}}{C}\!-\!O\!-\!R' + H_2O$$

Vinylacetate is considered an ester because it has the vinyl group from vinyl alcohol† and the acetate group from acetic acid. The polymer is polyvinylacetate. Sketch its structure. Would this be thermoplastic or thermosetting? Is it crystalline? (*Continued on page 344.*)

---

†Vinyl alcohol does not actually exist as a monomer, but polyvinyl alcohol is a real material.

Table 9.4 SUMMARY OF IMPORTANT POLYMERS

Group I. Thermoplastics

Polymer	Percentage of Market	Monomer(s) Used
Polyethylene	28	
Polyvinylchloride	17	
Polystyrene	16	
Polypropylene	6	
ABS	2	
Acrylics (example: polymethyl methacrylate, Lucite)	2	
Cellulosics	1	

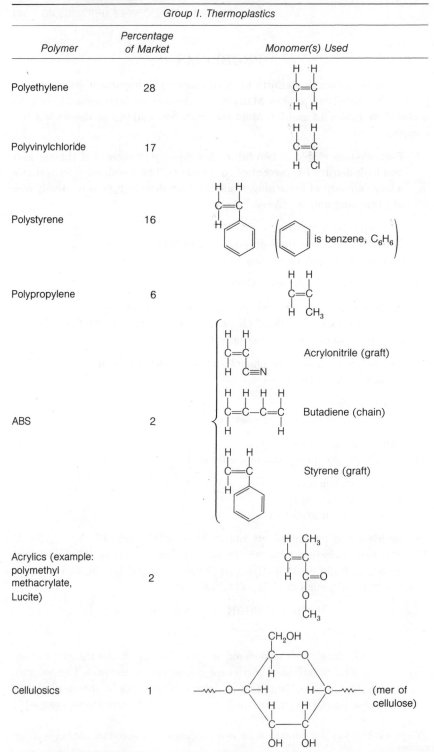

Table 9.4 SUMMARY OF IMPORTANT POLYMERS (*Continued*)

Group I. Thermoplastics (*Continued*)

Polymer	Percentage of Market	Monomer(s) Used
Acetals	<1	$-C H_2-O-$ (mer)
Nylons	<1	$H_2N-(CH_2)_6-NH_2$ $HO-C(=O)-(CH_2)_4-C(=O)-OH$
Polycarbonates	<1	$-O-C(=O)-O-C_6H_4-C(CH_3)_2-C_6H_4-$ (mer)
Fluoroplastics (example: polytetrafluorethylene)	<1	$CF_2=CF_2$
Polyester, thermoplastic type [Example: polyethylene-terephthalate (dacron)]	3	$HO-CH_2-CH_2-O-H$ $HO-C(=O)-C_6H_4-C(=O)-OH$

Group II. Thermosetting Polymers

Polymer	Percentage of Market	Monomer(s) Used
Phenolics (Example: phenol formaldehyde)	6	phenol $-H$ $O=CH_2$ $H-$ phenol
Amino resins (Example: urea formaldehyde)	4	$H_2N-C(=O)-NH-H$ $CH_2(=O)$ $H-NH-C(=O)-NH_2$
Polyesters, thermoset type	1	H_2C-OH, $HC-O-H$, H_2C-OH $HO-C(=O)-(CH_2)_x-C(=O)-OH$
Epoxies	1	epoxide$-C-R-C-$epoxide H_2N-R' (R and R' are complex polyfunctional molecules)

Table 9.4 SUMMARY OF IMPORTANT POLYMERS (*Continued*)

Polymer	Percentage of Market	Monomer(s) Used	
Polyurethane, also thermoplastic	1	OCN—R—NCO + HO—R′—OH (diisocyanate)	(R and R′ are complex polyfunctional molecules)
Silicones	1	$$\text{Cl—Si—Cl}$$ with CH_3 above and Cl below — Trichlorosilane; $$\text{H—O—Si—OH}$$ with CH_3 above and OH below — Trihydroxy silane	

In order to use the term polyester *resin*, the oxygen must be part of the carbon backbone. Therefore, the polyester resins are more complex. A diacid (two COOH groups) and a dialcohol (two OH groups) are reacted to give a linear polymer. However, if there are also double bonds, the polymer can be crosslinked, using the unsaturated bonds to give a network structure as shown in Fig. 9.14. Also, a trialcohol or polyalcohol may be used with a diacid to give a network structure. Thus a wide variety of materials are encountered in this polyester family.

9.6 The most common acrylic is polymethylmethacrylate, sold as Lucite or Plexiglas. The methylmethacrylate structure is given in Table 9.4. Sketch the structure of the polymer. Is it linear or network? Is it crystalline? Thermoplastic?

9.7 Another amino resin can be formed from melamine and formaldehyde. The structure of melamine is:

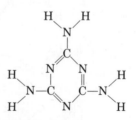

Show the polymerization reaction of melamine-formaldehyde and indicate the functionality of melamine if all hydrogens are replaceable.

9.8 The cellulose mer is shown in Table 9.4. Despite the complex structure, the regularity leads to crystallinity. Cellulose esters were the most important group of plastic materials before 1950. Remembering that an ester is an alcohol (OH group) plus an acid, sketch the formation of the ester cellulose acetate.

9.9 After over 50 million dollars in research a method was found for making acetal resin by polymerization of formaldehyde,

$$\underset{\text{H—C—H}}{\overset{\text{O}}{\|}}$$

Show how this is formed from formaldehyde. Is the material thermoplastic or thermosetting? Is it crystalline? Compare the structure with polyethylene (see Table 9.4).

9.10 The most important nylons are 6/6, 6, 6/10. The numbers refer to the number of carbon atoms in the mers which are joined by a condensation reaction. Let us take the case of 6/6 nylon, which is formed from adipic acid and hexamethylene diamine (Fig. 9.8). Actually the condensation takes place by combining the H of the NH_2 group with an OH to form water. Calculate the energy of the bonds broken and that of the new bonds formed. Compare the alternative possibility of breaking off an NH_2 group and the H of the OH group to form NH_3 gas as a condensation product. Is the nylon thermoplastic or thermosetting?

9.11 The structure of a polycarbonate is shown in Table 9.4. The molecule is regular in structure and can be crystallized by evaporation or slow cooling. Would you expect it to be thermoplastic?

9.12 Polytetrafluorethylene (TFE, Teflon) is an important plastic. What is the structure of the mer? Is the polymer thermoplastic or thermosetting? Suggest how it is attached to frying pans.

9.13 The basis of the *silicone* structure is silane, SiH_4, which is analogous to methane, CH_4. Silicon hydrides are named according to the number of silicon atoms in the chain; thus Si_3H_8 is trisilane. When chlorine replaces hydrogen, we have chlorosilane. It is possible to produce a backbone structure consisting of alternating silicon and oxygen atoms analogous to the acetal structure. These are the silicones. They are prepared by first reacting chlorosilanes with water to form the trihydroxy silane in Table 9.4. Show how these hydroxyl compounds can undergo condensation polymerization, giving a silicone.

9.14 Fiberglass is used in boat hulls and automobile bodies. Everyone is interested in this application of glass plus plastic, and although a number of basic steps are involved, we have built up the background to understand them.

First the glass must be chosen. There are two types of glass: low-alkali and high-alkali (15 percent) borosilicate. Which gives better weathering characteristics and is more expensive (see the discussion of ceramics)?

Next the glass has to be prepared for later blending with the resin. A pretreatment with a silane is used. Vinyl trichlorosilane, $Cl_3SiCH=CH_2$, is hydrolyzed in the presence of glass fiber. The OH groups of the $(HO)_3SiCH=CH_2$ react with the OH groups on the glass surface, leading to condensation of water and an —O— bond to the glass. The vinyl group is on the other side of the —O— bond, ready to bond with the polyester. Sketch how these reactions give the structure shown in Fig. 9.21.

Now let us consider a typical polyester formula, which appears complex at first:

> 0.2 mole phthalic anhydride
> 0.2 mole maleic anhydride
> 0.2 mole propylene glycol
> 0.2 mole ethylene glycol
> 0.2 mole styrene
> trace hydroquinone (0.02 percent)

This mixture is compounded carefully, giving a ready-to-use liquid. It is really made of two parts, the polyester and the monomer. We recall that an ester is made of two parts, an acid and an alcohol. In this case two acids (dehydrated to give the anhydrides) are present. One is an unsaturated acid so that crosslinking into a network structure is obtained later, and the other is saturated to avoid an excess of crosslinking, which would lead to brittleness. The formation of the polyester is shown in Fig. 9.21, expressing the anhydrides as the acids. The hydroquinone stabilizes the mixture until it is to be used. When the glass mat is ready, an initiator (peroxide) and accelerator are added to the polyester-styrene mixture. The mixture polymerizes, at first gelling and then hardening. Sketch how the styrene can

a. Polymerize
b. Bond to the polyester
c. Bond to the glass-silicone face

9.15 If all the isoprene in Fig. 9.11 is crosslinked with sulfur as shown, what percent by weight sulfur would have been added?

9.16 In Fig. 9.21 calculate the pounds of water released per pound of polyester formed [refer to part (*b*); maleic acid + ethylene glycol + phthalic acid provide one mer].

9.17 What would be the degree of polymerization if the average molecular weight of a styrene polymer was 73,000 g/mol. wt.?

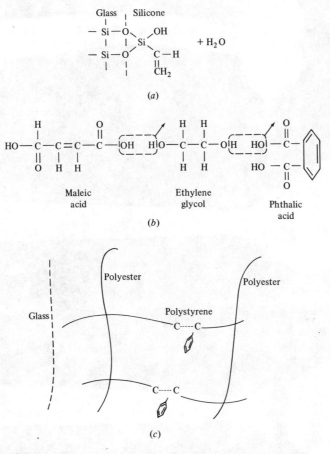

Fig. 9.21 *Formation of bonds in polyester-styrene-fiberglass.* (a) *Glass treated with silicone.* (b) *Formation of polyester.* (c) *Final structure.*

9.18 Which processing method would you use to make the following?

  a. Saran Wrap sheet
  b. Polypropylene rope
  c. Polyethylene squeeze bottle
  d. Dish of melamine
  e. Nylon fishing leader (0.007 in. diameter, clear)

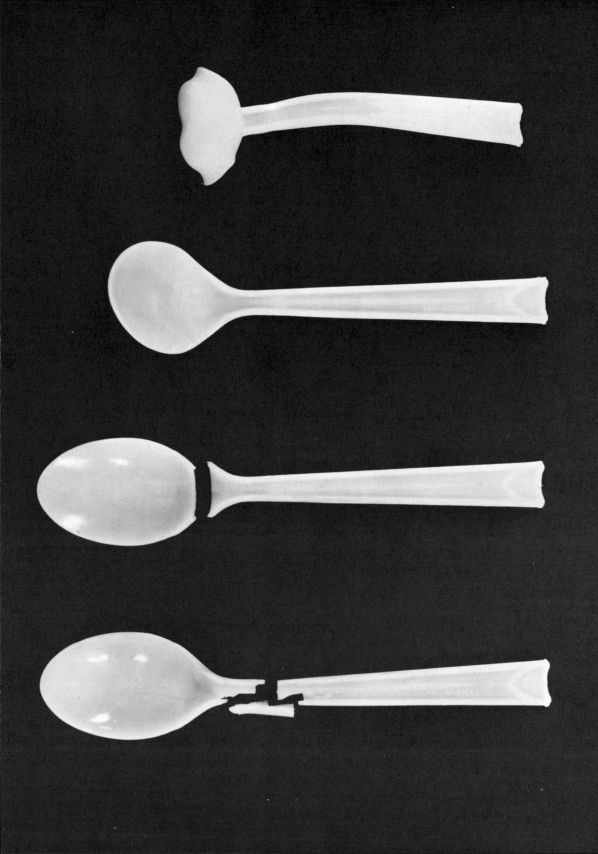

10

PROPERTIES AND APPLICATIONS
OF PLASTICS

THE spoons in the illustration were bent after immersion in water at different temperatures to show the very important effect of temperature on the mechanical properties of a thermoplastic material. The spoon at the top was immersed in boiling water and curled up readily when lightly stressed. The next spoon bent less readily at 150°F (66°C). The third spoon fractured at 120°F (49°C), while the bottom spoon splintered at 70°F (21°C).

In this chapter we will compare mechanical and thermal properties of high polymers and take up the reasons for transparent and translucent structures.

10.1 General

In this chapter we will correlate the mechanical properties of the plastics, such as strength and elongation, with the structures we have just discussed, and we will find that there are predictable relationships, just as there are in metals and ceramics. Corrosion resistance and the electrical properties will be discussed later, when we deal with special features of all the materials (Chaps. 12 to 14).

We have arranged the structures of the representative thermoplastic and thermosetting materials (Tables 10.1 and 10.2) in the same order in which they were summarized in the preceding chapter (Table 9.4), with the addition of certain important properties. Before taking up these specific properties, it is helpful to consider in a general way what we would expect from the different structures.

Just as in the metals and ceramics we were able to anticipate and interpret the properties in terms of the nature, amount, size, shape, distribution, and orientation of the phases, in the plastics it is important to follow the changes in the molecules and molecular arrangements to understand the properties.

In the thermoplastics we begin with the simple polyethylene structure, a carbon backbone with hydrogen atoms. The variations in properties are due to the degree of polymerization (i.e. the size of the molecule) and to grafting, a shape effect.

Advancing through polypropylene, polystyrene, polyvinylchloride, and Teflon, we see that we are changing the nature of the molecule by replacement of hydrogen atoms at the sides of the carbon backbone, which strengthens and even changes the chemical reactivity, as for example in Teflon.

Next we graft on additional molecules and add copolymers in the chain, as in acrylonitrile-butadiene-styrene (ABS). Then we stiffen the chain, as in nylons, acrylics, acetals, cellulosics, and polycarbonates. With the addition of polar groups as in nylon, hydrogen bonding gives further strengthening. Finally, with the use of crosslinking, as in polyesters and rubber, we can approach the rigidity of the thermosetting group.

In the thermosetting group the dominating effect is the network structure, giving essentially one giant molecule. The properties can, of course, be modified with fillers or plasticizers.

10.2 Effects of temperature and time

First let us consider the change in a typical property such as the modulus of elasticity E with falling temperature for an amorphous thermoplastic material (Fig. 10.1). As we decrease the temperature from the liquid state, reading from right to left in the figure, at first the material is *plastic,* flow is easy, and

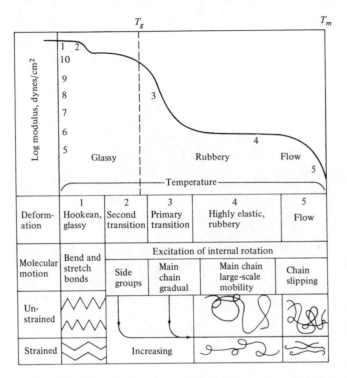

	1	2	3	4	5
Deform-ation	Hookean, glassy	Second transition	Primary transition	Highly elastic, rubbery	Flow
Molecular motion	Bend and stretch bonds	Excitation of internal rotation			
		Side groups	Main chain gradual	Main chain large-scale mobility	Chain slipping

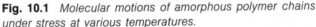

Fig. 10.1 *Molecular motions of amorphous polymer chains under stress at various temperatures.*

("Modern Plastics Encyclopedia," McGraw-Hill, Inc., New York, 1968.)

the molecular chains slip by each other. Next we encounter the *rubbery* range, a plateau where the extension is very great but elastic. The chains uncoil but go back to their original positions after the stress is removed. At lower temperatures we pass T_g and now the thermal agitation of the chains is less. When we are well into the glassy range, only small elastic movements are possible, and fracture takes place with very little elongation although the modulus is higher. The region between the glassy portion and the rubbery plateau is usually referred to as the *leathery* region.

Now let us consider the effects of speed of testing or stressing, again using the elastic modulus as a measure of bond strength. We realize that when strain takes place, molecules must move and such movement takes time. Therefore, Fig. 10.1 applies to one strain rate, and it is possible to modify the results by a change in the strain rate. Although we will treat such effects more fully in Chap. 11, we should point out here that the effects noted at higher temperatures in Fig. 10.1 would be obtained at lower temperatures with lower strain rates. Conversely, at a given temperature a short-time test results in a higher modulus and brittle or glassy behavior for a given polymer. On the other hand, the same

Table 10.1 PROPERTIES OF THERMOPLASTIC RESINS

Name and Structure[a]	Use, percent	Price,[b] dollars/lb	Tensile Strength,[c] psi	Percent Elongation	Rockwell Hardness, R	Impact,[d] izod, ft-lb	Modulus,[c] psi × 10³	Specific Gravity	Coefficient of Expansion, $(°F)^{-1} \times 10^{-6}$ $[(°C)^{-1} \times 10^{-6}]$	Heat Distortion,[e] °F (°C)	Burning Rate,[f] in./min	Typical Applications
Polyethylene												
High density	8	0.18 to 0.20	4,000	15 to 100	40	1 to 12	120	0.95	120 (216)	120 (49)	1	Clear sheet, Bottles
Low density	20	0.13 to 0.18	2,000	90 to 800	10	16	25	0.92	100 (180)		1	
Polypropylene	5	0.20 to 0.26	5,000	10 to 700	90	1 to 11	200	0.91	170 (308)	150 (66)	1	Sheet, pipe, coverings
Polystyrene	16	0.15 to 0.18	7,000	1 to 2	75	0.3	450	1.05	38 (68.5)	180 (82)	1	Containers, foams
Polyvinylchloride (rigid)	17	0.12 to 0.24	6,000	2 to 30	110	1	400	1.40	30 (54)	150 (66)	<1	Floors, fabrics
Polytetrafluoroethylene (Teflon)	<1	3.25	2,500	100 to 350	70	4	60	2.13	55 (99)	270 (132)	0	Chemical ware, seals, bearings, gaskets
ABS, acrylonitrile-butadiene-styrene copolymer	2	0.32 to 0.36	4,000 to 7,000	20 to 80	95	1 to 10	300	1.06	50 (90)	210 (99)	1	Luggage, telephones

Material	Structure											Uses
Polyamides (6/6 nylon)	$-\!\!\!\backslash\!\!\!-N-C-$ (with H on N, O double bond on C)	<1	0.75	11,800	60	118	1	410	1.10	55 (90)	220 (104)	Low	Fabric, rope, gears, machine parts
Acrylics (Lucite)	$-C-C-O-CH_3$ (with O double bond, CH₃ branch)	2	0.32 to 0.36	8,000	5	220	0.5	420	1.19	40 (72)	200 (93)	1	Windows
Acetals	$-C-C-O-O-C-$	<1	0.65	10,000	50	120	2	520	1.41	44 (79)	255 (124)	1	Hardware, gears
Cellulosics	(ring structure with O and C)	1	0.40 to 0.62	2,000 to 8,000	5 to 40	50 to 115	2 to 8	500 to 4,000	1.25	75 (135)	115 to 190 (46 to 88)	1.4	Fibers, films, coatings, explosives
Polycarbonates	$-O-C-O-R-$ (with O double bond)	<1	0.80	9,000	110	118	14	350	1.2	25 (45)	275 (135)	<1	Machine parts, propellers
Polyesters	$-C-O-$ (with O double bond)	3		8,000	300	117	1	340	1.3	33 (60)	130 (56)	Low	Magnetic tape, fibers, films

[a] A continuing bond is indicated by —~~—. R may be any complex molecule. In general, hydrogen atoms are not shown.
[b] 1974 prices.
[c] Multiply by 6.9×10^{-3} to obtain MN/m² or by 7.03×10^{-4} to obtain kg/mm².
[d] Multiply by 0.138 to obtain kg-m.
[e] Loaded at 264 psi (1.82 MN/m²).
[f] Multiply by 2.54 to obtain cm/min.

Table 10.2 PROPERTIES OF THERMOSETTING RESINS

Name and Structure[a]	Use, per-cent	Price,[b] dollars/lb	Tensile Strength,[c] psi	Percent Elonga-tion	Rockwell Hardness, R	Impact,[d] izod, ft-lb	Modulus,[c] psi × 10³	Specific Gravity	Coefficient of Expansion, (°F)⁻¹ × 10⁻⁶ [(°C)⁻¹ × 10⁻⁶]	Heat Dis-tortion,[e] °F (°C)	Burning Rate,[f] in./min	Typical Applications
Phenolics (phenolformaldehyde)	6	0.22	7,500	0	125	0.3	1,000	1.4	45 (81)	300 (149)	<1	Electrical equipment
Urea-melamine	4	0.37	7,000	0	115	0.3	1,500	1.5	20 (36)	265 (129)	0	Dishes, laminates
Polyesters	<1	0.22	4,000	0	100	0.4	1,000	1.1	42 (75.5)	350 (177)	1.4	Fiberglass composite, coatings
Epoxies	1	0.50	10,000	0	90	0.8	1,000	1.1	40 (72)	350 (177)	1	Adhesives, fiberglass composite, coatings
Urethanes	In change		5,000					1.2	32 (57.5)	190 (88)	<1	Sheet, tubing, foam, elastomers, fibers
Silicones	<1	1.00	3,500	0	89	0.3	1,200	1.75	20 (36)	360 to 900 (177 to 482)	<1	Gaskets, adhesives, elastomers

[a] A continuing bond is indicated by —◊◊◊—. R may be any complex molecule. In general, hydrogen atoms are not shown.
[b] 1974 prices.
[c] Multiply by 6.9×10^{-3} to obtain MN/m² or by 7.03×10^{-4} to obtain kg/mm².
[d] Multiply by 0.138 to obtain kg-m.
[e] Loaded at 264 psi (1.82 MN/m²).
[f] Multiply by 2.54 to obtain cm/min.

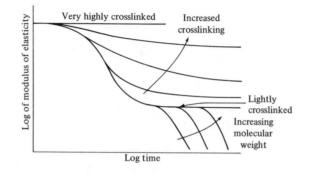

Fig. 10.2 *Effect of crosslinking on the modulus of elasticity.*

specimen would be rubbery or even flow if sufficient time were allowed for the material to react to the applied stress. Silly Putty is an excellent example of this varied behavior. This familiar material bounces like a ball (short-time stress) but flows like a liquid if left for a long time on a table top. In an extreme case, with very rapid stress application (a hammer blow) it will shatter.

To proceed further with our discussion of general properties, consider the variation with crosslinking, as shown in Fig. 10.2. With an increasingly crosslinked network there is a disappearance of the flow and rubbery ranges. In the case of natural rubber there is a great elasticity with only a small percentage of crosslinking, but when 70 to 80 percent of the available crosslinks are used, we have the hard, rather brittle material used in some combs and battery cases.

The effect of crystallinity is shown schematically in Fig. 10.3 for a thermoplastic material. Again the polymer becomes stronger but more brittle. The toughness can be increased by having partial crystallinity at a sacrifice in strength.

There is quite a variation in the degree to which the modulus and strength fall off with temperature for a given strain rate, as shown in Fig. 10.4. The chains with the simplest carbon skeleton such as polyethylene fall off severely, while the complex structures such as polycarbonate and ABS hold up quite well with increasing temperature.

10.3 Review of mechanical properties of polymers

With these qualifications in mind, let us take an overall view of the properties. In each case we will discuss the thermoplastic group (Table 10.1) first and then the thermosetting group (Table 10.2).

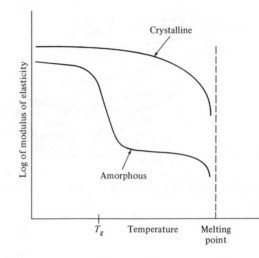

Fig. 10.3 *Effect of crystallinity on the modulus of elasticity.*

Tensile Strength. In the thermoplastics the highest tensile strengths are obtained in groups with stiffened carbon chains, such as the nylons, acrylics, acetals, and polycarbonates. The general level of 9,000 to 10,000 psi (62 to 69 MN/m²) for this group is well above the polyolefins at 2,000 to 7,000 psi (13.8 to 48.3 MN/m²). Also, the heat-distortion temperature is higher for the chain-stiffened polymers. [The heat-distortion temperature is the temperature at which considerable deflection takes place under a light stress (264 psi, 1.82 MN/m²).]

The thermosetting resins show higher levels of tensile strength than the simple olefins, as would be expected from the network structure. The strengths of the epoxies and urethanes equal those of the best thermoplastics, and these materials also have much higher heat-distortion temperatures. This is because the network structure is so strong that these materials do not melt but finally decompose at high temperatures.

Percent Elongation. In the polyolefins there is a sharp difference between the ductile polyethylene and polypropylene and the brittle polystyrene. This is due to the large styrene groups on the side of the carbon chain which prevent slippage. A more ductile modification is produced by blending styrene with rubber. The low elongation of the acrylics is due to a similar effect.

The network structure of the thermosetting materials, on the other hand, prevents uncoiling of the molecules because they are anchored by the network links. Therefore, all show little elongation at all temperatures.

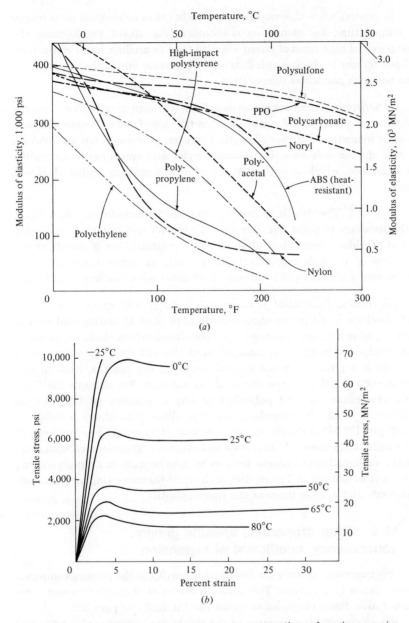

(a)

(b)

Fig. 10.4 (a) *Effects of temperature on properties of various resins.* (b) *Effect of temperature on the stress-strain curve of cellulose acetate.*

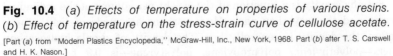

[Part (a) from "Modern Plastics Encyclopedia," McGraw-Hill, Inc., New York, 1968. Part (b) after T. S. Carswell and H. K. Nason.]

In general, when thermoplastics are tested at or below their glass transition temperature, low elongation is obtained (Fig. 10.4*b*). Furthermore the introduction of high rates of strain will also tend to produce lower elongation. The basic feature is whether uncoiling or slip can occur during the time interval of the test at a particular temperature.

Hardness. Hardness is customarily determined by a Rockwell instrument but on a special scale for plastics. As in the case of metals, higher hardness is roughly indicative of higher strength. However, hardness is not an accurate indicator of wear resistance. For example, nylon and acetal have outstanding wear resistance, which is not shown by a correspondingly higher Rockwell reading.

Impact. The details of impact tests will be included in Chap. 11. It is only necessary to point out here than an impact test is a high-strain-rate test and provides a good illustration of how a material that is ductile at the strain rate in a tensile test may become brittle at higher rates. Note, for example, that ABS shows 20 to 80 percent elongation but low impact strength.

Modulus of Elasticity. In general, in metals and ceramics the modulus of elasticity is fairly constant regardless of time of testing and over a reasonably wide temperature range near room temperature. However, as noted earlier, in plastics the modulus changes drastically with time and temperature. Also there is a great difference in moduli among the plastics. Even in one material, such as polyethylene, the modulus can vary. For example the high-density crystalline form of polyethylene has a modulus of 120×10^3 psi (828 MN/m^2), while the branched, less crystalline form shows a value of 25×10^3 psi (172 MN/m^2). We should note that these values are hundreds of times smaller than those of the metals and ceramics. Therefore, care must be taken if a substitution of a plastic for a metal is to be made in a closely mating part subject to deflection. The moduli of the rigid thermosetting materials are considerably higher than those of the thermoplastics.

10.4 Other properties: specific gravity, transparency, coefficient of expansion

It is important to note the low specific gravity of the plastics compared to other classes of materials. This leads to favorable strength-to-weight and stiffness ratios. Some examples are shown in the table on page 359.

Specific gravity is a function of the weight per volume of the individual molecules and the way they pack. Hydrocarbons are made of light atoms. Therefore, the density is generally low. The density of simple hydrocarbon polymers—polyethylene, polypropylene, polystyrene—is 0.9 to 1.1 g/cm^3.

Material	Tensile strength,[†] psi	Strength/weight,[‡] psi/(lb/in.^{3})
Cold-drawn SAE 1010 steel	53,000	156
ABS	6,500	170
6/6 nylon	11,800	290
Polyethylene (high density)	4,400	127

Substitution of a relatively heavy atom such as chlorine or fluorine for hydrogen gives polyvinylchloride with a density of 1.2 to 1.55 g/cm^3 and Teflon with 2.1 to 2.2 g/cm^3. The acetals are denser because of the C—O—C chain packing.

The crystalline form of a plastic is always denser than the amorphous form because of more efficient packing. The difference in density is important in determining transparency since the index of refraction is proportional to density. If the densities of the amorphous and crystalline materials are close, there is little dispersion as the light passes through a mixture of the two forms and the material is transparent. On the other hand, if there is a substantial difference between the densities, the material is opaque.

As an example, let us observe the series polyethylene, polypropylene, polypentene. In polyethylene the difference is great (specific gravity is 0.85 for the amorphous form and 1.01 for the crystalline). Therefore, where substantial crystallization takes place, the material is opaque. With polypropylene the difference is smaller (0.85 amorphous vs. 0.94 crystalline) and parts are at least translucent. With polypentene ($-C_5H_{10}-$) the densities are similar and the moldings are transparent.

It should be added, however, that a crystalline material may be transparent if it is cooled rapidly so that the crystallites are very small, shorter than the wavelength of the light to be transmitted.

Also, while amorphous polymers are transparent, fillers such as asbestos and carbon black are often used and these give an opaque product.

The most important point concerning the coefficient of expansion is that in plastics it is from 2 to 17 times as great as it is in a typical metal such as iron. This means that if a plastic is substituted for a metal in a part with close tolerance, due allowance should be made. Furthermore, if a composite metal-plastic part is molded of a brittle plastic, separation and cracking may take place. This is often noted on automobile steering wheels that have been through severe temperature changes.

[†]Multiply by 6.9×10^{-3} to obtain MN/m^2 or by 7.03×10^{-4} to obtain kg/mm^2.

[‡]Yield strength of metals or tensile strength of plastics (in psi) divided by density (in lb/in.^{3}).

10.5 Elastomers (rubbers)

In the preceding chapter we saw that rubberlike polymers are amorphous polymers with low glass transition temperatures so that the molecules can uncoil under the influence of stress. This behavior is illustrated in Fig. 10.5, which shows how an amorphous material with a high T_g (polystyrene) has a high modulus of elasticity compared with natural rubber.

EXAMPLE 10.1 Why is it usual to find a thermoplastic with a *higher* modulus of elasticity than an elastomer (rubber)?

ANSWER It is important to remember that the modulus of elasticity is defined as the load divided by the strain within the "elastic region." In rubbers a small load provides a high elastic strain and therefore a low modulus. Confusion usually arises when we emphasize the word "elasticity" rather than "modulus" in the definition. It is also well to appreciate that highly crosslinked elastomers show an increase in the modulus of elasticity with a decrease in elasticity (the maximum amount of elastic strain).

Typical commercial rubbers are listed in Table 10.3. Natural rubber, polyisoprene, and butadiene-styrene copolymer are in general use but have poor resistance to oil and gasoline and a limited temperature range. The more complex types such as nitrile have extended temperature ranges and better resistance to oil and gasoline.

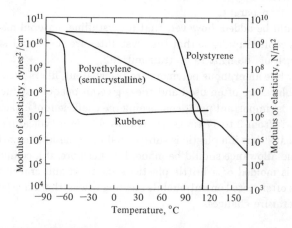

Fig. 10.5 *Comparison of the effect of temperature on the modulus of elasticity of rubber vs. thermoplastics.*

("Modern Plastics Encyclopedia," McGraw-Hill, Inc., New York, 1968.)

Table 10.3 PROPERTIES OF DIFFERENT RUBBERS (ELASTOMERS)

Common Name	Chemical Name	Tensile Strength,* psi	Percent Elongation	Resistance to Oil, Gas	Useful Temperature Range, °F (°C)
Natural rubber	Cis-polyisoprene	3,000	800	Poor	−60 to 180 (−51 to 82)
GR-S or Buna S	Butadiene styrene copolymer	250	3,000	Poor	−60 to 180 (−51 to 82)
Isoprene	Polyisoprene	3,000	400	Poor	−60 to 180 (−51 to 82)
Nitrile or Buna N	Butadiene acrylonitrile copolymer	700	400	Excellent	−60 to 300 (−51 to 149)
Neoprene (GR-M)	Polychloroprene	3,500	800	Good	−40 to 200 (−40 to 93)
Silicone	Polysiloxane	700	300	Poor	−178 to 600 (−117 to 315)
Urethane	Diisocyanate polyester	5,000	600	Excellent	−65 to 240 (−54 to 115)

*Multiply psi by 6.9×10^{-3} to obtain MN/m² or by 7.03×10^{-4} to obtain kg/mm².

10.6 Wood

Although the wood industry is of great importance in itself, at this point we can logically discuss the structure and properties of wood, considering it a high polymer. Let us begin with the familiar macrostructure of wood, then the microstructure, and finally the properties.

The characteristics of a typical cross-section are quite familiar and are illustrated in Fig. 10.6. The following aspects are particularly important in affecting mechanical properties.

The annual rings are made up of springwood and summerwood. The summerwood is denser and of higher strength. Wood rays, G, are encountered which can lead to radial cleavage. The central regions consist of heartwood, E, which conveys little sap, and softer sapwood, D, near the bark. Other features that are not shown include checks, which are lengthwise separations, and shakes, which are marks between annual growth rings. The two major types

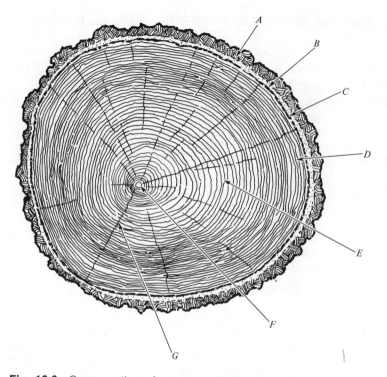

Fig. 10.6 *Cross-section of a tree trunk. A = bark, B = inner bark, C = cambium (cell former), D = sapwood, E = heartwood, F = pith, G = wood rays.*

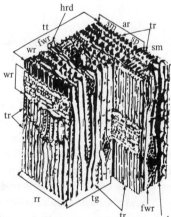

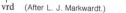

Fig. 10.7 *Microstructure of wood. Magnified three-dimensional diagrammatic sketch of a softwood.* tt = *end surface,* tg = *tangential surface,* rr = *radial surface,* tr = *fibers or tracheids,* wr = *wood ray,* fwr = *fusiform wood ray,* vrd = *vertical resin duct,* hrd = *horizontal resin duct,* sp = *springwood,* sm = *summerwood, and* ar = *annual ring.*

(After L. J. Markwardt.)

of wood are hardwood and softwood. The hardwoods, in general, are broadleafed trees, and the softwoods are the conifers or needle-bearing trees. Each group has a characteristic microstructure.

The microstructure of the softwoods is simpler and is shown in Fig. 10.7. Ninety percent of the structure is made up of longitudinal cells called "tracheids." These are about 0.1 in. (0.254 cm) long, 0.001 in. (0.00254 cm) in diameter, and with wall thicknesses of about 0.0003 in. (0.000762 cm). They are cemented together with another organic substance called "lignin," and recent work has shown that this is made up of a repeating structure of two complex molecules. The denser summerwood is due to the smaller rectangular-shaped cells as compared to the springwood. The rays form a horizontal cell structure, as shown. Resin ducts are also present. The chief differences in the hardwoods are smaller tracheids and the addition of open pores which are larger than the tracheids and are scattered among them. These form the chief sap-conducting structure and are a permeable network in the final wood product.

The mechanical properties are closely related to the structural features listed and to the water content of the structure, as shown in Table 10.4. The strength is markedly lower when measured perpendicular to the grain. The modulus of elasticity is higher in the hardwoods and increases with drying.

From these data we can see that the wood structure is similar to a honeycomb of high polymers. The directional weakness can be minimized by bonding wood layers properly with a plastic, as in plywood.

10.7   Special plastic products

Before considering the specific fields, we should discuss a few important special plastic products. These include coatings such as paint and adhesives, fibers, films, and foams.

Coatings. There are two broad classifications of plastic coatings, depending on whether the plastic is used alone or dissolved in a solvent that subsequently volatilizes. In fluid bed coating, an unheated bed of powdered plastic is kept in motion with air, i.e. fluidized, and then previously heated metal parts are dipped into the bed. A layer of plastic adheres to the part and spreads over the surface. It is possible to use the same process for applying a thermosetting plastic if the temperature of the part is adjusted to produce the thermosetting reaction.

A good example of the other type of coating, i.e. coating containing a solvent, is typical automobile touch-up paint. This material contains an acrylic resin in a solvent plus a very volatile solvent, such as CCl_2F_2, which acts as a propellant. In selecting a plastic we want T_g to be somewhat above room temperature so that the film will not be too soft but not so high that

Table 10.4 TYPICAL PROPERTIES OF CLEAR WOOD

Wood	Green					Air-dried (12 percent H_2O)			
	Water, percent	Specific Gravity	Yield Strength,* psi $\times$ 10^3	Modulus of Elasticity,* psi $\times$ 10^6	Tensile Strength Perpendicular to Grain,* psi $\times$ 10^3	Specific Gravity	Yield Strength,* psi $\times$ 10^3	Modulus of Elasticity,* psi $\times$ 10^6	Tensile Strength Perpendicular to Grain,* psi $\times$ 10^3
Hardwoods									
Yellow birch	67	0.55	4.2	1.50	0.43	0.62	10.0	2.01	0.92
Shagbark hickory	60	0.64	5.7	1.57		0.72	10.7	2.16	
Oregon white oak	72	0.64	4.6	1.51	0.94	0.72	6.6	2.28	0.83
Black walnut	81	0.51	5.4	1.42	0.57	0.55	10.5	1.68	0.69
Softwoods									
Northern white cedar	55	0.29	2.6	0.64	0.24	0.31	4.9	0.80	0.24
Douglas fir	38	0.40	3.6	1.18	0.33	0.43	6.3	1.40	0.34
Eastern white pine	68	0.34	3.1	1.02	0.30	0.36	6.0	1.28	0.30

* Multiply psi by 6.9 $\times$ 10^{-3} to obtain MN/m² or by 7.03 $\times$ 10^{-4} to obtain kg/mm².

the film will not be able to adjust to the movement of the metal underneath as the temperature changes.

Adhesives. These can be classified as permanently liquid, solidifying by physical processes, and solidifying by chemical processes. In Scotch tape the bond is a permanent liquid which sticks to the cellophane backing and with pressure to the paper or other surface. For this type of adhesive to work, the liquid must wet the surface and form strong bonds.

Examples of the second type are sealing wax or tar which are applied hot and then solidify. The modern electric glue gun utilizes a thermoplastic which hardens by this technique. As another example, airplane cement or rubber cement hardens by evaporation of a solvent. The cement for polyvinylchloride (PVC) pipe is in reality a solvent.

A good example of the third type is a thermosetting material such as epoxy. The liquid resin and the amine hardener are mixed and applied before setting, resulting from chemical action, can occur.

Possibly the most important adhesives are hybrids involving solidification by both physical and chemical means.

Fibers. These materials are produced from acrylics, rayon (derived from cellulose), cellulose acetate, spandex (polyurethane), Teflon, polyester, polyacrylonitrile, nylon, the olefins (polyethylene, polypropylene), cotton (α cellulose), and wool (a protein structure). In one method of making synthetic fibers a polymer melt is extruded through a die and put in filament form by cooling. After solidification the yarn is drawn at a temperature between T_g and T_m to orient the crystals in a favorable direction. To produce fabric with a memory, i.e. wash and wear, the cotton is treated with melamine, which tends to make the fibers regain their original position after washing or creasing because the deformation was elastic rather than permanent.

Films. Films are made by extrusion through a die [as wide as 10 ft (3.05 m)] and then the still-plastic mass is led through a series of polished rolls which orient the molecules by drawing.

Foams. These products are important for insulation and padding. Closed-cell foams are more desirable for life jackets, while open-cell foams are better for furniture. The polymer from which a foam is made can be a thermosetting plastic or a thermoplastic.

Foamed polyurethane, rubber, and vinylchloride are very important in the furniture market. The foam is made of polyurethane which is compounded to form CO_2 bubbles, and then the mixture is poured into a moving slab mold. The rigid foam is made by increased crosslinking and can be injected (foamed into place).

10.8 Industrial applications of polymers

Since the field of plastics is growing rapidly, it is interesting to review the applications in general and then to study one specific area, automotive applications, in detail. The approximate quantities used by various industries in 1969 are given in Table 10.5.

Appliances. In addition to serving as shapes for doors, cabinets, housings, and baskets, a good many parts are used where high corrosion resistance is needed, as in impellers, gears, and nozzles. The limited service demands and small size of parts lead to many good applications.

Construction. One million tons of plastics are used in paints, coatings, and adhesives and another million in structural items such as glazing, panels, pipe, and flooring.

Electronics. Polyethylene and vinyl coatings for cable and special Teflon-impregnated glass cloth for circuitry are important uses of plastics.

Furniture and Housewares. In furniture, polyurethane foam and polyvinylchloride coverings are important. In housewares, polypropylene and polyethylene are commonly used for containers.

Table 10.5 INDUSTRIAL USES OF POLYMERS, 1969

Industry	Polymers Used, tons $\times 10^3$ ($kg \times 10^9$)
Construction	3,500 (3.170)
Packaging	2,000 (1.814)
Automobile	500 (0.453)
Furniture, bedding	300 (0.272)
Appliances	250 (0.277)

Table 10.6 AUTOMOBILE INDUSTRY'S USE OF POLYMERS
COMPARED WITH OVERALL INDUSTRIAL CONSUMPTION, 1968

Material	Overall Market, percent	Cost, dollars/lb	Automobile Industry, percent
Polyolefins	~33	0.11 to 0.26	13
Polystyrene	16	0.14 to 0.26	
Polyvinylchloride	17	0.21 to 0.31	19
ABS	2	0.30 to 0.40	11
Acrylics	2	0.45 to 0.36	4
Cellulosics	1	0.40 to 0.80	3
Nylons	1	0.75 to 1.20	2
Acetals	1	0.65	2
Polycarbonates	1	0.80	1
Polyurethane foam			19
Polyester	3		18
Phenolics	6		4
Urea melamine	4		
Miscellaneous	15		4

Packaging. In 1969 approximately 2 million tons (1.814×10^9 kg) of plastics were used in this industry, with more than half going into sheet less than 10 mils (0.0254 cm) thick. Polyethylene, polyvinylchloride, and polypropylene were the most frequently used. Specialties such as polyester were used in "boil-in-bag" applications. Bottles use polyethylene, polyvinylchloride, and acrylic copolymers.

The Automobile Industry. This industry deserves close scrutiny because of its special engineering demands. To show the importance of the better-strength materials, let us compare the percentage consumption by the automobile industry in 1968 with industry in general in the same year (Table 10.6).

Most parts in automobiles have service requirements from -40 to $180°F$ (-40 to $82°C$) and higher. This eliminates many plastics which would either embrittle or become too soft.

EXAMPLE 10.2 The application of polymeric materials in automobiles is summarized in Table 10.7. Of course, in a number of applications plastics are in competition with other materials. Assuming the competitive material to be metal, either steel or aluminum, cite the advantages and disadvantages of the competing parts for door handles, carburetor components, and impact-bumper parts. (*Continued on page 374.*)

Table 10.7 SUMMARY OF APPLICATIONS OF PLASTICS IN TYPICAL AUTOMOBILES

Generic Name	Typical Trade Names	Typical Uses	Advantages	Disadvantages	Average Weight per Automobile, U.S. Production, 1970, lb (kg)
		Thermoplastics			
Vinyl	Geon Elvax Vygen Vinoflex	Wire insulation, hoses, heel pads, seat trim, door panel trim, armrest skins, headlining convertible tops, landau tops, exterior and interior moldings, crash-pad skins, door-locking knobs, coat hooks, horn pads, bumper fillers, seat-belt-retractor housings, steering wheels,* exterior door handles,* vinyl-covered seats and interior trim, vinyl-coated metal	Stable colors, weather-resistant, flexibility	Stiff at low temperatures, fogging (unless specially compounded)	18 (8.16)
Polyolefins Polyethylene	Marlex Alathon Hi-fax Petrothene	Splash shields, spring inter-liners, ducts, floor-pan plugs, water shields, protective seat covers, fuel tanks,* windshield washer bottles	Low cost, low friction, low-temperature impact	Not paintable, tendency to warp, tendency for stress cracking, low stiffness	17 (7.71)
Polypropylene	Tenite Marlex Pro-fax	Fender liners, accelerator pedals, interior trim, door panels, steering wheels, fan shrouds, fresh-air down spouts, cowl-trim panels, seat backs, wheelhouse and spare-tire covers, fuse blocks, electric connectors, battery cases,* cowl screens, lamp housings	Low cost, colorability, low friction, heat-resistant	Not paintable, poor low-temperature impact (Homo-polymer only)	

Material	Trade names	Applications	Properties	Limitations	Value
ABS	Cycolac Kralastic Lustran Abson	Instrument bezels, post covers, armrest bases, consoles, knobs, radiator grilles, air-conditioner outlets, instrument-panel trim plates, seat backs, push buttons, heater push-button housings, lamp housings,* interior trim (plated), deck-lid spoilers	Paintable, adequate impact, electroplatable, vacuum-metalizable heat-resistant, compounding latitude	Opaque, not weatherable	11 (5)
Polyvinyl Butyral	Saflex Butacite	Interlayer for safety glazing			2 (0.908)
Acrylic	Plexiglas Lucite Perspex	Lamp lenses, medallions, instrument facings, side markers	Clarity, weather-resistant	Heat-resistant; impact not quite adequate, except for some special compounds	3.5 (1.59)
Acrylic fiber bundles	Crofon	Interior signals, instrument lights*			
Cellulosic Acetate Butyrate Propionate	Tenite Forticel	Knobs, trim strips, steering wheels, push buttons	Good impact, clarity, colorability	Heat-resistant, odor	2.25 (1.02)
Nylon	Zytel Ultramid	Bushings, slides, gears, fasteners, timing chain sprocket, fuel filter housing, cable liners, fender extensions*	Low friction; heat-resistant; fuel-, oil-, and chemical-resistant; good toughness	High water absorption, brittle when dry	2.25 (1.02)
Acetal	Delrin Celcon	Same as nylon. Also housings, shear pins for energy-absorbing steering column, gears, carburetor components*	Low friction, heat-resistant (but lower than nylon), moisture-resistant, rigid, high pull-out strength of threaded studs, impact below nylon	Problems with paintability	2 (0.908)
Polycarbonate	Lexan Merlon	Interior lamp bases (indicator), lenses, exterior impact-bumper parts,* housings	Clarity, very high impact, heat-resistant	High cost, tendency to solvent crazing, poor weatherability	1 (0.454)

Table 10.7 SUMMARY OF APPLICATIONS OF PLASTICS IN TYPICAL AUTOMOBILES (*Continued*)

Generic Name	Typical Trade Names	Typical Uses	Advantages	Disadvantages	Average Weight per Automobile, U.S. Production, 1970, lb (kg)
Thermoplastics (Continued)					
Glass fiber-filled thermoplastic					
Polystyrene ABS Polyolefin Nylon Acetal		Air-conditioner heater push-button housing cover (nylon), instrument panels (styrene*), glove box door (styrene*), transmission cover for under-pan (propylene) housings,* switch parts	Increased strength, stiffness, heat-resistant, dimensional stability, low thermal expansion and contraction	High cost	1 (0.454)
Thermosetting Plastics					
Phenolic	Bakelite Durez Genal	Terminal blocks, coil towers, fuse retainers, insulators, carburetor spacers, switch components, brake components, distributor caps*	Temperature-resistant, electrical insulator, good strength and impact (in some special filled compounds)	Colorability, tracking	3 (1.36)
Phenolic laminate	Norplex Formica	Terminal blocks, electrical insulators, spacers, bushings, transmission thrust washers, printed circuit boards	Temperature-resistant, electrical insulator, high strength, high impact	High cost, must be machined	0.5 (0.227)

Material	Applications	Advantages	Limitations	Quantity, lb (kg)
Alkyd plastic	Brush holders (starters, wipers, heater blowers), slip-ring insulators, distributor caps	Arc- and tracking-resistant, heat-resistant, fast molding cycles	High cost, relatively short shelf life, limited colorability	2 (0.908)
Glaskyd Duraplex Reinforced polyesters Mineral-filled pre-mix (Bulk molding compound: polyester plus mineral fillers plus catalyst)	Exterior body components, fender extensions, fender skirts, bumper closures, hood scoops, door scoops, instrument panels,* molded semistructural components, rear-window frames, tail-light assembly*	Freedom from warping, good stud retention, excellent molded surface, dimensional accuracy, paintable without sanding, minimum property variation with temperature, flexural strength—14,000 psi	Material cost, limited production capacity per mold, mold flow limitations, subject to creep	5 (2.27)
Glass mineral-filled (sheet molding system: polyesters plus fillers plus glass fibers plus catalysts)	Air vanes, truck body panels,* upper grille panels (front-end sheet metal replacement)*	Freedom from warping, good stud retention, excellent molded surfaces, minimum property variation with temperature, dimensional accuracy, paintable without sanding, high physical properties, flexural strength—20,000 psi	Material cost, limited production capacity per mold, mold flow limitations, subject to creep	1 (0.454)

Foamed-in-place Trim Components

Material	Applications	Advantages	Limitations	Quantity, lb (kg)
Flexible ABS film (vacuum-formed skins)	Instrument-panel trim pads, door trim panels, trim bolsters	Capability to produce complex trim shapes, wide selection of skin materials, can be foamed to a defined impact performance	Skin shrinkage, poor edge retention, loss of grain for deep draws	Skins: 1.5 (0.681) Foam: 1.5 (0.681)

Table 10.7 SUMMARY OF APPLICATIONS OF PLASTICS IN TYPICAL AUTOMOBILES (Continued)

Generic Name	Typical Trade Names	Typical Uses	Advantages	Disadvantages	Average Weight per Automobile, U.S. Production, 1970, lb (kg)
		Foamed-in-place Trim Components (Continued)			
Plastisol and Drysol (slush-molded skin)		Instrument-panel trim pads, horn pads, headrests,* armrests, console covers	Nonshrink skins, good grain, good color match, semi-self-locking skins	Uneven skins, delicate tooling	Skins: 2.25 (1.02)
Rotocast skin		Horn pads, headrests, armrest pads, trim pads*	Nonshrink skins, good grain, good color match, self-locking skins	Uneven skins, delicate tooling	Foam: 3 (1.36)
Vinyl skin (Injection-molded. May also be assembled.)		Horn pads, armrests, trim pads*	Uniform skin thickness, more flexible skin, good color match and pattern reproduction, rapid production rate	Requires parting lines, not economical for low-volume parts	0.3 (0.136)
Urethanes Self-skinning foam		Horn pads, armrests,* small trim pads, steering wheels, bumpers, side scoops	Capable of producing grained and embossed surfaces; paintable; soft feel; freedom of styling in shape, fit, and body match	Not economical for high-volume parts, cannot mold undercuts, solid-metal engraved tools required	1 (0.454)
Other		Seats	Complex shapes are feasible, economical		12 (5.45)

Polyester film	Mylar†	Dial faces, trim components
Polyvinylidene chloride	Saran	Fuel filters
Epoxy potting compounds	Epon, Araldite, Bakelite, Epi-Rez	Voltage regulators
Solid urethane	Texin, Vibrathane, Estane	Bushings, bearings
Cellular poly-olefins	Ethafoam	Gaskets, visors
Polytetrafluoroethylene	Teflon	Seals, bushings, cable liners
Polyphenylene oxide	Noryl	
Ethylene-vinylacetate copolymer	Ultrathene	
Polysulfone	Astrel	

* Plastics in competition with other materials.
† Vacuum-metalized externally.

ANSWER

Application	Material	Advantages	Disadvantages
Door handles	Vinyl	Formability	Poor low-temperature impact
	Chrome-plated steel	Impact and wear resistance	Cost
Carburetor components	Acetals	Formability, toughness	Wear resistance
	Aluminum die casting	Strength and wear resistance	Cost
Impact-bumper parts	Polycarbonates	Toughness, corrosion resistance	Cost
	Chrome-plated steel	Strength, ease of repair	Corrosion resistance

SUMMARY

In discussing the properties of polymers it is helpful to separate the materials into thermoplastics and thermosetting plastics.

The thermoplastics vary in tensile strength from the polyethylenes at 2,000 to 5,000 psi (13.8 to 34.5 MN/m^2) to polyamides (6/6 nylon) at 11,800 psi (81.5 MN/m^2) as chain stiffening and hydrogen bonding are increased. The elongation of this group is generally good (above 5 percent) except in the case of polystyrene. The materials with amorphous structures such as Lucite and polystyrene are transparent, while the crystalline materials are translucent to opaque. On cooling from the melt most of the thermoplastics pass through five stages: liquid, viscous, rubbery, leathery, glassy.

The thermosetting plastics are stronger than the polyolefins and retain their rigidity to higher temperatures than the thermoplastics. They finally degrade rather than melt when heated and do not exhibit plastic or rubbery behavior. The impact strength is lower and the modulus higher than for the thermoplastics.

Wood may be considered a honeycombed natural high polymer. The basic differences between softwoods and hardwoods lie in differences in the honeycombed structure.

Elastomers are polymers with molecular structures which uncoil readily, exhibiting very high values of elastic elongation.

Adhesives are divided into three groups, based on whether they operate by permanent liquid films, application of a liquid which solidifies, or polymerization in place.

Fibers are produced by extrusion, filament formation, and drawing at controlled temperatures.

Key industrial applications of the polymers are discussed for the appliance, construction, electronics, furniture, packaging, and automobile industries.

DEFINITIONS

Rubbery range   The temperature range in which very great elastic elongation is obtained. The chains uncoil but return to their original positions.

Leathery range   The temperature range in which chains are more rigid and fracture takes place at less than 10 percent elongation.

Chain stiffening   The inclusion of foreign atoms in or along the carbon chain resulting in increased bonding.

Impact   The testing of specimens with high strain rate.

Elastomer   A high polymer with coiled structure resulting in rubberlike characteristics.

Springwood   The portion of wood that grows in the spring season. It is of lower density and strength than *summerwood*.

Softwood   Wood made up of 90 percent long-cell tracheids cemented with lignin. *Hardwood* exhibits smaller tracheids and scattered open pores.

Heat-distortion temperature   The temperature at which a polymer softens severely.

Foam   A polymer containing gas bubbles which add to insulating value and also "sponginess," as in a car seat.

Adhesive   A molecular structure that bonds two other materials by either physical or chemical means.

Fiber   A polymer in a finely drawn form.

PROBLEMS

10.1 Listed below are several automotive applications involving competitive materials, as discussed in Example 10.2. Comparing the polymers with the other materials indicated, cite the advantages and disadvantages of each.

a. Steering wheels: PVC vs. polypropylene
b. Fuel tanks: steel stamping vs. polyethylene
c. Battery cases: polypropylene vs. hard rubber
d. Lamp housings: polypropylene vs. sheet steel
e. Fender extensions: nylon vs. sheet steel

10.2 Compare Tables 9.4, 10.1, and 10.2. The strengthening mechanisms include chain stiffening, through the addition of other atoms and side groups, crosslinking, network structuring, and crystallization, to mention a few of the more important ones. Make a table and compare each material in the figures listed with low-density polyethylene. Indicate which of the strengthening mechanisms are most important in each use.

10.3 Rubber products sometimes undergo degradation with extended time. Explain why the following phenomena occur in rubber.

a. A rubber band fails after having been wrapped very tightly around an object for several months.
b. An automobile heater hose bursts.
c. A rubber washer in a water faucet will no longer seal.

10.4 Anyone who has ever owned an automobile will recognize the following situations. Suggest what might be occurring in the structure of the material.

a. The insides of the windows develop a haze which resembles, but is not, a smoke haze.
b. Vinyl seats crack with time. Those areas exposed to sunlight are especially susceptible.

10.5 Polyethylene as used in containers is normally translucent. If a piece of this material is pulled in tension, it transmits less light (becomes chalky white). Explain.

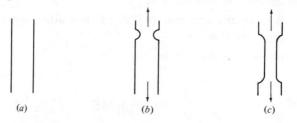

(a) (b) (c)

Fig. 10.8 *Behavior of polyethelene. (a) Unstressed. (b) Necking down begins. (c) Necking down continues with more stress.*

10.6 Most of the adhesives cannot be used for all types of environments. Explain. What type of glue could be used where it would be exposed to weathering?

10.7 Explain the behavior of polyethylene in the tensile specimen shown in Fig. 10.8. Why doesn't failure take place at the reduced section in (*b*) instead of the necking down continuing?

10.8 Some of the polymers develop a potential toxicity if they are burned and the fumes allowed to become airborne. How much HCl could be obtained per pound of combusted PVC?

11

MECHANICAL PROPERTIES OF MATERIALS; COMPOSITE STRUCTURES

THE catastrophic failure of certain steel ships, often while moored quietly at dockside, has led to extensive investigation of the "fracture toughness" of materials. A tensile specimen cut from the steel plate of these ships will invariably exhibit good strength and ductility. The brittle fracture of the ship results from a combination of constraints and lowered temperature, as discussed in the text.

In this chapter we will take up mechanical properties which have not been discussed in any detail to this point and which are important in the selection of materials. These include fatigue, creep, impact, and wear.

11.1 General; the beginning of materials selection

We have discussed the basic structures of metals, ceramics, and plastics and some of their properties. We can now approach the major problem of materials engineers—the selection of a structure that will perform safely, with adequate service, at the lowest total cost in a given application. They must take into account not only the mechanical properties of materials but also electrical, magnetic, thermal, and optical properties, as well as corrosion resistance. Finally, they must know whether the best material can be cast or fabricated into the shape of the component needed.

In this chapter we will begin studying this decision-making procedure by discussing the ranges of *mechanical* properties that are available in existing materials. We will start with a survey of the conventional tensile- and compressive-test properties which have already been discussed. Next we will study the testing for impact, fatigue, and wear resistance at low and high temperatures. We cannot hope to be comprehensive in this effort, but it is important to show how mechanical properties that are measured by methods other than the tensile test can be quite different from tensile strength and are often more important in determining service life. Finally, we will describe the design freedom made possible by composite structures.

11.2 Summary of mechanical properties

To illustrate the variations in properties, we have assembled a representative group of materials in Table 11.1. The aluminum and magnesium alloys are shown because of their light weight and, as we shall discuss later, the importance of their conductivity and corrosion resistance. The copper alloys are included because of their exceptional ductility and conductivity. A variety of steels from the low-strength plain carbon to the high-strength and stainless types are given. Gray iron and ductile iron are included because of castability, cost, and machinability.

In the ceramics several representative materials are shown. Although there is intensive development in fibers of graphite, boron, and other materials, their usage is not general enough for detailed discussion here and these are taken up briefly under composite materials in the latter part of the chapter.

In the plastics section four thermoplastics—polyethylene, polymethylmethacrylate (Plexiglas), acrylonitrile-butadiene-styrene (ABS), nylon—and two thermosetting plastics—phenolic and epoxy—as well as one rubber are included.

11.3   Tensile strength and yield strength

It is important to distinguish between tensile strength and yield strength because in ductile materials there may be quite a difference between these two properties, while in brittle materials there is practically no difference. For example, in annealed aluminum the tensile strength is 10,000 psi (69 MN/m^2) and the yield strength 4,000 psi (27.6 MN/m^2). In class 40 gray iron, however, both strengths are 40,000 psi (276 MN/m^2). Another point in the metals is the large difference between annealed, cold-rolled, and heat-treated conditions, and values have been included in the table to recall this. This is a unique advantage of the metals because a part can be formed in the soft condition and then hardened later or even hardened by the forming operation.

We see that, in general, the yield strength of the metals is well above that of other materials. Ceramics have higher tensile strength in some cases, but because of extreme brittleness the strength is difficult to exploit except in composites.

In many cases it is more appropriate to use the specific strength or strength-to-weight (yield strength-to-density) ratio as a basis for comparison. On this basis the light metals, graphite, and certain plastics show favorable properties. Similarly, specific stiffness or modulus-to-density ratio is used as an index of selection where deflection is important.

When these factors are taken into consideration along with cost per pound or cost per cubic inch, we see that there can be close competition between cast iron, aluminum, and plastic for the same part. In many cases the manufacturing cost rather than the material cost is the main consideration.

In most metals compressive yield and ultimate strength are the same as the tensile values. However, in low-ductility materials such as cast iron and concrete the compressive strength is several times greater. Therefore, in service these materials are often loaded in compression and tensile loading is minimized.

11.4   Percent elongation

There are two principal uses of the value of percent elongation. A high value (>20 percent) indicates that the material can be cold-worked and formed to shape. Second, the elongation, together with the yield strength, is a measure of toughness. In many cases a designer will consider high values of elongation as a safety factor. It should be realized, however, that most failures involve fatigue, as discussed later, in which ductility is not directly involved. Furthermore, if a complex structure undergoes distortion of say 5 percent, it is usually of little value afterward.

Table 11.1 REPRESENTATIVE PROPERTIES AND COSTS OF DIFFERENT ENGINEERING MATERIALS

Classification	Family of Material[a]	Chemical Analysis, percent	Structure[a]	Tensile Strength,[b] psi × 10³	Yield Strength,[b] psi × 10³	Per-cent Elongation[b]	Hardness	Modulus,[b] psi × 10⁶	Brittle Temperature,[c] °F (°C)	Creep Temperature,[d] °F (°C)	Cost,[e] dollars/lb	Specific Strength[f] × 10³	Specific Stiffness[g] × 10⁶
							Metals						
1060	Aluminum	99.6 Al	Ann. FCC	10	4	43	BHN 19	10	None	220 (104)	0.25	41	105
1060	Aluminum	99.6 Al	CR FCC	19	18	6	BHN 35	10	None	220 (104)	0.25	184	105
2014-T6	Aluminum	4.5 Cu	FCC + θ	70	60	13	BHN 135	10	None	220 (104)	0.25	610	105
AZ31	Magnesium	3 Al, 1 Zn	HCP	38	28	9	BHN 55	6		220 (104)	0.40	451	97
110	Copper	99+	Ann. FCC	32	10	45	40 R_F	17	None	300 (149)	0.50	31	53
110	Copper	99+	CR FCC	50	40	6	85 R_F	17	None	300 (149)	0.50	124	53
260	Brass	70 Cr, 30 Zn	Ann. FCC	46	11	65	54 R_F	16	None	300 (149)	0.45	36	52
260	Brass	70 Cu, 30 Zn	CR FCC	76	63	8	82 R_B	16	None	300 (149)	0.45	204	52
1020	Steel, HR	0.2 C	BCC + carbide	55	30	25	BHN 111	30	0 (−18)	800 (427)	0.03	106	106
1020	Steel, CR	0.2 C	BCC + carbide	61	51	15	BHN 121	30	0 (−18)	800 (427)	0.03	180	106
1080	Steel, HR	0.8 C	BCC + carbide	112	62	10	BHN 229	30	0 (−18)	800 (427)	0.03	220	106
4340	Steel, ann.	0.4 C + alloy	BCC + carbide	100	80	25	BHN 200	30	0 (−18)	800 (427)	0.03	286	106
4340	Steel, Q and T	0.4 C + alloy	BCC + carbide	30	290	8	BHN 580	30	0 (−18)	800 (427)	0.03	1,040	106
302	Stainless steel	18 Cr, 8 Ni	CR FCC	110	75	35	BHN 240	30	None	1000 (538)	0.22	286	106
Class 40	Gray iron	3 C, 2 Si	BCC + graphite	40	40	0.4	BHN 220	17	None	800 (427)	0.03	154	65
80-55-06	Ductile iron	3.5 C, 2 Si	BCC + graphite	80	55	6	BHN 190	23	−50 (−46)	800 (427)	0.04	211	88

Ceramics

Material	Structure	Specific Gravity					Temperature °F (°C)			Hardness			
Glass	Soda	Fiber amorph.			0	10	900 (482)		0.02				111
Glass	Borosilicate	Fiber amorph.			0	9	1200 (649)		0.03				113
Carbon	Carbon	Hexagonal	1.2	1.2	0	2.7	2000 (1093)		0.30			20	50
Titanium carbide	Titanium carbide	Cermet	130	130	0	50	2000 (1093)		2.00			500	248

Plastics

Specific Gravity	Material	Structure			Temperature °F (°C)	Hardness						
0.95	Polyethylene	Linear poly.	4	100	75 (24)	40 R[h]	0.12	0.16	118	3		
1.19	Polymethacrylate	Linear poly.	8	5	45 (7)	220 R	0.42	0.34	185	11		
1.06	ABS	Graft poly.	6	20		95 R	0.3	0.35	158	8		
1.1	6/6 nylon	Linear poly.	11	60		118 R	0.41	0.80	274	10		
1.4	Phenolic	Network	7	0		125 R	1	0.22	138	20		
1.1	Epoxy	Network	10	0		90 R	1	0.50	250	25		
0.9	Natural rubber	Crosslinked	3	700	−65 (−54)		Low		94	4		
0.72	Wood (oak)	Fiber	0.8	6.6	None		2.3		31	88		

[a] Abbreviations: HR = hot-rolled, CR = cold-rolled, ann. = annealed, Q and T = quenched and tempered, amorph. = amorphous, poly. = polymer.

[b] Multiply by 6.9×10^{-3} to obtain MN/m² or by 7.03×10^{-4} to obtain kg/mm².

[c] Temperature at which failure caused by brittleness is a design factor.

[d] Temperature at which creep is a design factor.

[e] Costs are for alloys not finished materials.

[f] Specific strength = $\dfrac{\text{yield strength (psi)}}{\text{density (lb/in.}^3)}$; tensile strength (psi) is used for plastics in place of yield strength.

[g] Specific stiffness = $\dfrac{\text{modulus of elasticity (psi)}}{\text{density (lb/in.}^3)}$

[h] These are Rockwell hardness values measured on a special scale for plastics.

11.5   Hardness

There is a tremendous range in hardness, with the ceramics containing the highest-hardness materials such as diamond and tungsten carbide, the metals next, and the plastics lowest. A different hardness scale, Rockwell R, is used for the plastics which is below the common scales for metals. In the metals there is good correlation between hardness and strength, but in the ceramics the true strength which should accompany high hardness is not evident because of brittleness and the existence of small flaws. In thin fibers such as fiberglass high strength is evident.

11.6   Modulus of elasticity

This property is evidence of interatomic bonding, and as a result the ceramics, which have covalent and ionic bonds, have the highest modulus. The metals are next with metallic bonding, and the plastics lowest with van der Waals bonding. While there are strong covalent bonds within the molecules in the plastics, the forces between molecules are only van der Waals attractions. Also, as noted in Chap. 10, the moduli of plastics vary greatly with temperature and time of testing.

It is particularly important to note that in the metals the modulus is the same within a given alloy family. For example, all steels, whether heat-treated or annealed, show a modulus of approximately 30×10^6 psi $(2.07 \times 10^5 \text{ MN/m}^2)$.

11.7   Effects of notches, defects, temperature, fatigue

Up to this point we have used the conventional tensile and hardness tests to show how mechanical properties are determined by structure. Now that we are summarizing mechanical properties for use in design, we must point out that great errors, resulting in catastrophic failures, have resulted when only these conventional properties were taken into consideration. The presence of notches and defects can cause failure of a component at a fraction of the yield strength; elevated temperatures produce easy flow as a function of time (creep); and low temperatures can lead to brittle fracture. Cyclic loading as in a spring can produce fracture at much lower stresses than indicated by the simple tensile test. We should emphasize again that the structure of a material is the fundamental variable. For example, we may have two alloys of the same tensile strength but of different structures. If the tensile test is conducted at high temperatures, we will probably have widely different values; in other words, similar mechanical properties at room temperature are no guarantee of similar properties at other temperatures. Similarly, performance in a test does not ensure performance in the presence of notches. To discuss this important field of mechanical testing, let us consider the following factors:

1. Stress concentration
2. Brittle and ductile fracture, low-temperature effects
3. Fatigue (cyclic loading)
4. Elevated-temperature testing (creep)

11.8 Stress concentration

If we drill a hole of radius r through a bar, we reduce the cross-section of the bar (Fig. 11.1), and we might calculate that the effective stress in the bar at the hole would be

$$S_{nom} = \frac{P}{(w - 2r)t}$$

where S_{nom} = nominal stress
P = load
w = width
t = thickness

(The normal area of the bar would be wt, and we reduce this by $2rt$ to take care of the area of the cross-section lost by the hole.)

Calculations such as this led to the disastrous failure of many highly stressed components, especially in the presence of cyclic stresses. The error is that the stress measured at the sides of the hole is not this nominal stress but is much higher. This is because the stresses that would normally be supported by the material where the hole was drilled are *concentrated* at the edge of the hole (Fig. 11.1). It is found, furthermore, that the smaller the ratio of the radius of the hole to the width of the bar, the *higher* the *stress concentration factor K*, reaching a value of about 2.8 at a radius-to-width ratio of 0.05. Measurements beyond this point are difficult to make.

EXAMPLE 11.1 Given P = 20,000 lb$_{force}$, r = 0.1 in., w = 1 in., t = 1 in., calculate the maximum stress in a bar similar to that in Fig. 11.1.

ANSWER

$$S_{nom} = \frac{20,000}{(1 - 0.2)1} = 25,000 \text{ psi} \quad (172.5 \text{ MN/m}^2)$$

but since K = 2.7 (r/w = 0.1),

Stress at edges of hole = 2.7 × 25,000 = 67,500 psi   (466 MN/m²)

(This is why so many fatigue failures start at holes and other "stress raisers" such as fillets, notches, etc.)

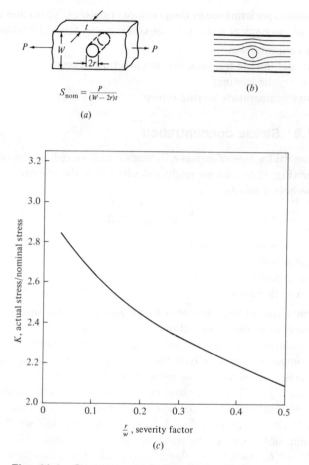

$$S_{\text{nom}} = \frac{P}{(W - 2r)t}$$

(a)

(b)

(c)

Fig. 11.1 *Stress concentration factor K due to a transverse hole in the tension plate member.*

(C. Lipson, G. C. Noll, and L. S. Clock, "Stress and Strength of Manufactured Parts," McGraw-Hill Book Company, New York, 1950.)

11.9   Brittle fracture, fracture strength

Many years ago Griffith found that if glass was drawn into thin fibers, the tensile stress required to cause failure was much higher than that for rods of larger cross-sections. He postulated that the larger sections always contained some cracks (now called "Griffith cracks"), and that the probability of finding defects in fine fibers was lower and, therefore, the tensile strength was higher. He also pointed out that the greater the length of the crack, the lower the tensile strength, even after the reduction in cross-sectional area due to the crack was taken into consideration.

In recent years it has been found that this theory applies to high-strength metals and plastics and is important in preventing catastrophic failures. The reason these failures are called "catastrophic" is that once a crack starts to grow, it propagates rapidly, so that failure is sudden. This does not mean that if a crack exists, there will automatically be sudden failure. For example, railroad crossovers are made of a very ductile austenitic steel (12 percent manganese) in which cracks develop as a result of continuous loading by train traffic. The cracks grow very slowly, and the specification calls for repair welding after they attain a specified length. By contrast, every chemistry student is familiar with the easy fracture of a glass tube after a small notch has been made in it with a file.

A general formula has been developed in engineering mechanics which relates the length of a crack to the stress required for fracture:

$$\sigma_f = \sqrt{\frac{EG_c}{\pi a}}$$

where σ_f = fracture stress required to propagate the crack and cause failure
    E = Young's modulus of elasticity
    G_c = fracture energy (work required to produce the fracture surface)
    a = one-half the length of the crack

Estimates of G_c can be made for brittle materials in which the work required to produce the fracture surface is calculated from the bonding forces between the atoms. For a ceramic such as MgO the calculated value of G_c is close to the value obtained by measuring σ_f for a given crack length and solving for G_c. [G_c (calculated) = 0.085 lb/in. vs. 0.10 lb/in. measured (0.0152 kg/cm vs. 0.0179 kg/cm measured).] However, for alloys such as high-strength steel the agreement is poor. [G_c (calculated) = 0.13 lb/in. vs. 300 lb/in. (0.0234 kg/cm vs. 53.6 kg/cm measured).] The explanation is that the work required for fracture is not simply the work to create the fracture surface but includes a good deal of plastic deformation in the volume around the crack. The practical result is that the high-strength steel is less notch-sensitive. We can analyze this point further by rearranging the formula:

$$\sigma_f \sqrt{\pi a} = \sqrt{EG_c}$$

For a given material tested at a given temperature to failure, in many cases we find that $\sigma_f \sqrt{\pi a}$ = constant. The fracture toughness K_{Ic} is defined as $K_{Ic} = \sigma_f \sqrt{\pi a}$.

EXAMPLE 11.2 If the value of K_{Ic} is 95,000 psi (in.)$^{1/2}$ for a given steel, and we can only inspect for cracks as long as 0.2 in., what is the highest design stress we may use safely?

ANSWER

$$\sigma_f \sqrt{\pi \times (0.1)} = 95,000 \qquad \text{since } 2a = 0.2 \text{ in.}$$
$$\sigma_f = 170,000 \text{ psi} \quad (1,173 \text{ MN}/\text{m}^2)$$

This is considerably below the yield strength of 250,000 psi (1,725 MN/m²) obtainable in these steels, and this value would ordinarily be used in design.

11.10   Temperature effects

Let us take up now the effects of the temperature of testing on the fracture strength. One of the most common tests is the so-called "Charpy impact test," Fig. 11.2a. A square bar with a V notch is struck by a calibrated swinging arm, and the energy absorbed is measured. It is relatively simple to examine the effects of temperature by immersing several specimens in advance in liquids at different temperatures and then transferring them quickly to the test fixture.

The types of data obtained are shown schematically in Fig. 11.2b. FCC metals show high impact values and no important change with temperature. However, BCC metals, polymers, and ceramics show a transition temperature below which brittle behavior is encountered. It should be emphasized that there is great variation in the actual transition temperatures for different materials. For the metals and polymers it is between −200 and 200°F (−129 and 93°C), while for the ceramics it is above 1000°F (538°C).

There is a distinct difference in the appearance of the fractures of low-carbon steels, depending on whether the specimen was tested and broken below or above the transition temperature. As indicated in Fig. 11.3b, the appearance of Charpy V notch specimens varies from ductile to brittle as the specimen temperature is reduced from 200 to −321°F (93 to −196°C). Careful observation of the specimens shows that a shear type fracture as shown by the presence of a shear lip is characteristic of the specimens tested at higher temperatures, while shear is absent in the specimens tested at the lowest temperatures. Photomicrographs taken with the scanning electron microscope provide additional evidence (Fig. 11.3c, d, and e). The specimen with highest impact strength shows a dimpled surface with myriads of cuplike projections of deformed metal. In contrast, the brittle sample shows principally cleavage fractures, almost like split platelets of mica. The specimen at 68°F (25°C) shows a mixture of both types of fracture, as expected.

One of the most striking examples of the importance of transition temperature is the failure of the Liberty ships produced during World War II. These ships were made of low-carbon steel of good ductility in the usual tensile test. However, 25 percent of these ships developed severe cracks, often while anchored in harbor, and many broke in two. The failures were analyzed as the result of constraint (stress concentration) caused by square hatches in the deck,

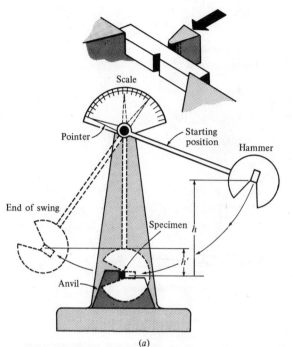

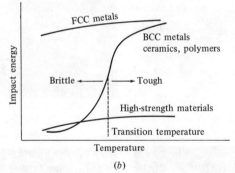

Fig. 11.2 (a) *Operation of a Charpy impact test.*
(b) *Effect of temperature on the impact strength of various materials (schematic).*

[Part (a) from H. W. Hayden, W. G. Moffatt, and J. Wulff, "The Structure and Properties of Materials," vol. 3: Mechanical Behavior, John Wiley & Sons, Inc., New York, 1965.]

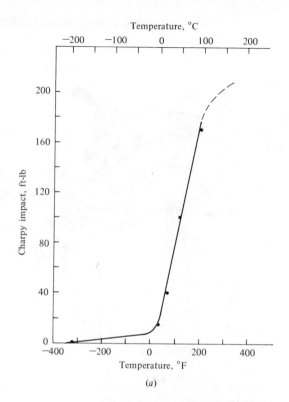

(a)

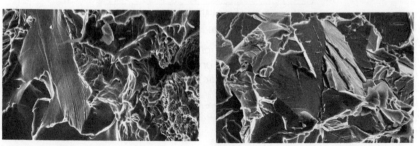

(b)

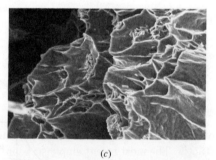

(c)

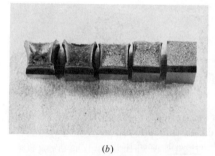

(d)

(e)

coupled with the use of steel with a transition temperature (Charpy) near the operating temperature. However, it has been found that the Charpy test does not give a wholly satisfactory answer, as shown by the recent failures of oil tankers, and a new test called the "dynamic tear test" has been developed by the Naval Research Laboratory. In this test notched samples of varying sections can be tested with a drop weight, and the transition temperatures indicated by this test are more reliable indicators of steel quality.

In the use of polymers the temperature for brittle fracture (elongation less than 1 percent) can be found by tensile testing as well as by impact. In the case of the slower-speed tensile test, brittle behavior is encountered at temperatures well below T_g, the glass transition temperature, as shown by the following data from normal tensile tests.

Material	Glass transition temperature, $°F(°C)$	Brittle temperature, $°F(°C)$
Polymethylmethacrylate	220 (104)	113 (45)
Rigid polyvinylchloride	165 (74)	−4 (−20)

The structural explanation for the development of a brittle temperature in the polymers is that above this temperature the stress field ahead of a crack does much more work in uncoiling and displacing molecules, analogous to plastic deformation in a metal. Below the brittle temperature rupture takes place with a minimum of such displacement, and the total energy required for rupture is lower.

11.11 Reduction in impact strength by neutron radiation

Considerable experimental work has been done to measure the effects of exposure of materials to radiation such as that encountered in a nuclear reactor. The most recent emphasis is on structural materials such as types 304 and 316 stainless steels which are subjected to neutron flux densities charac-

Fig. 11.3 (a) Charpy impact vs. temperature for SAE 1020 steel, hot-rolled. (b) Charpy impact fractures after testing at (right to left) −196, 0, 25, 50, 93°C. (c) to (e) are scanning electron micrographs of Charpy specimens showing shear failure above transition temperature, cleavage below transition temperature, and mixed shear and cleavage at intermediate temperature; 1020 steel, hot-rolled, 1500X. (c) Tested at 93°C, ductile (shear) fracture, 170 ft-lb. (d) Tested at 25°C, mixed fracture, 40 ft-lb. (e) Tested at −196°C, cleavage fracture, <1 ft-lb.

teristic of liquid-metal-cooled, fast-breeder reactors. In general, the yield strength increases and the elongation decreases, but the most concern is with the decrease in toughness as shown by the Charpy V notch impact specimen and dynamic tear tests. As an example, type 304 plate decreased from 230 ft-lb (V notch Charpy) before radiation to 70 ft-lb after radiation. The effect is due to actual atomic transformations and movement due to the neutron bombardment. However, just as the effects of cold working can be removed by annealing, the effects of neutron bombardment can be ameliorated by subsequent heat treatment. In view of these substantial effects, the resistance of a material to change under these conditions is an important consideration in reactor design.

Fig. 11.4 *Typical fatigue failure of a crankshaft. The crack began at the stress concentration at the hole in the front of the picture. The steps in the fracture caused by successive cycles of loading lead to an oyster-shell effect. The final failure took place suddenly as shown by the small fracture surface at the rear.*

(Courtesy of H. Mindlin, Battelle.)

11.12 Fatigue

A survey of the broken parts in any automobile scrap yard reveals that the majority failed at stresses below the yield strength. This is not due to imperfections in the material but due to the phenomenon called "fatigue." If we take a bar of steel and load it a number of times to a stress, at 80 percent of the yield strength for example, it will ultimately fail if stressed through enough cycles. Furthermore, even though the steel would show 30 percent elongation in a normal tensile test, no elongation is evident in the appearance of the fatigue fracture. A typical crankshaft failure is shown in Fig. 11.4.

Fatigue testing in its simplest form involves preparing test specimens with carefully polished surfaces and testing at different stresses to obtain an *S-N* curve which relates *S* (stress required for failure) to *N* (number of cycles) (Fig. 11.5). As would be expected, the lower the stress, the greater the number of cycles to failure.

The curve for ferrous materials exhibits what is called the "fatigue strength" or "endurance limit" (Fig. 11.6). In other words, beyond 10^6 cycles, further cycling does not cause failure. On the other hand, for nonferrous materials such as aluminum, the curve continues to decline and it is necessary to test the material for the number of cycles to be encountered in service.

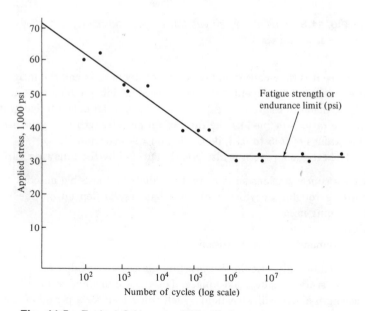

Fig. 11.5 *Typical S-N curve (fatigue) for a ferrous material showing an endurance limit (schematic).*

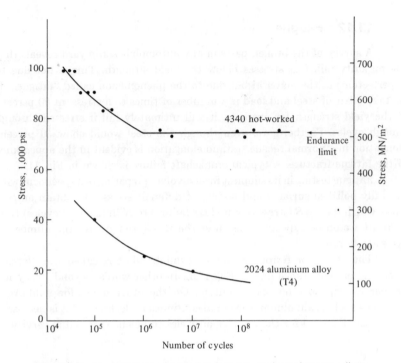

Fig. 11.6 *Typical S-N curves for ferrous and nonferrous alloys.*
(After Garwood and Alcoa.)

Since it is time-consuming to conduct fatigue tests and the fatigue data require statistical treatment because of reproducibility difficulties, attempts have been made to relate the tensile test to fatigue data. The fatigue ratio or endurance ratio is defined as the ratio of fatigue strength to tensile strength and has values of 0.45 to 0.25, depending on the material.

The fatigue strength is also greatly affected by the following variables:

1. Stress concentrations such as radii at fillets and possible notches
2. Surface roughness, which indicates that results depend on the type of machining used
3. Surface residual stress
4. Environment such as corrosion

In ceramics, particularly in certain glasses and oxides, a phenomenon known as "static fatigue" takes place. These materials and even some ultra-high-strength alloys will withstand a high static load for a period of time and then fail suddenly. This type of failure does not take place in dry air or *in vacuo*, so it is related to a chemical reaction between the water in the atmosphere and the highly stressed surface.

11.13 Behavior at elevated temperatures, creep

In contrast to the brittleness encountered in some materials at low temperatures, softness and ease of plastic flow are characteristic of many materials at elevated temperatures. We recall how brass can be hot-worked and is relatively plastic above its recrystallization temperature. Polyethylene and other thermoplastic resins are plastic at the temperature of boiling water. Fused silica must be heated to incandescence with an oxygas flame to soften.

While the ability to flow plastically is a great advantage from the point of view of formability of a part, it is a distinct handicap if the part is to be used at elevated temperatures. Polyethylene, for example, cannot be used for hospital containers that are to be sterilized in boiling water. Aluminum cannot be used in engine parts that reach 600°F (315°C).

To measure the suitability of an alloy for service at a given temperature, creep testing, as described in Chap. 6, is performed. An indication of the temperature range at which creep becomes important is given in Table 11.1. It is evident that creep is a factor to be considered in designing with plastics even at room temperature.

We may now summarize the preceding sections on impact, brittle fracture, fatigue, and creep as design considerations necessary to special types of mechanical or stress environments. One important point is that tensile test data represent performance for a particular rate of testing and test temperature which, although very useful in the establishment of allowable design stresses, may not truly represent performance in service. Therefore, if a material in service will have a minimum time to adjust to an applied stress, we must be concerned with high-strain-rate and low-temperature tests such as impact, fatigue, and the drop-weight tests. In the other extreme, where we have plenty of time for stress adjustment (relaxation) and diffusion, creep test data are necessary to the design.

11.14 Wear and abrasion

The most important characteristic of wear is its unpredictability. As examples, a gray cast iron is an excellent engine-block material because it retains lubricant and avoids seizing of the piston. On the other hand, the graphitic areas crumble under high stresses and cannot be present in a component such as a railroad car wheel or grinding ball. In another case a general "rule" is to avoid contact of similar metals. However, gray iron piston rings perform well against the gray iron engine block and gray iron tappets† do well against a gray iron camshaft.†

---

† These contain controlled amounts of massive iron carbide.

In spite of this apparent confusion, wear problems can be approached reasonably if the mechanism of wear is known. In many cases a careful observation of worn parts will disclose which structures failed and what conditions developed at the interface. Before discussing specific cases, let us consider the fundamentals of wear in the following order:

1. The nature of wearing surfaces
2. Effects of pressing surfaces together (static contact)
3. Interaction between sliding surfaces
4. Effects of lubrication
5. Wear resistance of different material combinations (Sec. 11.15)

The Nature of Wearing Surfaces. Even the best polished surface has two types of inhomogeneity. By using an accurate diamond stylus we can find pits and scratches of the order of 1000 Å or hundreds of times greater than the size of the unit cell. In addition to these physical differences, we know that variations in grain orientation, in the nature of phases and inclusions, can occur at the wearing surface. In other words, even with the best machining we will have two relatively rough inhomogeneous mating surfaces. Furthermore, we know that it is possible to have two very different microstructures of the same overall macrohardness. For example, one structure might contain martensite at BHN 600 microhardness plus graphite flakes which would lower the overall hardness to BHN 500. On the other hand, another structure might be homogeneous martensite at BHN 500. Under one type of wear, involving high loads and good lubrication, the second structure would probably perform better, particularly if the load crushed the iron at the edges of the graphite. However, with moderate loads and intermittent lubrication the first structure would resist galling (sticking) better than the second.

Static Contact. Let us consider the simple case of a ball bearing on a flat plate. This is incorrectly called "point contact" because the ball and plate both deform, giving a circular area of contact. It can be shown by engineering mechanics that the shear stress is a maximum not at the surface but rather occurs beneath it. If a is the radius of the circle of contact, the maximum shear stress is at $0.6a$ beneath the surface.

It is possible to calculate the maximum stress at this level as a function of the radius of the ball or other "point" contact. From this value we can calculate the load required to produce flow for different radii of the ball and different plate materials (Table 11.2).

In conventional terms, a high load is required to deform tool steel with a ball. However, the point of the table is that when we consider that a large ball will have tiny rough points of radii as small as 10^{-4} cm, then we must expect some deformation even with very light loading (1.4×10^{-3} g).

Table 11.2 LOAD REQUIRED TO PRODUCE FLOW IN DIFFERENT MATERIALS USING INDENTERS (STEEL BALLS) OF DIFFERENT RADII

Plate Material	Load, g, for Ball Radius Indicated		
	10^{-4} cm	10^{-2} cm	1 cm
Copper	2.5×10^{-6}	2.5×10^{-2}	250
Mild steel	4.7×10^{-5}	0.47	4,700
Tool Steel	1.4×10^{-3}	14	140,000

Interaction between Sliding Surfaces. When surfaces slide over each other the projections will produce very high stresses and flow in the areas of contact. If we accept this we can understand the inevitable breaking-in period better. In many cases the high stresses result in high friction, and welding, followed by shearing takes place between the parts.

1. If the weld is weaker than either metal, little wear takes place. This is the case, for example, when a tin-base-alloy bearing wears against steel.
2. If the junction is stronger than one of the metals, then shear takes place in the weaker metal. For example, when steel wears on lead, the fracture takes place in the lead.
3. If the junction is stronger than both metals, tearing will occur in both metals and wear will be rapid.

The effects of surface temperature are also important. Under many simple conditions of sliding wear, lead actually melts and covers the surface, while steel is hardened by local heating above the austenite transformation temperature. The presence of freshly formed martensite is often noted in steel or iron parts subjected to wear, as in brake drums.

Effects of Lubrication. When a lubricant is present, there will be either hydrodynamic lubrication or boundary lubrication. In hydrodynamic lubrication the attempt is to preserve a fluid film so that the surfaces do not touch. Many factors operate against this, including intermittent operation, breakdown of the lubricant molecule, and heavy loads with slow speeds. Under these conditions welding or galling can take place.

With boundary lubrication the attempt is to coat the surface with lubricant molecules. The metal reacts with the lubricant to form a metal soap. In other cases extreme pressure lubricants containing sulfur, phosphorus, or chlorine react with the metals to reduce seizing. Molybdenum disulfide and graphite are also used where extreme pressure exists, to serve as solid fragments in the bearing area with easy basal cleavage.

11.15 Wear-resistant combinations

It is possible to dissolve lead in the liquid copper alloys, and then on solidification the lead precipitates in fine globules. These alloys are often used against steel in slow-speed bearings with intermittent lubrication. The copper is hardened by the addition of tin and zinc, as in the alloy 85% Cu, 5% Sn, 5% Zn, 5% Pb. For some applications a higher-strength alloy is needed and aluminum bronze (89% Cu, 11% Al) is used. For gears a favorite combination is hardened steel wearing against a nickel-tin bronze (11% Sn, 87% Cu, 2% Ni). The tin provides a hard δ phase and the nickel gives solid solution hardening of the copper.

There are many applications involving ceramic and plastic bearing materials. Sapphire bearings have long been used in precision equipment and watches, and it is possible to obtain 1-in.- (2.54-cm-) diameter sapphire balls. Nylon composites are used in applications where previously metal was required, such as fishing reel gears.

11.16 Abrasion resistance

In the usual wear application the introduction of gritty foreign material is carefully avoided. However, in some cases, such as grinding ore and conveying sand, the abrasive is part of the system. The chief variables in abrasion are the hardness of the abrasive and the type of loading of the equipment, whether impact or steady compression.

Unfortunately, the hardest and most abrasion-resistant alloys, such as tungsten carbide, can be used in only a few installations because the carbides will crack out. The most useful abrasion-resistant alloys in mining and other grinding are martensitic white cast iron, hardened steel, and austenitic manganese steel. The white irons are the most wear-resistant, while the manganese steel is the toughest.

In some instances, where the part is not heated or cut by the abrasive, rubber-coated parts are very successful. Conveyor belts and gloves for sandblast operators are examples.

11.17 Composite and laminated materials

There are many cases in which an outstanding combination of properties can be obtained by using two or more materials together. Perhaps the most common example is a fiberglass automobile fender. The plastic used alone would have insufficient strength and excessive deflection, and the glass alone would obviously be too brittle.

Fiberglass is a true composite; that is, one material (glass) is completely surrounded by a matrix (polyester plastic). Let us look at the entire range of duplex materials first and then return to analyze the factors influencing the strength of a composite.

Surface coatings may have either a mechanical, adhesive type bond or a diffusion bond in which there is a continuous change in structure. Examples of adhesive bonds are painted surfaces, rubber coating, and plating. In contrast, there are diffusion bonds in carburized and nitrided surfaces, bimetal castings, hot galvanized pipe, and brazed and welded assemblies.

Laminated construction may also involve either mechanical or diffusion bonds. In plywood the weakness and warpage of the wood in one direction are overcome by gluing pieces in different orientations. In Alclad sheet an age-hardened aluminum, a 4.5 percent copper alloy core with poor corrosion resistance, is covered with a layer of pure aluminum. The heating during and after bonding must be controlled so that the copper from the core does not diffuse to the surface, thereby lowering the corrosion resistance of the sheet.

Turning now to the true composites, we find two important classes (Fig. 11.7):

1. Composites in which the fibers of the strengthener are continuous
2. Composites in which the fibers of the strengthener are discontinuous and may be merely chopped-up pieces

We will analyze only the factors that make up the overall strength and modulus of elasticity of the composite in the first case, which gives the strongest combination.

CASE 1. *Composites with the continuous fibers in a matrix.* As an example, let us start with the simple case of a bar 0.5 in.^{2} (3.23 cm^2) in cross-section in which we have continuous glass fibers in a matrix of polyester, a typical fiberglass.

Let us analyze the effect of increasing the amount of glass on the strength and modulus of elasticity of the composite.

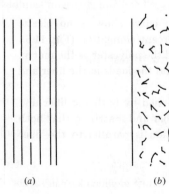

(a) (b)

Fig. 11.7 *Variations in fiber distribution in a composite material such as fiberglass. (a) Continuous or practically continuous fibers. (b) Chopped-up, discontinuous fibers.*

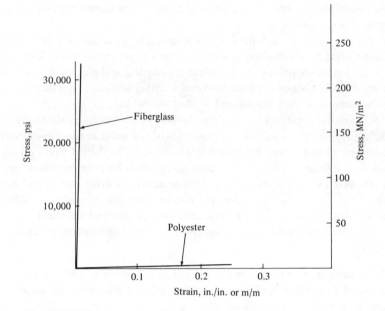

Fig. 11.8 *Stress-strain curves for fiberglass and polyester. At 0.003 strain, the stress in the fiberglass is 30,000 psi (207 MN/m²). At 0.003 strain, the stress in the polyester is 30 psi (0.207 MN/m²).*

Let us assume the following typical values for the material:

Material	Tensile strength,† psi × 10³	Yield strength,† psi × 10³	Modulus,† psi	Percent plastic elongation
E fiberglass	250	250	10 × 10⁶	0
Polyester	4	4	10 × 10³	25

The glass will have a stiff stress-strain curve showing no plastic elongation, while the polyester will have 25 percent elongation (Fig. 11.8).

We will consider the glass as the fiber and the polyester as the matrix surrounding the fiber. To enable us to understand the loads in the fiber and matrix, let us conduct the following analysis.

First, we apply a load P that will be carried by both the fiber and the matrix. We assume a good bond between fiber and matrix so that both stretch the same amount and the load direction is parallel to the fiber alignment direction.

———

†Multiply by 6.9 × 10⁻³ to obtain MN/m² or by 7.03 × 10⁻⁴ to obtain kg/mm².

We can write

$$P_c = P_f + P_m \tag{1}$$

where P_c = total load on composite
P_f = load carried by fiber
P_m = load carried by matrix

Also, we know that the stress in the fiber σ_f is the load divided by the cross-sectional area of the fiber:

$$\sigma_f = \frac{P_f}{A_f} \quad \text{or} \quad P_f = \sigma_f A_f \tag{2}$$

Writing similar equations for the composite and the matrix, and substituting in Eq. (1), we have

$$\sigma_c A_c = \sigma_f A_f + \sigma_m A_m \tag{3}$$

Dividing by A_c, the total cross-sectional area of the composite,

$$\sigma_c = \sigma_f \frac{A_f}{A_c} + \sigma_m \frac{A_m}{A_c} \tag{4}$$

We note that A_f/A_c is the fraction of the total area occupied by the fiber. In this case it is also the volume fraction of the fiber, since the lengths of the fiber, composite, and matrix are all the same. If we let C_f denote the area (volume) fraction of fiber and C_m the area (volume) fraction of matrix, then $C_f + C_m = 1$, and rewriting Eq. (4), we obtain

$$\sigma_c = \sigma_f C_f + \sigma_m C_m \tag{5}$$

Now let us look at the strain produced by the load. If we call the overall strain ε_c, the strain in the fiber ε_f, and that in the matrix ε_m, then

$$\varepsilon_c = \varepsilon_f = \varepsilon_m$$

because we assumed a good bond and all are strained equally under load P_c.

Let us divide the terms of Eq. (5) by ε_c, ε_f, and ε_m, separately:

$$\frac{\sigma_c}{\varepsilon_c} = \frac{\sigma_f}{\varepsilon_f} C_f + \frac{\sigma_m}{\varepsilon_m} C_m \tag{6}$$

We can now substitute the overall modulus of the composite E_c for σ_c/ε_c, the modulus of the fiber E_f for σ_f/ε_f, and the modulus of the matrix E_m for σ_m/ε_m:

$$E_c = E_f C_f + E_m C_m \tag{7}$$

This allows us to determine the deflection or strain.

EXAMPLE 11.3 Suppose we have a 0.5-in.^{2} (3.23-cm^2) bar that is 60 percent glass and 40 percent polyester by volume. What are the stresses and strains in the individual components?

ANSWER

$$E_c = (10 \times 10^6 \text{ psi})(0.6) + (10 \times 10^3 \text{ psi})(0.4)$$
$$= 6.004 \times 10^6 \text{ psi} \quad (0.414 \times 10^5 \text{ MN}/\text{m}^2)$$

The strain will be σ_c/E_c or, since the bar is 0.5 in.^{2} (3.225 cm^2),

$$\varepsilon = \frac{P_c}{0.5\, E_c} = \frac{2\, P_c}{E_c}$$

Next let us find the factors affecting the loads carried by the glass and the plastic.

The load carried in the fiber will be the stress in the fiber times its area, and similarly in the matrix. We can write

$$\frac{\text{Load carried in fiber}}{\text{Load carried in matrix}} = \frac{\sigma_f A_f}{\sigma_m A_m} \tag{8}$$

or, using the relation $\sigma = E\varepsilon$,

$$\frac{P_f}{P_m} = \frac{E_f \varepsilon_f A_f}{E_m \varepsilon_m A_m} \tag{9}$$

But $\varepsilon_f = \varepsilon_m$ and $A_f/A_m = C_f/C_m$, so

$$\frac{P_f}{P_m} = \frac{E_f}{E_m} \frac{C_f}{C_m} \tag{10}$$

From this expression we can calculate the loads carried by the fiber and the matrix and predict the overall strength of the composite.

Let us take our 60 percent glass–40 percent plastic mixture and assume a total load of 10,000 lb (4,540 kg). From Eq. (10) the ratio of the loads will be

$$\frac{P}{P_m} = \frac{10 \times 10^6}{10 \times 10^3} \frac{0.6}{0.4} = 1,500$$

The load on the plastic will be

$$P_m = P_c - P_f$$
$$P_m = 10,000 \text{ lb} - 1,500P$$
$$P_m = 6.6 \text{ lb} \quad (3.0 \text{ kg})$$

$$\sigma_m = \frac{P_m}{A_m} = \frac{6.6 \text{ lb}}{(0.5 \text{ in.}^2)(0.4)} = 33 \text{ psi}$$

$$P_f = 9{,}993 \text{ lb} \quad (4{,}537 \text{ kg})$$

$$\sigma_f = \frac{P_f}{A_f} = \frac{9{,}993}{(0.5 \text{ in.}^2)(0.6)} = 33{,}300 \text{ psi} \quad (230 \text{ MN/m}^2)$$

Checking that $\varepsilon_m = \varepsilon_f$, we have

$$\varepsilon_m = \frac{\sigma_m}{E_m} = \frac{33}{10^4} = 3.3 \times 10^{-3}$$

$$\varepsilon_f = \frac{\sigma_f}{E_f} = \frac{33{,}300}{10^7} = 3.3 \times 10^{-3}$$

We have therefore accomplished our objective for Case 1, composites with continuous fibers. It is important to note that the modulus of the composite is merely the weighted sum of the moduli of the fiber and matrix, and that the load carried in the fiber is affected by the *ratio* of the moduli:

$$E_c = E_f C_f + E_m C_m \qquad \text{from Eq. (7)}$$

$$\frac{\text{Load in fiber}}{\text{Load in matrix}} = \frac{E_f}{E_m} \frac{C_f}{C_m} \qquad \text{from Eq. (10)}$$

In the second case, composites with discontinuous fibers, the critical factors are the length-to-diameter ratio of the fiber, the shear strength of the bond between the fiber and the matrix, and the amount of fiber. All these variables affect the strength of the composite. However, we find a good deal of fiberglass made with chopped-up pieces of glass because of the ease with which it is formed into complex shapes even when the length-to-diameter ratio is unfavorable.

Adherence of Fiber to Matrix. In the preceding section we assumed that the strength of the bond of the glass to the polymer was greater than the strength of the plastic. This is not always the case and considerable research is in progress on this point.

In Figs. 11.9 and 11.10 photomicrographs of a polycarbonate–glass fiber composite showing poor bonding are shown in contrast with a well-bonded composite. The properties are given in Table 11.3.

In the material molded at 190°C the tensile strength is lower in the material containing glass fibers, while after treatment at higher temperatures, which improves the bond, the strength and elongation are higher in this material.

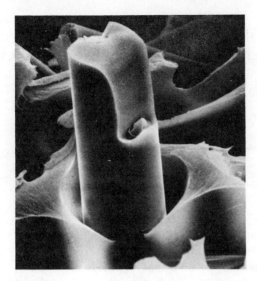

Fig. 11.9 *Tensile fracture as seen with scanning electron microscope of an epoxy glass–polycarbonate composite molded at 190°C (374°F). Note the void around the fibers. 9100X.*

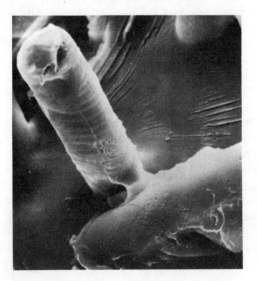

Fig. 11.10 *Same as in Fig. 11.9 but molded at 275°C (527°F) followed by annealing at 245°C (473°F) for 3 hr. 8600X.*

(Photographs courtesy of J. L. Kardos, F. S. Cheng, and T. L. Tolbert, Washington University/Monsanto Association, St. Louis.)

Table 11.3 PHYSICAL PROPERTIES OF TYPE E
GLASS–POLYCARBONATE COMPOSITE

Material	Tensile Strength,* psi × 10³	Percent Elongation	Modulus of Elasticity,* psi × 10⁵
Molded at 190°C			
Resin alone	9.0	5.1	3.25
Resin and glass	7.6	1.3	8.19
Molded at 275°C plus 3 hr at 245°C			
Resin alone	8.9	4.9	3.11
Resin and glass	11.2	2.3	9.28

* Multiply psi by 6.9 × 10⁻³ to obtain MN/m² or by 7.03 × 10⁻⁴ to obtain kg/mm².

Further examination indicates that treatment at higher temperatures results in the formation of tiny crystalline regions in the resin which are accompanied by improved bonding. The material with the higher-temperature treatment also wets the fibers. It may also provide a layer of intermediate modulus, improving stress transfer without fracture.

SUMMARY

While tensile test data are used widely in both design and the specification of materials, it is vital to realize that other mechanical properties can be more important in determining the safety of a given design. Even in applications involving simple tension, the specific strength or specific modulus may be more important than the tensile strength alone. Other factors are stress concentration, brittle fracture, notches, fatigue, temperature, and wear resistance.

Stress concentration may be planned or accidental, as in "stress raisers." Holes and sharp corners may lead to fracture even if a generous factor of safety has been taken into account (based on the average stress across a section).

Brittle fracture may take place at stresses below the yield strength in materials with low ductility, such as hardened steel. Small flaws and cracks exist in all materials, and the resistance to crack propagation varies greatly. The concept of fracture toughness is particularly important in designing with high-strength materials.

Notched impact tests such as the Charpy impact test can be used to indicate the brittleness of materials at low temperatures. In the transition temperature range the fracture changes from ductile shear to brittle cleavage.

Fatigue tests indicate the stresses which can safely be used in parts subjected to cycles of loading. The fatigue strength is the stress at which fracture takes place after a given number of cycles and varies from 0.2 to 0.45 of the tensile strength.

Elevated-temperature tests such as creep and stress-rupture tests indicate the reliability of a material in continued service at high temperatures. If long-time service is planned, a simple tensile test at the temperature involved will not indicate either the fracture strength or the deformation to be expected.

Wear resistance is not a function of hardness alone but involves the interaction between the mating structures and the lubrication conditions. Abrasion resistance is an extreme case of wear resistance and is extremely important in the mining and materials-handling industries.

In general, composite materials do not exhibit the average of the properties of the individual materials but may develop an advantageous balance. In fiberglass, for example, the high strength and high modulus of the glass lead to good overall strength, and the plastic adds toughness to the composite.

DEFINITIONS

High-temperature testing For practically every material there is a range of temperature in which it softens and deforms in increasing amounts as a function of time. When this occurs, it is necessary to obtain test data of stress vs. time. These are obtained from high-temperature tests, and the properties measured include creep and stress required for rupture.

Impact testing A bar, usually notched, is struck by a calibrated pendulum and the energy required to cause fracture is measured. Although the term "impact strength" is used, the results are expressed in terms of energy. The principal reason for the fracture is not the impact but the presence of the notch, which causes a brittle stress condition known as a triaxial stress condition. This is discussed in texts on strength of materials.

Transition temperature If the samples for the impact test are tested at different temperatures, some materials such as steel show a rapid fall in impact strength at a low temperature such as 0°C. This is called the "transition temperature," and the fracture changes from ductile to brittle at this point. (This brittleness with temperature usually does not exist in FCC metals.)

Fatigue testing If a specimen is subjected to cycles of loading and unloading either in tension, compression, or bending, the stress required for fracture is much lower than in a single stress application. The "fatigue strength" (also called the "endurance limit" in steels) is the stress required to produce failure in a specified number of cycles, usually 10^6.

Specific strength Yield strength (psi) divided by density (lb/in.^{3}) gives a relative measure of the strength of a given volume. (Tensile strength is used for polymers in place of yield strength.)

Specific stiffness Modulus of elasticity (psi) divided by density (lb/in.^{3}) gives a relative measure of the resistance to deflection.

Notch effects Accidental nicks produced during manufacture, sharp corners at changes in section, or highly stressed sharp threads can lead to fracture when the overall stress is only a fraction of the yield strength. These factors are also called "stress raisers."

Stress concentration factor, K The stresses around a smooth hole drilled through a stressed member are well above the average stress in the cross-section. K = actual stress/nominal stress.

Fracture toughness Resistance to propagation of a crack under stress once the crack has been initiated is $K_{Ic} = \sigma_f \sqrt{\pi a}$ (see test). The *fracture stress* for a crack of length $a/2$ is σ_f.

Hydrodynamic lubrication In this type of lubrication the motion of the bearing surfaces produces a continuous lubricant film between the surfaces.

Boundary lubrication In this condition the lubricant film is discontinuous in places and special lubricants are added to prevent galling.

Galling This occurs when two metal surfaces are in contact and sticking or welding takes place, resulting in surface roughening and increased wear.

Composite material A material composed of two materials bonded together with one serving as a matrix surrounding the particles or fibers of the other. Fiberglass is an excellent example.

PROBLEMS

11.1 In each of the applications described below, attempt to find a satisfactory material from those given in Table 11.1 and specify any unique structure, treatment, and properties desired. If other materials are more suitable, specify them from tables in other chapters. Tell briefly the shortcomings of other materials (by groups) for this application.

Example: For a hand chisel for masonry the best choice is 1080 steel, which can be austenitized, quenched, and tempered to provide tempered martensite, 50 R_C. The heat treatment should be applied only to the tip, or else the shank should be tempered at a higher temperature for greater toughness. Shortcomings of other materials, as groups, are that ceramics are too brittle under impact and plastics are too soft. High-carbon steel has greater hardness than the alloy grades listed and is less expensive.

Often several materials will be competitive, as for example for porch screening. Give the advantages and disadvantages of each.

a. The head of a carpenter's hammer

b. A $\frac{1}{8}$-in. (0.317-cm) drill for wood

    c. A $\frac{1}{8}$-in. (0.317-cm) drill for production machining of 1030 steel

    d. A toothpaste tube

    e. A container for beer

    f. The hood of an automobile

    g. The windshield of an automobile

    h. A conveyor belt for sand for loading trucks

    i. An aircraft wing girder

    j. A girder for a 20-story building

    k. The body and frame of an electric toaster

    l. The crankshaft of a V-8 engine

    m. The cylinder block of a V-8 engine

    n. The rear bumper of an automobile

    o. A 14-ft (4.27-m) power-boat hull

    p. The hull for a 600-ft (183-m) ore boat

    q. A 48-in. (1.22-m) diameter sewer pipe

    r. Transparent wrapping material

    s. Lining for a furnace in which steel is melted at 3000°F (1649°C)

    t. Porch screening

    u. Valves for domestic piping

    v. Trays for conveying steel parts through an austenitizing furnace at 1650°F (899°C)

11.2 A high-strength composite material is made by bonding silica glass fibers (all aligned in the same direction) with an epoxy. The composite contains 45 percent (by weight) glass. The data are as follows:

Epoxy: Density = 1.3 g/cm³;   Young's   modulus = 300×10^3 psi
$(0.207 \times 10^4 \text{ MN/m}^2)$

Glass: Density = 2.6 g/cm³;   Young's   modulus = 10×10^6 psi
$(6.9 \times 10^4 \text{ MN/m}^2)$

    a. Calculate Young's modulus for the composite measured parallel to the fibers.

    b. In order to make a more rigid glass-epoxy composite, we could increase the volume fraction of glass. If each fiber of glass were 0.0005 in. (0.00127 cm) in diameter, and if each fiber were coated with an epoxy film 0.0001 in. (0.000254 cm) thick before bonding with more epoxy, what would be the maximum volume fraction of glass that one could achieve in a unidirectionally aligned composite? Assume a cross-section shows the fibers in an HCP structure.

    c. What would Young's modulus be in part *b*?

11.3 Throughout the text we have said that the properties of a material depend on (1) nature, (2) amount, (3) size, (4) shape, and (5) distribution and orientation of the phases. As examples we could use:

1. Pure iron vs. iron alloyed with nickel
2. Effect of amount of spheroidal carbide on hardness of steel
3. Effect of particle size on age hardening

Continue this series for (4) and (5) for iron-base alloys and then repeat it for copper-base alloys and ceramic and polymeric materials.

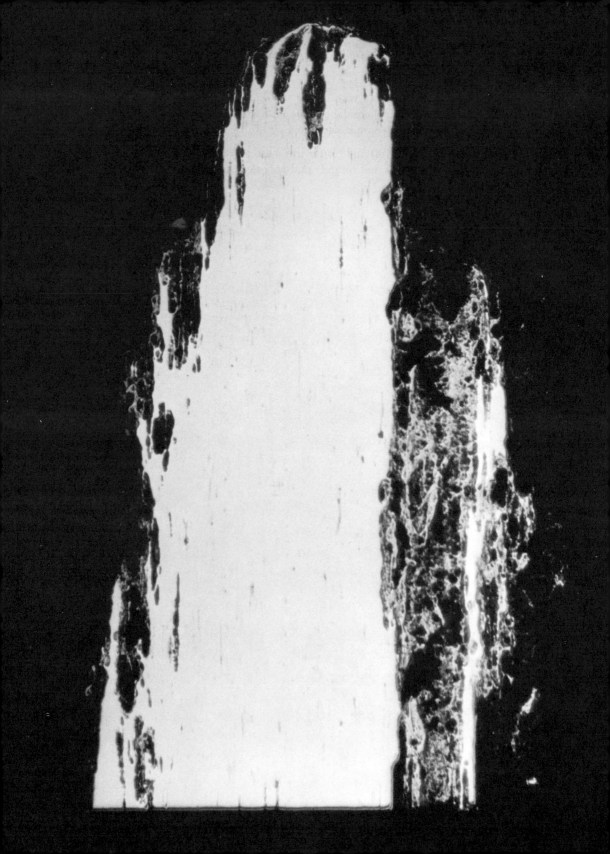

12

CORROSION OF MATERIALS

THIS photomicrograph at 8X shows the longitudinal section of a nail collected from a wreck over 100 years old on the shore of Cape Hatteras, North Carolina. The material is wrought iron, as shown by the large gray fibers of slag in a matrix of ferrite. This is an example of the most important cause of corrosion: the iron (in this case next to the slag fibers) dissolves to form iron ions. These migrate to react with OH⁻ ions which are liberated in the oxygen-rich regions to form rust.

In this chapter we take up the basic reactions occurring in corrosion and their relation to the behavior of materials in corrosive environments.

12.1 Introduction

Corrosion is not a favorite subject of engineers. Many a proud designer or project engineer has developed a new component or process with outstanding performance only to encounter premature failure due to corrosion. Furthermore, despite active research by corrosion engineers, a visit to the local scrap yard will show that large percentages of automobiles and domestic appliances still fail because of corrosion. As a result, the annual cost of corrosion and corrosion protection in the United States is of the order of $8 billion, far more than the annual profit of the nation's largest corporation.

On the positive side, there is real satisfaction in designing a component that can resist punishing service conditions under which other parts fail. A major area in the development of the gas turbine is a corrosion (oxidation) application. The key to the problem is an understanding of the nature of corrosion and how different structures react to it. We will see that the understanding of the nature of the different phases of materials developed in earlier chapters is of vital assistance in surmounting the corrosion problem.

We will find that in most cases corrosion is not completely prevented—we cannot afford platinum bodies for automobiles—but that the objective is to avoid obsolescence of the component due to corrosion. For example, for a long time the average automobile was scrapped because of body corrosion, but now the corrosion life has been doubled by a combination of coatings.

Finally, metals, ceramics, and plastics all suffer attack from different atmospheres and solutions. However, we will soon learn that the corrosion of metallic materials is electrochemical in nature, while in the other cases a simple solution is usually involved. Therefore, we shall first consider the corrosion of metallic materials.

12.2 Corrosion of metals

In discussing the corrosion of metals we will cover four areas:

1. *Chemical principles:* the driving force for corrosion, anode and cathode reactions in a corrosion cell, the solution tendency of metals and alloys, effects of concentration, inhibitors, passivity
2. *Corrosion phenomena* based on chemical principles: galvanic action, selective leaching, hydrogen embrittlement, oxygen corrosion cells, pit and crevice corrosion; and combined mechanical-corrosive effects: stress corrosion, corrosion fatigue, corrosion erosion, cavitation
3. *Corrosion environments:* the reaction of metals to different atmospheres, fresh and salt water, and chemicals
4. *Corrosion in gas* at elevated temperatures, scaling, and growth

Chemical Principles

12.3 Does the metal react?

The first question to ask in analyzing corrosion is: Does the metal react with its environment and, if so, what is the nature of the corrosion product? A copper wire placed in distilled water simply does not react; therefore, there is no problem. An aluminum wire reacts, but the corrosion product, aluminum oxide, is so adherent that there is no further reaction. An iron wire reacts more slowly than the aluminum wire at first, but reaction continues because the product, rust, is nonprotective. Therefore, in each case we must study the reaction involved and the types of products generated.

12.4 Anode and cathode reactions (half-cell reactions)

Whether the corrosion is spectacularly fast, such as zinc dissolving in hydrochloric acid, or quiet, such as rust insidiously forming on the back side of an automobile rocker panel, the basic types of reactions are the same.

1. There is an *anode* reaction at which point metal goes into solution as an ion; i.e. it corrodes. For example, in the reactions just mentioned,

$$Zn \longrightarrow Zn^{2+} + 2e$$

$$Fe \longrightarrow Fe^{2+} + 2e$$

where e = electron, or, in general,

$$M \longrightarrow M^{n+} + ne$$

where M = metal.

2. The electrons flow through the metal part until they reach a point where they can be used up (*cathode* reaction). Again we use the examples given above.

   In the case of zinc in acid the electrons combine with hydrogen ions at the surface. Atomic hydrogen is formed. Most of this combines to form molecular hydrogen which bubbles off, but some dissolves in the metal. This is important in cases of hydrogen attack, discussed later:

$$2H^+ + 2e \longrightarrow 2H \longrightarrow H_2 \text{ (gas)}$$

   In the case of iron the solution is neutral, and we have a reaction involving oxygen and water using up the electrons from the anode to form hydroxyl ions:

$$O_2 \underset{\text{(Dissolved)}}{} + 2H_2O + 4e \longrightarrow 4OH^-$$

Corrosion of iron in pure water will not occur in the absence of dissolved oxygen.

Table 12.1 POSSIBLE CATHODE REACTIONS
IN DIFFERENT GALVANIC CELLS

Cathode Reaction	Example
$2H^+ + 2e \longrightarrow 2H^0$	Acid solutions; see text.
$O_2 + 2H_2O + 4e \longrightarrow 4OH^-$	Neutral and alkaline solutions.
$O_2 + 4H^+ + 4e \longrightarrow 2H_2O$	Using both O_2 and H^+ in acid solutions.
$M^{3+} + e \longrightarrow M^{2+}$	This is encountered when ferric ions are reduced to ferrous.
$M^{2+} + 2e \longrightarrow M^0$	When iron is placed in a copper-salt solution, the electrons from solution of the iron reduce copper ions to metallic copper.

For corrosion to progress, it is essential to have both anode and cathode reactions; otherwise a charge builds up, stopping corrosion. The anode reaction is generally the simple case of metal going into solution, but a variety of cathode reactions are encountered, depending on the conditions (Table 12.1). Note that in all the cathode reactions electrons are *absorbed*.

Several points should be emphasized here: At the anode electrons are left in the metal as the metal ions leave. The metal in this region is negative. At the cathode electrons are absorbed by ions. This region is positive because positively charged ions accumulate to receive electrons.

12.5 Cell potentials

Now that we have defined corrosion cell action, we can go on to consider the driving force of a cell. Why, for example, will zinc corrode in nonoxidizing dilute acid such as HCl while copper will not? Each metal has a different driving force for solution which can be measured as a voltage (Table 12.2). On the one hand, we see that elements such as zinc have a strong negative voltage with respect to hydrogen, indicating that the metal will go into solution and the hydrogen will come out. On the other hand, with the noble metals such as gold, silver, and copper there is no driving potential to replace hydrogen. However, with an oxidizing acid, copper dissolves because of the potential difference between the reaction $O_2 + 4H^+ + 4e \rightarrow 2H_2O$ and copper.

It is useful to review the method of determining these potentials because it gives us another insight into the nature of corrosion.

12.6 Hydrogen half-cell

To establish standard conditions, an electrode of the metal to be tested is placed in a 1 molar solution of its ions (Fig. 12.1). A semipermeable membrane divides the cell. For the other half of the cell a platinum electrode is placed in a 1 molar solution of hydrogen ions (produced by an acid). A stream of

Table 12.2 STANDARD OXIDATION–REDUCTION POTENTIALS
FOR CORROSION REACTIONS*

Corrosion Reaction	Potential, volts vs. normal hydrogen electrode†
$Au \longrightarrow Au^{3+} + 3e$	+1.498
$O_2 + 4H^+ + 4e \longrightarrow 2H_2O$	+1.229
$Pt \longrightarrow Pt^{2+} + 2e$	+1.200
$Pd \longrightarrow Pd^{2+} + 2e$	+0.987
$Ag \longrightarrow Ag^+ + e$	+0.799
$2Hg \longrightarrow Hg_2^{2+} + 2e$	+0.788
$Fe^{3+} + e \longrightarrow Fe^{2+}$	+0.771
$O_2 + 2H_2O + 4e \longrightarrow 4OH^-$	+0.401
$Cu \longrightarrow Cu^{2+} + 2e$	+0.337
$Sn^{4+} + 2e \longrightarrow Sn^{2+}$	+0.150
$2H^+ + 2e \longrightarrow H_2$	0.000
$Pb \longrightarrow Pb^{2+} + 2e$	−0.126
$Sn \longrightarrow Sn^{2+} + 2e$	−0.136
$Ni \longrightarrow Ni^{2+} + 2e$	−0.250
$Co \longrightarrow Co^{2+} + 2e$	−0.277
$Cd \longrightarrow Cd^{2+} + 2e$	−0.403
$Fe \longrightarrow Fe^{2+} + 2e$	−0.440
$Cr \longrightarrow Cr^{3+} + 3e$	−0.744
$Zn \longrightarrow Zn^{2+} + 2e$	−0.763
$Al \longrightarrow Al^{3+} + 3e$	−1.662
$Mg \longrightarrow Mg^{2+} + 2e$	−2.363
$Na \longrightarrow Na^+ + e$	−2.714
$K \longrightarrow K^+ + e$	−2.925

(left margin top to bottom: More cathodic *... * More anodic*)*

* Measured at 25°C.

† In some chemistry texts the signs of the values in this table are reversed; for example, the half-cell potential of zinc would be +0.763 volt. This depends on whether we consider charge motion in the external circuit or within the cell. In any case it is agreed that zinc goes into solution with respect to the hydrogen half-cell, and we shall adopt the convention that corrosion is at the more negative electrode of the cell.

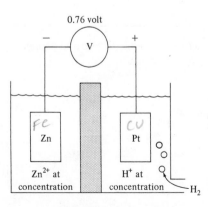

Fig. 12.1 *Diagram of a half-cell for measurement of potential.*

hydrogen gas is bubbled around the platinum electrode. Therefore, the platinum merely serves to adsorb the gas and does not take part in the reaction.

If the metal is more reactive than hydrogen, we have the following half-cell reactions when the electrodes are connected:

$$M - ne \longrightarrow M^{n+}$$
$$nH^+ + ne \longrightarrow nH^0$$

and the negative potentials in Table 12.2 are developed. On the other hand, if the metal is nobler (less reactive) than hydrogen, we have

$$nH^0 - ne \longrightarrow nH^+$$
$$M^{n+} + ne \longrightarrow M^0$$

and the positive potentials in Table 12.2 are developed.

With these examples in mind we should visualize the situation if the metal is not surrounded by a 1 molar concentration of its ions. If the concentration of the ions is less than 1 molar, the driving force to dissolve will be greater because the ions available for the reverse reaction to $M^+ + e \rightarrow M^0$ are fewer. The change in electrode voltage for each tenfold change in concentration, to 0.1 molar for example, is 0.0591 volt. This is for the case in which there is a transfer of one electron per ion going into solution. In the case of divalent ions the voltage change is 0.0295 volt and for trivalent ions 0.0197 volt. If the concentration is above 1 molar, the driving voltage is correspondingly less.

These effects are derived from the well-known Nernst equation, which can be simplified to suit our case at 25°C:

$$E = E_0 + \frac{0.0592}{n} \log C_{\text{ion}}$$

where E = new electromotive force (emf)

E_0 = standard emf of Table 12.2

n = number of electrons transferred as
$$M^0 - ne \longrightarrow M^{n+}$$

C_{ion} = molar concentration of ions

EXAMPLE 2.1 What would be the new half-cell voltages for zinc and copper if the ion concentrations were changed to 0.1 molar?

ANSWER

Zinc
$$E = -0.763 + \frac{0.0592}{2} \log 0.1$$
$$= -0.763 - 0.0296$$
$$= -0.793 \text{ volt}$$

Copper
$$E = +0.337 + \frac{0.0592}{2} \log 0.1$$
$$= +0.337 - 0.0296$$
$$= +0.307 \text{ volt}$$

The result for zinc shows a greater tendency to corrode. For copper we see a greater tendency to go into solution, but the potential of a copper-zinc cell would be less because the copper is the electrode at which ions plate out.

12.7 Cell potentials in different solutions

There are even more serious corrections to be made to the driving voltage of a given corrosion couple because the metals are rarely in solutions of their own salts. For example, the order of half-cell potentials in salt water is given in Table 12.3 for a group of commercial alloys. Note that aluminum and zinc have changed places compared with the standard potential sequence.

12.8 Corrosion rates

Now let us take up the rate of corrosion. We have introduced the concept of solution potential, and we find with further study that equilibrium and this emf are closely associated. That is, the emf indicates the *tendency to corrode* as well as the anode and cathode, but it does not directly govern the corrosion *rate*. For example, a battery is merely a case of controlled corrosion. If we were to ask for a 1.5-volt battery in a store, we could choose from dozens of sizes, all of which would provide 1.5 volts; the life in ampere-hours of the batteries would be decidedly different, however. Thus the corrosion rates would be different because of the different battery sizes and different materials in the batteries. In effect, the corrosion rate depends on Faraday's law:

$$\text{Weight of metal dissolving (g)} = kIt$$

where $k = \dfrac{\text{at. wt. of metal}}{\text{no. of electrons transferred} \times 96,500 \text{ amp-sec}}$

I = current (amp)

t = time (sec)

In corrosion we are usually interested in the loss in weight from a given area per unit time. Therefore, dividing both sides of the above equation by the area gives us the corrosion rate as equal to a constant times the current density [current (amp)/area (cm^2)]. In other words, the rate of weight loss from a given area is proportional to the current density.

Table 12.3 HALF-CELL
POTENTIALS OF VARIOUS ALLOYS
IN SALT WATER*

Metal or Alloy†	Potential,‡ volts, 0.1 N calomel scale
Magnesium	−1.73
Zinc	−1.10
A612	−0.99
7072, Alclad 3003, Alclad 6061, Alclad 7075	−0.96
220-T4	−0.92
5056, 7079-T6, 5456, 5083, 214, 218	−0.87
5154, 5254, 5454	−0.86
5052, 5652, 5086, 1099	−0.85
3004, B214, 1185, 1060, 1260, 5050	−0.84
1100, 3003, 615, 6053, 6061-T6, 6062-T6, 6063, 6363, Alclad 2014, Alclad 2024	−0.83
13, 43, cadmium	−0.82
7075-T6, 356-T6, 360	−0.81
2024-T81, 6061-T4, 6062-T4	−0.80
355-T6	−0.79
2014-T6, 113, 750-T5	−0.78
108, 108-F	−0.77
85, 380-F, 319-F, 333-F	−0.75
C113-F	−0.74
195-T6	−0.73
B195-T6	−0.72
2014-T4, 2017-T4, 2024-T3 and T4	−0.68 to −0.70§
Mild steel	−0.58
Lead	−0.55
Tin	−0.49
Copper	−0.20
Bismuth	−0.18
Stainless steel (series 300, type 430)	−0.09
Silver	−0.08
Nickel	−0.07
Chromium	−0.49 to +0.018

(left margin, vertical: ↑ More anodic)

*53 g NaCl + 3 g H_2O_2 per liter, 25°C.

†The potential of all tempers is the same unless temper is designated.

‡Data from Alcoa Research Laboratories.

§The potential varies with quenching rate.

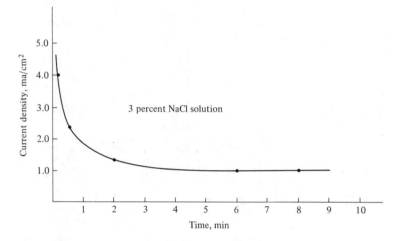

Fig. 12.2 *Corrosion current decrease in an aluminum-iron galvanic cell (25°C, 77°F).*

If we determine the current density for an aluminum-iron corrosion cell, we find the initial current is relatively high (4 ma/cm²) and gradually falls off until it reaches 1 ma/cm² after 6 min (Fig. 12.2). Furthermore, the emf changes during this period. Such a decrease in corrosion rate with time is caused by *polarization* and is very important to corrosion control.

As an example, consider the copper-zinc cell shown in Fig. 12.3a. The zinc will act as an anode, and its emf will tend to become more cathodic (positive) as the current or current density increases, as shown in Fig. 12.3b. Meanwhile, the copper is cathodic and will tend to become more anodic, as shown by the emf–current density relationship. The two curves will intersect, which gives us a corrosion emf of less than 1.10 volts and a corresponding current density which determines the corrosion rate.

Therefore, if we want to lower the corrosion rate, we must find a way to decrease the current density of the cell. This involves the concept of polarization. Polarization is really composed of two effects: *activation polarization* and *concentration polarization*. The latter is easier to describe first, since it follows directly from our discussion of the effect of concentration on cell potential.

Let us again consider the copper-zinc corrosion cell in operation. As zinc goes into solution, the electrons are delivered to the copper. In order for the electrons to be absorbed, positive ions are needed at the copper cathode. In Fig. 12.3 these are copper ions. After current has passed for a short time, there will be a scarcity of positive ions at the cathode because of the time needed for diffusion of ions within the solution. This is shown schematically in Fig. 12.4. The greater the requirement for positive ions, as with very high initial

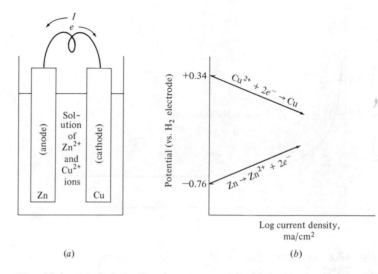

(a)　　　　　　　　　　　　　(b)

Fig. 12.3 (a) *Polarization in a copper-zinc galvanic corrosion cell (25°C, 77°F). (b) Potential vs. log current density.*

corrosion rates, the sooner the appearance of an ion-depleted layer at the cathode. The net effect, of course, is to decrease the corrosion rate.

Furthermore, we saw in Example 12.1 that the cathode voltage becomes more negative at lower ion concentrations. Therefore, depletion of ions at the cathode changes the cathode voltage, as shown in Fig. 12.5, and gives a lower corrosion rate.

We should also note that a high concentration of Zn^{2+} ions at the anode interface might also inhibit more zinc from going into solution and hence might cause concentration polarization at the anode. However, such concentration

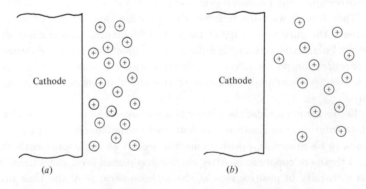

(a)　　　　　　　　　　　　　(b)

Fig. 12.4 *Concentration polarization (ion depletion) due to a high rate of reaction at the cathode. (a) Time = 0. (b) Time > 0.*

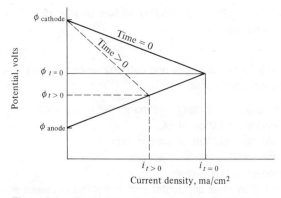

Fig. 12.5 *Effect of concentration polarization at the cathode in lowering corrosion rate (current density decrease in a cathode controlled polarization).*

polarization at the anode is usually negligible compared to the cathode, since a good deal of metal surface is exposed and can ionize.

Finally, it must be emphasized that concentration polarization is related to diffusion. Therefore, conditions that change diffusion rates also affect concentration polarization. As an example, stirring the liquid will reduce the concentration gradient of positive ions.

Similar effects leading to lower concentration polarization and increased corrosion rates can be predicted for an increase in temperature and an increase in the concentration of the ions in solution.

Now let us consider activation polarization, which is the common variety of polarization when the reaction rate is low. A good illustration of this effect is the evolution of hydrogen at the cathode. The hydrogen is first formed as atomic hydrogen; then it is necessary for two atoms to form molecular hydrogen; next the molecules must form a bubble; and finally the bubble rises. A slowdown in any step can retard the entire sequence. The term "activation polarization" is used to give the idea of the "activation energy" needed to overcome the resistance of the slowest step. There is also activation polarization at the anode. This is the barrier the metal atom or ion encounters in leaving the metal specimen and entering the solution. It is interesting that this value is larger for the transition metals iron, cobalt, nickel, and chromium than for silver, copper, and zinc.

Let us summarize these effects. The concentration polarization is more of a chemical barrier, while the activation polarization is more of a physical or electrical type barrier. Both reduce corrosion rate.

The polarization effects at anode and cathode are rarely equal. When the polarization results in a greater decrease in cathode potential, the reaction is said to be *cathodically controlled,* as shown in Fig. 12.5. The term *anodic control* is used similarly.

EXAMPLE 12.2 Write the anode and cathode reactions (half-cell reactions) for the following conditions:

a. Copper and zinc in contact and immersed in seawater
b. As in *a* with the addition of HCl
c. As in *a* with the addition of copper ions
d. Copper immersed in fresh water
e. Iron immersed in fresh water
f. Cadmium-plated steel scratched and immersed in seawater

ANSWER

a. Anode: $Zn \rightarrow Zn^{2+} + 2e$
   Cathode (copper): $O_2 + 2H_2O + 4e \rightarrow 4OH^-$ (see Table 12.1)
b. Anode: $Zn \rightarrow Zn^{2+} + 2e$
   Cathode (copper): $O_2 + 4H^+ + 4e \rightarrow 2H_2O$
   (Since H^+ ions can help use up electrons generated by corrosion of zinc, the corrosion rate is higher.)
c. Anode: $Zn \rightarrow Zn^{2+} + 2e$
   Cathode: $Cu^{2+} + 2e \rightarrow Cu$
d. Little or no corrosion because copper is noble. Compare the emf for a solution of copper with the equation for $O_2 + 2H_2O + 4e \rightarrow 4OH^-$.
e. Anode: $Fe \rightarrow Fe^{2+} + 2e$
   Cathode: $O_2 + 2H_2O + 4e \rightarrow 4OH^-$
f. Anode: $Cd \rightarrow Cd^{2+} + 2e$
   Cathode: $O_2 + 2H_2O + 4e \rightarrow 4OH^-$
   [*Note:* Cadmium is cathodic to iron in the standard oxidation-reduction potentials (Table 12.2) but anodic to iron in salt water.]

12.9 Inhibitors

The reason we have given this detailed account of polarization is that one of the most important methods for the control of corrosion is the use of inhibitors. The action of these chemicals is related to polarization. It is possible to reduce the corrosion rate drastically by inhibiting the reaction at *either* the anode or the cathode. In many cases the role of the inhibitor is to form an impervious, insulating film of a compound on either the cathode or the anode. A common case is the use of chromate salts in automobile radiators. The iron ions liberated at the anode surface combine with the chromate to form an insoluble coating. Another case is the use of gelatin which is adsorbed and limits ion reaction with the electrode.

12.10 Passivity

A logical extension of inhibition is the development of a passive film on the surface of a metal. A classic example is the corrosion of iron in nitric acid. If a piece of iron is placed first in concentrated nitric acid and then in dilute nitric acid, no appreciable corrosion occurs. A thin adherent *passive* film of iron oxide is produced in the concentrated acid. If the sample is then scratched, rapid corrosion takes place in the dilute acid because of the rupture of the film. When the sample is not immersed first in concentrated acid, both the initial and continuing corrosion are rapid in the dilute acid.

The importance of the experiment is that *self-repairing passive films* are formed when more than 10 percent chromium is present in the iron. In general, iron, chromium, nickel, titanium, and aluminum alloys can form passive layers.

Passivating can also be explained with the aid of a polarization curve obtained by measuring cell current for the material as a function of voltage. A material that passivates exhibits the anode polarization curve shown in Fig. 12.6. As the anode is driven to higher current densities (generally by exposure to an oxidizing acid), a maximum value called $i_{critical}$ is reached. Upon reaching this value the anode forms a protective layer and its corrosion decreases to a value called $i_{passive}$. It is important to realize that if the initial corrosion rate

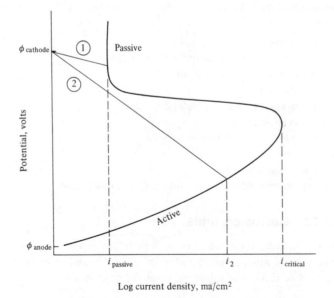

Fig. 12.6 *Schematic polarization curve of a passive metal. Cathode polarization results in high corrosion i_2, while polarization maintains a passive anode and lower corrosion rates ($i_{passive}$).*

is not great enough, $i_{critical}$ will not be exceeded, and this can cause some peculiar effects. For example, stainless steel will corrode more readily in a weak oxidizing acid ($i < i_{critical}$) than in a concentrated oxidizing acid (i reaches $i_{passive}$).

The polarization of the cathode can determine whether the passive condition is reached by the anode. In Fig. 12.6, for instance, two possible *cathode* polarization curves have been superimposed on the anode polarization curve. In the case of cathode 1 the intersection with the anode is in the passive region. Cathode 2 polarizes at a higher rate (such as through concentration polarization), and it intersects the anode polarization curve within the active region, giving a higher corrosion rate. Therefore, care must be exercised with coupling passive metals to highly polarizable cathodes because passivation may be destroyed and a high corrosion rate may be obtained.

Corrosion Phenomena

12.11 Types of corrosion

Practically any case of corrosion can be explained by using the basic principles we have just discussed. However, there are a number of corrosion situations or phenomena which are encountered frequently and which are given special names and deserve careful analysis. Also, these common cases serve well as illustrations of how to apply the basic principles. The phenomena to be covered are:

1. Corrosion units
2. Galvanic corrosion, macroscopic and microscopic cases
3. Selective leaching (dezincification, etc.)
4. Hydrogen damage
5. Oxygen-concentration cells, water-line attack
6. Pit and crevice corrosion
7. Combined mechanical-corrosive effects
   a. Stress corrosion
   b. Corrosion fatigue
   c. Liquid velocity effects: corrosion erosion, cavitation

12.12 Corrosion units

In the standard corrosion tests, which are discussed later in detail, samples of a metal are measured and weighed before and after immersion for a given period of time. The following common units are used:

mdd: milligrams lost per square decimeter per day
ipy: inches corroded per year
mpy: mils corroded per year (1 mil = 0.001 in.)

The last two terms are preferable because they allow us to visualize easily the long-range effect. These values apply to uniform corrosion and may only be used in design if nonuniform corrosion, such as pit corrosion, is not present.

12.13 Galvanic corrosion

In a general way all corrosion depends on galvanic action, but this term means specifically a type of corrosion that occurs because two materials of different solution potential are in contact. For our purposes here we shall also include galvanic effects on the microscopic scale after discussing the macroscopic case.

Macroscopic Cases of Galvanic Corrosion. Perhaps the best illustration of this phenomenon was given by an inspired sculptor who produced a figure for a central plaza in New York City. To express himself he used a bronze body, an aluminum crown, an iron sword, and a stainless steel base. The result was a fine collection of galvanic cells, and the original form did not last long.

Many less spectacular cases of galvanic action exist, and often the more reactive member of the couple is deliberately used as a *sacrificial anode*. In the common case of galvanized steel or wire, the steel is coated with zinc either by being dipped into molten zinc or by electroplating. The effects in Table 12.4 were obtained in an experiment in which zinc and steel samples of equal size were tested separately for the same time period in the attached or coupled condition.

Note that when uncoupled, both zinc and steel corrode in the solutions, but when coupled, the zinc protects the steel, corroding at an accelerated rate as a sacrificial anode.

Another common couple is found in the "tin" can, which is composed of steel covered with a thin layer of tin. If a cut section is allowed to corrode, the steel rather than the tin will usually corrode. This confirms a prediction that might be made from Table 12.2, since iron has a greater solution

Table 12.4 WEIGHT LOSS OF ZINC AND STEEL IN THE UNCOUPLED AND COUPLED CONDITIONS

| | Weight Change of Each Sample, g | | | |
| | Uncoupled | | Coupled | |
Solution	Zinc	Steel	Zinc	Steel
0.05 molar Na$_2$SO$_4$	−0.17	−0.15	−0.48	+0.01
0.05 molar NaCl	−0.15	−0.15	−0.44	+0.01
0.005 molar NaCl	−0.06	−0.10	−0.13	+0.02

potential. The question might then be asked: Why aren't cans made of galvanized steel? The answer is that while the zinc would protect the steel, the ions would be present in the food at an undesirable level. In the manufacture of the tin cans great care is taken to avoid contact of the steel with the food (through the use of soldered joints and lacquering). Therefore, the corrosion rate is only that of tin or lacquer, which is generally low.

As another example, rivets of a different material than that of the basic structure are often used, especially as an expedient in repair. In the case of copper rivets in steel sheet there is a *large anode area* (the steel), and the galvanic action is not serious. However, when steel rivets are used in a copper sheet, all the metal loss is concentrated in a small anodic region and there is a large area for cathode reactions to absorb electrons. The corrosion is catastrophic. Figure 12.7*a* shows the effect of anode-to-cathode area on the corrosion rate (current) of just such an iron-copper system. Therefore, galvanic coupling is to be avoided, especially in systems exposed to aqueous solutions. Another example is shown in Fig. 12.7*b*, in which a brass pipe and a cast-iron fitting were used as part of a steam condensate trap. Since the iron is anodic, it has almost completely corroded away.

The question may be asked: Suppose we have a sample of an alloy with several phases present. Will these not act like small galvanic couples and corrode? As a matter of fact, this effect is the reason we can distinguish different phases in the microstructure. Usually there is not much to be seen in the as-polished condition; it is only after etching that the details of structure are visible. There are many commercial cases where this behavior is important and we will now discuss some examples.

WELD DECAY OF 18% CR, 8% NI STAINLESS STEEL. Type 304 stainless steel contains 18% Cr, 8% Ni, and 0.08% C maximum. It is usually delivered in the single-phase austenitic structure obtained by rapid cooling from elevated temperatures. The Cr_4C phase can precipitate, however, if the steel is reheated in the two-phase field. During welding the weld zone is heated to the liquid state and cools rapidly enough to avoid carbide precipitation. However, there is a region adjacent to the weld which has just been heated high enough to precipitate Cr_4C. This usually takes place at grain boundaries. Because of the high chromium content of the carbide, the nearby regions are impoverished in chromium in the process of forming the carbide. The chromium level falls below 10 percent in these regions near the grain boundary. Hence, the low-chromium regions are not passive (<10 percent Cr), whereas the remainder of the matrix is passive. The result is galvanic action between the grain boundary region and the higher-chromium regions within the grain. It should be emphasized that it is not the Cr_4C particles that are corroded, but the low-chromium–iron matrix. The Cr_4C particles have been recovered after corrosion and under the electron microscope appear as platelike crystals. This type of corrosion can be avoided by using a solution heat treatment at 2000°F (1093°C) after welding.

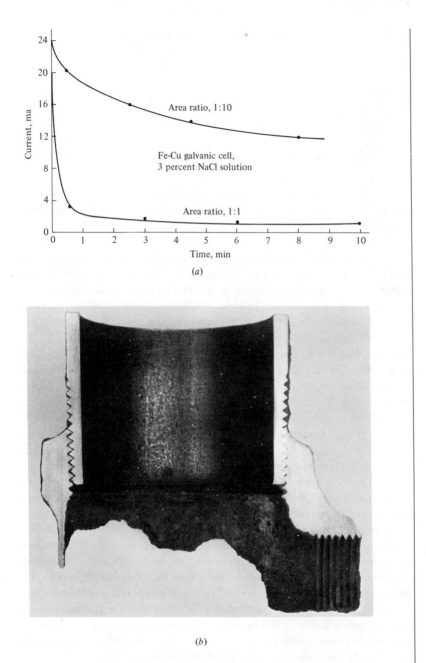

Fig. 12.7 (a) Corrosion rate (current) as related to anode-to-cathode area (25°C, 77°F). (The upper curve shows the higher corrosion with a large cathode and a small anode.) (b) Condensate trap of iron fitting in brass pipe (cross-section) with the iron fitting badly corroded.

Table 12.5 CHEMICAL COMPOSITIONS OF
SEVERAL STAINLESS STEELS

Stainless Steel	Chemical Analysis, percent			
	C	Cr	Ni	Other
Type 304	0.08 max.	18	8	
321	0.08 max.	18	8	Ti = 5 × percent C
347	0.08 max.	18	8	Nb = 10 × percent C
304L	0.03 max.	18	8	

This type of grain boundary corrosion also can be minimized by using 18% Cr, 8% Ni steels to which a stronger carbide-former than chromium (titanium or niobium) is added or by specifying a very low-carbon level, as in Table 12.5.

The use of the special steels is not always a safeguard against galvanic action because chromium carbide may still form in preference to other carbides in a certain temperature range. The special case called "knife-line attack" can be encountered in type 347. All carbides are dissolved in the weld and niobium carbide then precipitates in the range 2250 to 1450°F (1232 to 788°C) on cooling. However, if the material at the edge of the weld is cooled rapidly in the range for niobium carbide formation but more slowly from 1450 to 950°F (788 to 510°C), *chromium* carbide can precipitate, again forming an active-passive condition. The weld will be corroded in a narrow region: hence the name "knife line."

Cases of galvanic action are not limited to the high-alloy steels. An interesting case in plain-carbon steel piping is called "ringworm corrosion." This selective attack takes place near the end of the pipe which has been *especially heated* to forge the flange portion. This treatment results in a spheroidized iron carbide structure which has a different solution potential than the untreated balance of the pipe. A circular form of attack occurs near the junction of the two regions.

In aluminum alloys there is a marked difference between pure aluminum and some of the age-hardened alloys, particularly those containing copper in corrosive liquids such as salt water. In a 4 percent copper alloy, for example, the potential after solution heat treatment is uniform (-0.69 volt). After aging, however, there is a change in potential near the $CuAl_2$ particles that precipitate. The matrix near the grain boundaries when the precipitation is heavy is -0.78 volt, while the balance of the structure is still -0.69 volt. In this condition the alloy is subject to *intergranular corrosion*. For protection a pure aluminum coating is used.

Even in single-phase alloys galvanic effects will develop. A common source of potential difference is segregation during solidification (Chap. 4). An as-cast cupronickel alloy is shown in Fig. 12.8. By reheating the alloy

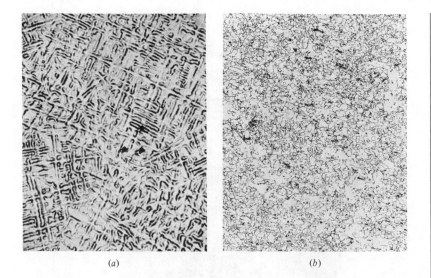

(a)   (b)

Fig. 12.8 (a) 70% Cu, 30% Ni *alloy as-cast, dendritic structure. 100X, chromate etch. (b) Same as (a), but after rolling and homogenization at 1700°F (927°C). 100X, chromate etch.*

to below the solidus and allowing diffusion to take place, the material is homogenized.

In a homogeneous alloy the material at the grain boundaries is at a high solution potential because of the dislocation concentrations leading to poorer bonding and higher strain energy. After cold working, the material tends to corrode more than annealed material. An interesting illustration of the use of this effect in criminal investigations is shown in Fig. 12.9. A series of identification numbers was stamped on an engine part (1). Next the identification was ground away (2). However, upon etching the ground surface the numbers reappeared (3). The reason is that the metal is cold-worked below the visible bottom of the impression, and therefore still etches differentially after the impression is ground off. To avoid this action in the corrosion of cold-worked parts, at least partial annealing is necessary (recovery).

12.14 Selective leaching (dezincification, etc.)

"Selective leaching" is a term that can be used in general to cover a number of mouth-filling classifications, such as "dezincification," "dealuminization," etc. The case of brass will illustrate this type of failure.

It is noted that after exposure to fresh or salt water, brass may develop copper-colored layers or plugs of material (Fig. 12.10a and b). Examination

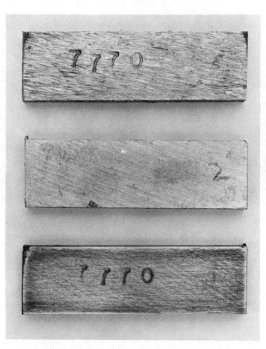

Fig. 12.9 *Revealing stamped numbers after they have been "ground away." At the top, numbers are stamped on the bar. In the middle, numbers are ground away. At the bottom, after 10 min in boiling 50 percent sulfuric acid, the numbers reappear because the severely cold-worked region beneath the stamped number etches more rapidly. This illustrates the higher solution potential of cold-worked metal.*

will show that these are in fact spongy low-strength regions of copper from which the zinc has leached out. The usually accepted mechanism is that the brass dissolves slightly, then the copper ions are displaced by more zinc going into solution, and the copper plates out. This is obviously a dangerous phenomenon when spongy plugs of copper form in a pipe under pressure. In these cases a copper-nickel alloy is used.

To avoid the problem, the zinc content of the pipe should be lowered. The most sensitive alloys contain 40 percent zinc and have a second phase (β) which aggravates the problem. In severe cases, where the 30 percent zinc alloys are troublesome, the addition of tin and arsenic (1% Sn, 0.04% As) or reduction of zinc to below 20 percent will alleviate the problem.

(a)

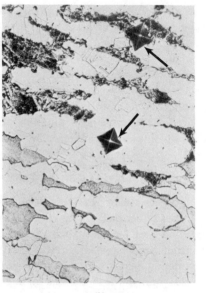

(b)

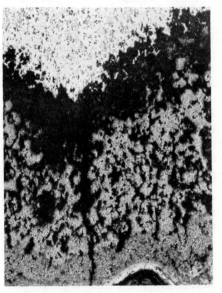

(c)

Fig. 12.10 (a) 60% Cu, 40% Zn brass. FCC α phase, light, VHN 100. BCC β phase, gray, VHN 130. Spongy copper, dark, VHN 52. 100X, chromate etch. (b) 60% Cu, 40% Zn brass. FCC α, light. BCC β phase, gray. Spongy copper, dark. Note that corrosion takes place in the β regions (higher zinc content and more active). The selective leaching is called dezincification. 500X, chromate etch. (c) Gray cast-iron pipe showing graphitic corrosion. The original diameter of the pipe included all the black and gray corrosion product in the lower two-thirds of the photomicrograph. Only a spongy mass of graphite and some silicate remain. The white area at the top is unattacked gray iron in which graphite flakes can be seen. 50X, unetched.

In gray cast-iron pipe the iron matrix may be slowly leached away, leaving the insoluble graphite behind. This is called "graphitic corrosion" or, often improperly, "graphitization." This is usually a very slow effect, and pipe showing this condition has been in service for centuries. On the other hand, it can occur in a few years under extreme soil conditions, such as where cinders might have been used for backfill. The microappearance of this type of corrosion is shown in Fig. 12.10c.

Selective leaching of aluminum from copper alloys and of cobalt from a cobalt-tungsten-chromium alloy has also been encountered.

12.15   Hydrogen damage

In discussing the basic atom movements in corrosion we said that hydrogen discharged at the cathode could form bubbles *or* dissolve in the metal and diffuse through it. There are cases involving either hydrogen evolution from applied currents, as in electroplating, or hydrogen from corrosion in acid solution. In *hydrogen blistering* the atomic hydrogen (H) diffuses through the metal and, finding a void, diffuses into it, forms molecular hydrogen (H_2), and exerts high pressure, tending to spread out the void. The pressure of molecular hydrogen in equilibrium with atomic hydrogen is over 10^5 atm (1.033×10^6 kg/m^2). In *hydrogen embrittlement* the dissolved interstitial atoms lead to low ductility and low impact strength, perhaps because of interaction with microcracks. These effects can be minimized by baking for a long time to remove hydrogen and by avoiding couples which produce it.

12.16   Oxygen-concentration cells, water-line corrosion

In contrast to the galvanic effects just discussed, corrosion can appear to occur without any starting potential. A drop of water will corrode a polished iron surface, a homogenized steel tank will corrode at the water line.

To explain these cases, let us refer to the equations shown in the emf series:

$$O_2 + 4H_2O + 4e \longrightarrow 4OH^- \qquad +0.401 \text{ volt}$$
$$Fe \longrightarrow Fe^{2+} + 2e \qquad -0.440 \text{ volt}$$

We see, therefore, that in an oxygen-concentration cell involving solution of iron at the anode and discharge of electrons by forming OH$^-$ at the cathode, a potential of 0.841 volt exists. In the apparently innocuous case of the drop of water we can predict the following sequence of events.

Some iron dissolves in the water, and oxygen and water react to absorb the electrons. The oxygen in the water can only be replaced rapidly at the edge of the drop. This region of high oxygen concentration develops as the cathode

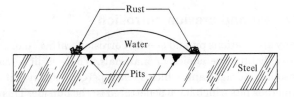

Fig. 12.11 *Corrosion of steel by a drop of water.*

(Fig. 12.11). Iron will dissolve inside the drop, but more will go into solution closer to the edge regions than at the center. This is because there is a higher resistance for electron travel from the center; i.e. anodes and cathodes are favored when close together. A ring of rust is formed at the edge of the drop because the iron ions migrate faster than the OH^- ions. The ferrous hydroxide is oxidized to hydrated ferric hydroxide (rust) in a secondary reaction.

The case of water-line corrosion is similar. We find that a tank that is kept only partly filled with water will corrode more rapidly than a completely filled one. Corrosion takes place at the water line because of the gradient in oxygen concentration which develops between the solution at the water surface and that at depth. The greatest attack is *just below* the water line, where there is a pronounced drop-off in oxygen concentration plus a short electron path to the water line, where hydroxyl ions are formed.

There are, of course, other examples of oxygen affecting the corrosion rate. Home hot-water-heating systems are usually "closed systems." The initial oxygen is very rapidly used up, and the corrosion rate becomes very low because of the low amount of oxygen. On the other hand, if an "open system" is used (fresh water is continuously added), the high amount of dissolved oxygen results in a continued high corrosion rate.

Stains on the bottom of cooking utensils from "cooked-on food" are usually the result of concentration cells. The liquid under barnacles on a ship's hull will develop a different concentration of ions than in the normal salt water and hence a concentration corrosion cell will form.

EXAMPLE 12.3 Water allowed to dry on stainless steel will usually result in "water spots." How do these spots occur?

ANSWER This is an oxygen-concentration-cell effect which can be increased by other ions. If a large area of water, say several inches in diameter, is followed during its process of evaporation, the spot occurs in the very last portion to evaporate. This may be explained by the higher oxygen and other ion concentrations in the last liquid. (Dissolved salts and gases remain in the liquid, and since the volume of liquid becomes smaller, these concentrations become greater.)

12.17 Pit and crevice corrosion

In many applications pit and crevice corrosion, rather than the overall corrosion rate, dictate the choice of materials. There is little comfort in owning a gasoline storage tank which is 99.9 percent intact but in which there are numerous pits which have penetrated to the outside. This unhappy situation can be avoided by an understanding of the causes of pitting and the realization that certain well-known combinations of materials and environments are prone to this phenomenon.

Until recently the formation of a pit was considered merely a special situation of an oxygen-concentration cell, but this did not explain the important role of ions such as chlorides. The most recent concept is that a pit begins at a surface discontinuity such as an inclusion or grinding mark. An oxygen-concentration cell develops between the discontinuity and the surrounding material. The feature that involves the chloride ions is that within the incipient pit positive metal ions dissolve and accumulate. These attract chloride ions. The metal chloride concentration begins to build up in the pit. If the chloride is iron chloride, for example, this hydrolyzes to give HCl: $M^+Cl^- + H_2O = MOH + H^+Cl^-$.

The combination of chloride and hydrogen ions then accelerates the attack. In proof of this mechanism, it has been found that the fluid within crevices exposed to an overall neutral dilute sodium chloride solution contains 3 to 10 times as much chloride ion as the bulk solution and a pH of 4 rather than 7.

Crevice corrosion follows the same mechanism as pit corrosion, since the crevice serves as a ready-made pit in which the oxygen concentration is low.

In combating pitting the most important point is to avoid combinations of materials and environments that are known to be susceptible. For example, many of the materials with passive surfaces exhibit pitting because of the large potential difference between the passive and active regions of the pit. Furthermore, chloride ions are known to destroy passive layers locally. Among the stainless type materials the following sequence, from the most to the least susceptible, is found for salt water:

Type 304 stainless steel (18% Cr, 8% Ni)
Type 316 stainless steel (18% Cr, 8% Ni, 2% Mo)
Titanium

It is also interesting that titanium forms a very stable passive layer under oxidizing conditions. Therefore, it has excellent corrosion resistance and finds extensive use in the chemical industry.

12.18 Combined mechanical-corrosion effects

In many cases a component will fail because of the combined effect of mechanical or hydraulic factors and corrosion. These cases are of three types: stress corrosion, corrosion fatigue, liquid velocity effects (corrosion erosion, cavitation).

The typical test for *stress corrosion* is to take a sample of metal in the form of a small beam, apply a permanent bending load, and place the beam in a corrosive liquid. Failure will occur far more rapidly than in an unstressed part. Another method is to take a deformed sample such as a stamping and observe the onset of cracking. Classic cases are the *season cracking* of drawn-brass cartridge cases and the *caustic embrittlement* of steel in boilers. In both these cases we have a highly deformed structure with high residual stress and a suitable environment. In the case of the brass cartridge cases the cracking is associated with the presence of ammonia from decaying organic matter plus the high humidity of the tropics. In the boiler tubes the elastic strain is produced by cold-rolling the ends and the presence of caustic in the environment.

Stress corrosion is usually accompanied by an intergranular fracture in which the material shows no ductility even though the tensile test results may indicate a high plastic elongation capability for the material. Although many examples are due to residual stresses from processing, the stresses can also result from the service conditions. Thermally induced stresses resulting from different coefficients of expansion have caused stress-corrosion failure in some bimetal plumbing fixtures.

EXAMPLE 12.4 Where does corrosion normally occur first in an automobile bumper?

ANSWER The portions that have been most cold-worked are usually the first to corrode. Just as on a microscopic scale grain boundaries are anodic to the surrounding grain, the macroscopic area of high residual stress will be anodic to the nonworked portion. Of course, a corrosive medium such as salt spray from roads is necessary to cause the corrosion. Chromium (cathodic to steel) does not offer any galvanic protection for the steel and, once ruptured, makes the problem worse because of the cathode-to-anode area relationship.

Fatigue failures or cyclic stress failures were discussed in Chap. 11. Here it was pointed out that such failures are very sensitive to surface conditions. Pits caused from corrosion cause a surface stress concentration that easily propagates as a crack under a cyclic stress; hence the name *corrosion fatigue*.

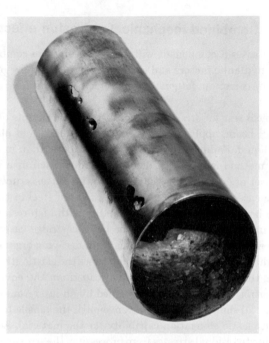

Fig. 12.12 *Copper sewer pipe with holes eroded by dripping waste fluids. Although copper has excellent corrosion resistance, waste fluids that dripped and concentrated on standing finally penetrated the pipe.*

Normally an increase in liquid velocity is accompanied by an increase in corrosion rate. Therefore, a spinning disk in water would have a higher corrosion rate at its periphery. (Recall that high velocities decrease concentration polarization.) A normal manifestation is a gouging out or *erosion*. However, all erosion need not result from high liquid velocities. A dripping action has also been known to "wear a hole" in materials such as copper sewer pipe, as shown in Fig. 12.12. *Cavitation,* on the other hand, results from bubbles of vapor popping against a surface such as a ship propeller.

Corrosive Environments

12.19 General

The most important corrosion problems occur in three types of environments:

1. The atmosphere
2. Water, fresh and salt
3. Chemicals: acids, alkalis, salts

In each case we will discuss the types of reactions encountered.

12.20 Atmospheric corrosion

Although atmospheric corrosion is not spectacular, the cost of it is. The annual bill in the United States is $2 billion. After much experimentation, corrosion engineers have found three distinctly different corrosion rates in industrial, marine, and rural environments.

The problems in the industrial environment arise from SO_2, which leads to H_2SO_4 and H_2SO_3. Salt and other contaminants from roads also lead to accelerated rates.

In marine atmospheres the chief problem is salt spray.

In rural atmospheres rain and dust cause the principal problems. In contrast, in the Atacama Desert in Chile automobiles wrecked 50 years ago appear just as they did at the time they were discarded.

Recently the change in atmosphere has changed the life of galvanized eave troughs. In the past, in highly industrialized areas it was not uncommon to have to replace such items in as little as 5 years. The present concern with air pollution has resulted in a life extension of more than 20 years for these troughs, which compares favorably with their life span in a rural atmosphere.

In selecting a material for atmospheric exposure the first decision to be made is whether a shiny metallic surface such as that given by stainless steel is necessary. If not, there are two principal alternatives.

1. Use an alloy that forms a protective coating, such as a steel with small amounts of copper and nickel, which develops a brown surface. The extreme case is the use of copper alloys that develop an attractive green patina. Styles change, and while shiny metal surfaces used to be in fashion, the muted surfaces have recently found favor.
2. Apply paint or plastic coatings.

12.21 Water

The attack from fresh water varies widely, depending on the dissolved salts and gases. The principal contaminants are chloride ions, sulfur compounds, iron compounds, and calcium salts. There is little difference between plain and low-alloy steels. Cast iron and ductile iron are widely used for water pipe. At critical junctions such as valves the mating surfaces are generally specified as copper alloys. In general, in cases of dezincification or dealuminization, alloys with over 80 percent copper are used. Monel, aluminum, some stainless steels, and cupronickel are also employed, depending on the application.

Seawater attacks ordinary steel and cast iron fairly rapidly, and protection by painting or a sacrificial anode is used. For example, for ocean-going vessels zinc sacrificial anodes are bolted at intervals to the hull. Pitting is encountered in stainless steel, and brass with less copper may dezincify. Titanium has excellent resistance.

12.22 Chemical corrosion

The petroleum and chemical industries offer the most severe problems. In the petroleum industry salt water, sulfide, organic acids, and other contaminants accelerate corrosion. Stainless steel, Stellite (a cobalt-base alloy), and Monel are used.

In chemical corrosion there are a few specific cases of success:

Corrosive chemical	Resistant material
Nitric acid	Stainless steels
Hot oxidizing solutions	Titanium
Caustic solutions	Nickel alloys
Concentrated sulfuric acid	Steel
Dilute sulfuric acid	Lead
Pure distilled water	Tin

As a typical example of the effects of concentration and temperature, the materials used to resist sulfuric acid are shown in Fig. 12.13. Note that the temperature-concentration graph is divided into 10 zones of increasing severity of attack. As the attack becomes more severe, the number of materials that can be used to give a corrosion rate of less than 0.020 in./yr dwindles rapidly until only gold, glass, and platinum are left. The behavior of 316 stainless steel is interesting in that it can resist higher acid concentrations better than intermediate concentrations. This is related to passivation, whereby a high concentration of oxidizing acid is required to form a *stable* passive layer. Passivity is not achieved at low concentrations, and therefore the corrosion rate is higher.

Corrosion in Gas

12.23 General

At first it may seem out of place to discuss a topic such as oxidation in the same chapter with corrosion. However, as we delve into the actual mechanisms it will be evident that oxidation is closely related to other forms of corrosion, since ions and electrons must be transferred in both processes. We will consider first the types of scales or oxides encountered, then how they form, next the rate of oxidation, and finally the scale-resistant alloys and special cases such as catastrophic oxidation and internal oxidation.

12.24   Types of scales formed

When a metal is exposed to air, for example, it is important to determine first whether the scale occupies a smaller or larger volume than the metal it came from. For instance, if we oxidize the outer layer of a piece of magnesium, the volume of the scale formed will be less than the volume of metal from which it was formed. In the case of iron the volume of the scale will be greater than that of the parent metal, and therefore the scale will be protective. Magnesium oxidizes rapidly because cracks appear in the scale, giving ready access to the metal beneath.

EXAMPLE 12.5 What is the ratio of oxide volume to metal volume for the oxidation of magnesium? (The specific gravity of magnesium is 1.74 and that of MgO is 3.58.)

ANSWER Assume 100 g of Mg is oxidized to MgO:

$$Mg + \tfrac{1}{2}O_2 \rightarrow MgO$$

$$\text{Volume of Mg} = \frac{100 \text{ g}}{\text{density of Mg}} = \frac{100 \text{ g}}{1.74 \text{ g/cm}^3} = 57.5 \text{ cm}^3$$

Letting x = MgO produced, we have

$$\frac{100 \text{ g}}{\text{mol. wt. Mg}} = \frac{x \text{ g}}{\text{mol. wt. MgO}}$$

$$x = \frac{100 \times 40.32 \text{ g}}{24.32} = 167 \text{ g}$$

$$\text{Volume of MgO} = \frac{167 \text{ g}}{\text{density of MgO}} = \frac{167 \text{ g}}{3.58 \text{ g/cm}^3} = 46.6 \text{ cm}^3$$

or   57.5 cm^3 Mg → 46.6 cm^3 MgO

The ratio is then 46.6/57.5 = 0.810.

The ratio of the volume of the scale to the volume of the parent metal is called the "Pilling-Bedworth ratio" and may be calculated from the formula

$$\text{P-B ratio} = \frac{Wd}{Dw}$$

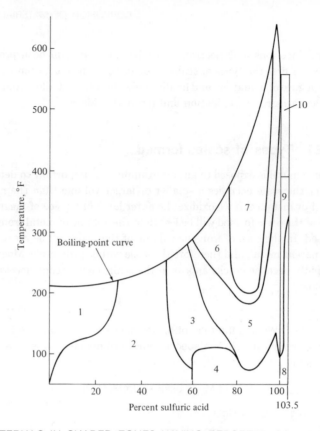

Boiling-point curve

Temperature, °F

Percent sulfuric acid

103.5

MATERIALS IN SHADED ZONES HAVING REPORTED CORROSION RATES LESS THAN 20 MPY

Zone 1	
10 percent aluminum bronze (air-free)	Gold
Glass	Platinum
Lead	Silver
Copper (air-free)	Zirconium
Monel (air-free)	Tungsten
Rubber (up to 170°F)	Molybdenum
Impervious graphite	Type 316 stainless (up to 10% H_2SO_4,
Tantalum	aerated)

Zone 2	
Glass	Tantalum
Silicon iron	Gold
Lead	Platinum
Copper (air-free)	Silver
Monel (air-free)	Zirconium
Rubber (up to 170°F)	Tungsten
10 percent aluminum bronze (air-free)	Molybdenum
Austenitic ductile iron	Type 316 stainless (up to 25% H_2SO_4 at
Impervious graphite	75°F, aerated)

MATERIALS IN SHADED ZONES HAVING REPORTED CORROSION
RATES LESS THAN 20 MPY (*Continued*)

Zone 3	
Glass	Tantalum
Silicon iron	Gold
Lead	Platinum
Monel (air-free)	Zirconium
Impervious graphite	Molybdenum

Zone 4	
Steel	Impervious graphite (up to 96% H_2SO_4)
Glass	Tantalum
Silicon iron	Gold
Lead (up to 96% H_2SO_4)	Platinum
Austenitic ductile iron	Zirconium
Type 316 stainless (above 80% H_2SO_4)	

Zone 5	
Glass	Tantalum
Silicon iron	Gold
Lead (up to 175°F and 96% H_2SO_4)	Platinum
Impervious graphite (up to 175°F and 96% H_2SO_4)	

Zone 6	
Glass	Gold
Silicon iron	Platinum
Tantalum	

Zone 7	
Glass	Gold
Silicon iron	Platinum
Tantalum	

Zone 8	
Glass	Gold
Steel	Platinum
18% Cr, 8% Ni stainless steel	

Zone 9	
Glass	Gold
18% Cr, 8% Ni stainless steel	Platinum

Zone 10	
Glass	Platinum
Gold	

Fig. 12.13 *Corrosion resistance of materials when exposed to sulfuric acid.*

(M. B. Fontana and N. D. Green, "Corrosion Engineering," McGraw-Hill Book Company, New York, 1967.)

where W = molecular weight of the oxide

$\quad d$ = density of the metal

$\quad w$ = atomic weight of the metal

$\quad D$ = density of the oxide

When this ratio is less than 1, the scale is nonprotective because of cracking. On the other hand, if the ratio is much greater than 1, there is danger that the scale will crack off because the volume difference (P-B ratios greater than 2.3) is so great. This ratio, therefore, serves as a rough screening test, but it must be realized that other factors such as coefficient of expansion, melting point, vapor pressure, and high-temperature plasticity are also important characteristics.

The P-B ratio is not just an experimental observation but is based on the concept that ratios less than 1 result in tensile strains in the oxide scale. As pointed out in Chap. 7, oxides and ceramics, in general, do not possess high tensile strengths, and therefore scale cracking can occur. However, their compressive strengths are also not infinite, and large P-B ratios can result in spalling off of the protective scale.

12.25 Mechanism of scale formation

Assuming we have a relatively adherent scale, how does it grow? There are two possibilities: The oxygen or oxygen ion can diffuse through the scale to the metal, or the metal or metal ion can diffuse through the scale to the surface and react with oxygen. This is a rather important distinction because in the first case the scale would grow at the metal-oxide interface, while in the second it would grow at the oxide-air interface. Wagner suspected that because metal ions are generally smaller than oxygen ions the growth occurred at the outer surface, and he performed the experiment illustrated in Fig. 12.14. Instead of oxidation with air the reaction of silver with liquid sulfur in a cell was chosen. Two weighed cakes of silver sulfide were placed on a specimen of silver, and then liquid sulfur was placed over the cakes. At a given temperature two reactions were possible. Ag^+ ions could travel

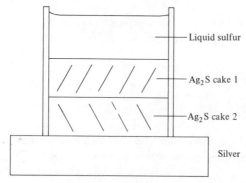

Liquid sulfur

Ag_2S cake 1

Ag_2S cake 2

Silver

Fig. 12.14 *Diffusion experiment to show whether metal ions (Ag^+) or anions (S^{2-}) diffuse more rapidly. The results can be compared to oxide scale formation in metal systems.*

through the sulfide and react at the upper surface with sulfur, or sulfide ions could migrate to the metal-sulfide interface. After the cell was disassembled, it was found that the upper cake had gained much more weight than the lower cell, indicating that the metal ion migration was most important. The usual oxidation mechanism, therefore, is the migration of metal ions and electrons to the outer surface to react with oxygen. Since the mobility of electrons is usually an order of magnitude higher than that of the ions, this is usually not a limiting factor except under very special circumstances.

Oxide Defect Structures. As discussed later in the section on semiconductors, many oxide structures do not follow the exact chemical formula but may have an excess of one ion or the other. This leads to vacancies at certain lattice points in order to attain electroneutrality. The presence of vacancies increases the diffusion rate. However, if an alloy element is added which reduces the vacancies, the oxidation rate is lowered. A rather startling case is that the addition of a small amount of lithium, an easily oxidized element, lowers the oxidation rate of nickel by occupying vacancies in the nickel oxide lattice.

12.26 Oxidation rates

When the scale is adherent, build-up usually follows the *parabolic law.* If the rate of oxidation is proportional to the thickness of the scale (see the problems), then we obtain the relation

$$W^2 = k_p t + C$$

where W = weight of scale
   t = time of exposure
   k_p, C = constants for the particular reaction

If the scale cracks, as at P-B ratios less than 1 or greater than 2.3, the rate is *linear:*

$$W = kt$$

In cases of fairly thick scale containing voids or when a limiting thickness is approached (such as aluminum in air), an empirical *logarithmic* relation is obeyed: $W = k \log (Ct + A)$, where k, C, and A are constants.

12.27 Scale-resistant materials

The most important needs for scale-resistant materials are at temperatures from 800 to 2000°F (427 to 1093°C). Iron, nickel, and cobalt oxidize appreciably above 1000°F (538°C). The addition of chromium, silicon, and aluminum leads to the formation of protective, adherent scale. Although these elements oxidize more readily than iron, they form with iron new structures, such as the spinel types discussed in Chap. 7. The importance of forming the proper scale is especially apparent when attempts are made to use pure molybdenum at elevated temperatures. Despite the excellent creep resistance, the volatility of MoO_3 leads to reduction of section thickness at a rapid rate. Similar problems are encountered with tungsten and niobium.

12.28 Special cases

Certain alloys oxidize very rapidly under special conditions, and this phenomenon is called "catastrophic oxidation." The most insidious type is the effect of small amounts of vanadium pentoxide or lead oxide in the gas phase. These oxides can combine with the normal metal scale to provide a low-melting-point phase in the scale which provides a liquid with rapid oxygen transport to the metal so that attack is accelerated.

Another interesting case is called "internal oxidation." In this case the scale is more permeable to oxygen than to the metal ion. The oxygen dissolves in the metal, migrates, and forms oxide particles in the metal itself. While this causes some hardening, the embrittling effect is usually undesirable in high-temperature alloys.

Corrosion of Ceramics and Plastics

12.29 General

Ceramics and plastics, often in conjunction with metals, offer alternative solutions to corrosion problems. The principal drawback in the ceramics is the lack of ductility, and in the plastics it is the effects of low and elevated temperature and in some cases water absorption and flammability. On the other hand, none of the effects associated with metal corrosion, such as pitting and galvanic action, is encountered. In many cases plastic parts can be used to insulate metal parts from interaction as long as oxygen-concentration cells do not form under the plastic. We will consider the resistance of ceramics and plastics to the same group of corrosive environments that attack the metals: atmosphere, water, and chemicals.

An important general rule in predicting the performance of an organic material in a solvent is that *like dissolves like*. Structures that contain polar groups such as OH, C≡N, COOH will be attacked or swelled by polar solvents such as acetone, alcohol, and water. (Remember that a polar group is one in which the electrons are shifted to one of the elements or are unbalanced around the carbon atom because of the stronger attractive forces.) These polar groups resist solution by the balanced nonpolar solvents such as carbon tetrachloride, gasoline, and benzene.

On the other hand, polymers with nonpolar groups such as methyl (CH_3) and phenyl (C_6H_5) are resistant to the polar solvents, but they are swollen or dissolved by the nonpolar solvents.

Furthermore, the straight-chain (aliphatic) polymers tend to dissolve in straight-chain solvents such as ethyl alcohol, while those with benzene rings (aromatic polymers) tend to dissolve in aromatic solvents such as benzene.

As molecular weight increases, solubility decreases. Also, with greater crystallinity and consequently denser packing and intermolecular forces, solubility decreases.

12.30   Atmosphere

Ceramics are only slowly affected by the atmosphere, as shown by the many structures made of brick and cement that have stood for centuries. The principal dangers in weathering are the effects of water entering cracks or joints and expanding on freezing. Salt in water aggravates the problem.

Plastics are affected slowly by the atmosphere and particularly by sunlight. There are wide differences in this effect on various plastics, as shown in Table 12.6.

12.31   Water

Ceramics are used widely for containers and piping when they are non-porous. Glass-lined and enameled steel tanks have been used for many years with great success.

Plastics are also generally resistant to water and make excellent protective coatings. It should be noted, however, that there is a small percentage of water absorption in all cases except Teflon and a very slight amount for polyethylene and polypropylene. This absorption can aggravate problems such as creep in fibers that are used for clothing; thus we get wrinkles in summer weather.

12.32   Chemicals

There are wide differences in the resistance of ceramics to chemicals. In the glasses pure silica and borosilicate are very resistant, but the soda-lime glasses are slowly attacked by alkalis. Basic refractories such as magnesia are attacked by acids. Organic solvents have no effect on the typical ceramic.

Plastics also show a great deal of variation in resistance to chemicals. Most are resistant to weak acids and alkalis. However, strong acids decompose cellulose acetate; some oxidizing acids decompose melamines and phenol-formaldehyde. Strong alkalis and organic solvents also attack certain plastics. The most resistant materials are Teflon, polyethylene, and vinyl.

Sometimes the corrosive media come from unexpected sources. As an example, we often have considerable difficulty trying to maintain eyeglass frames. A combination of body fluids, exposure to airborne hydrocarbons common in many laboratories, and stress results in failure in plastic eyeglass frames in 12 to 24 months, as shown in Fig. 12.15.

Table 12.6 CORROSION RESISTANCE OF PLASTICS*

Material	Acids		Alkalis		Organic Solvents	Water Absorption, percent/24 hr	Oxygen and Ozone	High Vacuum	Ionizing Radiation	Temperature Resistance, °F	
	Weak	Strong	Weak	Strong						High	Low
Thermoplastics											
Fluorocarbons	Inert	Inert	Inert	Inert	Inert	0.0	Inert		P	550	G, 275
Polymethylmethacrylate	R	A-O	R	A	A	0.2	R	Decomp.	P	180	
Nylon	G	A	R	R	R	1.5	SA		F	300	G, 70
Polyether (chlorinated)	R	A-O	R	R	G	0.01	R			280	G
Polyethylene (low density)	R	A-O	R	R	G	0.15	A	F	F	140	G, 80
Polyethylene (high density)	R	A-O	R	R	G	0.1	A	F	G	160	G, 100
Polypropylene	R	A-O	R	R	R	<0.01	A	F	G	300	P
Polystyrene	R	A-O	R	R	A	0.04	SA	P	G	160	P
Rigid polyvinylchloride	R	R	R	R	A	0.10	R		P	150	P
Vinyls (chloride)	R	R	R	R	A	0.45	R	P	P	160	P
Thermosetting Plastics											
Epoxy (cast)	R	SA	R	R	G	0.1	SA		G	400	L
Phenolics	SA	A	SA	A	SA	0.6			G	400	L
Polyesters	SA	A	A	A	SA	0.2	A		G	350	L
Silicones	SA	SA	SA	SA	A	0.15	R		F	550	L
Ureas	A	A	A	SA	R	0.6	A		P	170	L

* Abbreviations: R = resistant, A = attacked, SA = slight attack, A-O = attacked by oxidizing acids, G = good, F = fair, P = poor, L = little change, decomp. = decomposes.
Source: M. B. Fontana and N. D. Green, "Corrosion Engineering," McGraw-Hill Book Company, New York, 1967.

Fig. 12.15 *Plastic eyeglass frames that failed because of applied stress and corrosive media. Although thermoplastics generally have good corrosion resistance, the combination of stress and body fluids caused this failure at normal temperature.*

12.33 Summary of corrosion prevention

In summary, the methods commonly available to inhibit corrosion are:

1. Materials selection, also used to prevent galvanic cells
2. Alteration of environment—temperature, velocity, ion concentration—and the use of inhibitors
3. Design changes, such as removing pockets that may hold corrosive fluids
4. Cathodic protection, which can be applied to almost any metal
5. Anodic protection, which can only be used for materials that passivate
6. Application of coatings—metallic, ceramic, or organic—including metallic coatings as sacrificial anodes

Finally, as was pointed out at the beginning of the chapter, corrosion is seldom totally prevented. Economics plays a most important role, and an optimization of both costs and required life span is the normal ultimate solution.

SUMMARY

Corrosion takes place because of the solution or other reaction of the surface of a component with its environment. In the case of metals each instance of corrosion can be studied as an electrochemical cell in which the metal goes into solution, i.e. corrodes at the anode. For the process to continue, the electrons left in the solid at the point of solution must migrate through the structure and be consumed at the cathode. By far the most common reaction, that seen in the rusting of iron, is the reaction with water and dissolved oxygen to produce OH^- ions: $2H_2O + O_2 + 4e \rightarrow 4OH^-$.

A starting point in estimating the tendency of a metal or alloy to corrode is its half-cell potential, the voltage it develops when connected to a standard electrode. However, while this shows that elements such as aluminum and magnesium have a high tendency to corrode, it does not tell us anything about the nature of the corrosion product, which is important. If a passive or un-reactive film forms, corrosion will stop. A great many specialized terms, defined below, have developed, and practically all involve applications of simple elec-trochemical cell theory.

Oxidation of metals in gases is a specialized case of corrosion and is governed by a combination of the tendency of a metal to react and whether a protective scale is formed.

Finally, corrosion in both ceramic and polymeric materials is possible, although usually it is not as prevalent as it is in metals.

DEFINITIONS

Corrosion　The solution or harmful reaction of the structure of a material with its environment.

Anode　The electrode of a cell at which metal goes into solution; the negative pole in the external circuit.

Cathode　The electrode of a cell at which metal ions are plated out or negative ions are created, as in the reactions $2H_2O + O_2 + 4e \rightarrow 4OH^-$ and $Cu^{2+} + 2e \rightarrow Cu^0$.

Cell potential　The electromotive force (emf) developed by a cell.

Hydrogen half-cell　A reference half-cell involving hydrogen adsorbed on a platinum wire in a standard solution of hydrogen ions.

Half-cell potential of an element　The emf developed when an element is coupled with a hydrogen half-cell.

Concentration polarization　A decrease in the emf of a cell due to build-up of ions around the electrodes.

Activation polarization　A decrease in cell potential due to intermediate steps necessary to complete the reaction.

Inhibitor　A chemical added to produce a film or coating which slows down the corrosion reaction.

Passivating　The formation of a film of reaction product that inhibits further reaction.

Galvanic corrosion　Corrosion that is accelerated by the presence of electri-cally connected dissimilar metals.

Sacrificial anode   An active metal which, when attached to the object to be protected, makes the object cathodic and therefore causes solution to take place only on the active metal.

Cathodic protection   Plating or attachment of a more active metal which dissolves, thereby preventing solution of the component to be protected.

Weld decay   Corrosion at or adjacent to a weld as a result of galvanic action due to differences in structure produced by the welding.

Intergranular corrosion   Preferential corrosion at grain boundaries due to the formation of an anode because of precipitation of a second phase at the grain boundaries.

Selective leaching   Preferential solution of one element in an alloy. In brass the zinc dissolves and a spongy copper deposit remains (dezincification).

Graphitic corrosion   Selective corrosion of gray cast iron in which only a structure of graphite and some oxides remain.

Hydrogen blistering   The development of blisters due to the diffusion of hydrogen into voids.

Oxygen-concentration cell   A galvanic cell caused by differences in oxygen concentrations. An example is the water line in a partially filled tank.

Pit corrosion   Corrosion due to differences in oxygen and ion concentrations at the base of a pit compared to the surface.

Season cracking   The acceleration of corrosion due to high residual stresses. *Caustic embrittlement* is a similar effect in an alkaline environment.

Corrosion fatigue   Fatigue failure aggravated by corrosion.

Erosion   The wearing away of a surface because of combined mechanical-corrosion effects.

Scale   A reaction product formed on a surface that is attacked by gas or liquid.

Pilling-Bedworth ratio   The ratio of volume of scale to volume of parent metal. Values much different from 1 may lead to high scaling rates.

PROBLEMS

Let us be perfectly candid about the problems in this section. We could put together a group of electrochemical calculations that would be a great exercise in slide-rule manipulation. But this might give the erroneous impression that we can go out and calculate such things as the corrosion rate of an automobile fender in the spring mush of Michigan roads. There has been some excellent work in the application of thermodynamics to corrosion by M. Pourbaix, but this cannot yet be applied directly to the average complex situation. Also, quantitative calculations of potentials necessary for protection of pipes have been made.

Therefore, in this problem section we will illustrate the application of the general principles set forth in the sections on chemical principles of metallic corrosion, corrosion phenomena, and gas corrosion.

12.1 Write the ion-electron equations for the anode and cathode reactions (half-cell reactions) in the following cases. (If you believe corrosion will not occur, write "no reaction.")

    a. An opened (punctured) tin can at the bottom of a fresh-water lake
    b. An opened (punctured) tin can in the ocean. Why would the rate be more rapid here than in part *a*?
    c. A stainless steel (302) piece of trim on an automobile that is exposed to intermittent splashing from salt solutions
    d. An automobile fender (1010 steel) covered on the inside with a coat of asphalt with a few holes through the asphalt
    e. A copper heating coil brazed to a steel pipe with a 70% Cu, 30% Zn braze. The junction is exposed to dripping from a hot-water tank.
    f. The effect of attaching a "copper ground" from a "live" electric stove to a steel water pipe which sweats in the summer
    g. A punctured Alclad aircraft wing (pure aluminum covering alloy 2014) exposed to salt water

12.2 Compare the merits of using tin and zinc

    a. As a protective coating for cans for food
    b. For outdoor fencing

12.3 Large amounts of copper are obtained from copper mine water by immersing iron scrap in the solution and later collecting fine copper powder (cement copper). Write the ion-electron equations and discuss the economics of this technique compared to other possible methods such as electrolysis, evaporation, and reduction of the salt.

12.4 A student argues that dead spots in the circulation of a tank should not be harmful because corrosion products could build up and protect the region from corrosion. Discuss.

12.5 With most alloys the reduction of oxygen concentration in the solution retards corrosion. For what group of alloys is this not true?

12.6 Why are zinc sacrificial anodes preferred to magnesium sacrificial anodes for protection of steel ship hulls? A number of years ago it was found that the zinc had to be quite pure to be effective and, in particular, had to be free from second phases. Why?

12.7 Describe two ways in which buried ductile-iron pipe lines can be protected from corrosion.

12.8 One device for protecting boat hulls uses a platinum electrode protruding from the steel hull. How does this work? To which electrode of a battery should it be connected, and to which should the steel be connected?

12.9 A nail is immersed completely in oxygenated water. At which locations would corrosion occur? Write the anode and cathode reactions.

12.10 Antique brass pots and kettles may not be great "finds," since they often leak because of a corrosion phenomenon. What would the corrosion be called if it had the following appearances?

a. Fine cracks that appear to follow grain boundaries
b. Copper rather than the normal yellow brass color

12.11 Assume that concentration polarization is present in a certain cell. Modify Fig. 12.7 to show (schematically) the effect of

a. Increased temperature
b. Increased electrolyte concentration

12.12 How would Fig. 12.7 appear if current density rather than current was used?

12.13 Under what conditions might you expect concrete to corrode? (Recall the type of bonding in concrete.)

12.14 Medicine has advanced to the point where more and more synthetic components have been considered as replacements for natural bone and tissue. What are some of the requirements of the material in order to resist corrosion?

12.15 The adherence of Al_2O_3 to aluminum involves a close match of (*a*) the aluminum-to-aluminum distances in the (111) planes of the metal with (*b*) the close-packed planes of Al_2O_3 (Fig. 7.11). What is the percentage difference in spacing?

12.16 Magnesium parts are heat-treated in an atmosphere containing 1 percent SO_2, forming a sulfate coating. Why is this preferable to air? (The density of $MgSO_4$ is 2.66 g/cm³ and of MgO is 3.58 g/cm³.)

12.17 Why does the addition of chromium, a rapidly oxidized element, improve the oxidation resistance of austenitic nickel steels?

12.18 Derive the equation for the parabolic oxidation rate. (Let the rate of oxidation dW/dt be proportional to $1/x$, where x is the thickness of the scale.)

12.19 Show by calculation whether the nitrification of titanium might be expected to be linear or parabolic (related to oxidation). The reaction is

$$2Ti + N_2 \longrightarrow 2TiN$$

The specific gravity of titanium is 4.50 and that of TiN is 5.43.

12.20 There is a current interest in the development of biodegradable plastics so that "plastic trash" will decay if thrown along the roadside. What would be some of the important characteristics of these corrodible plastics?

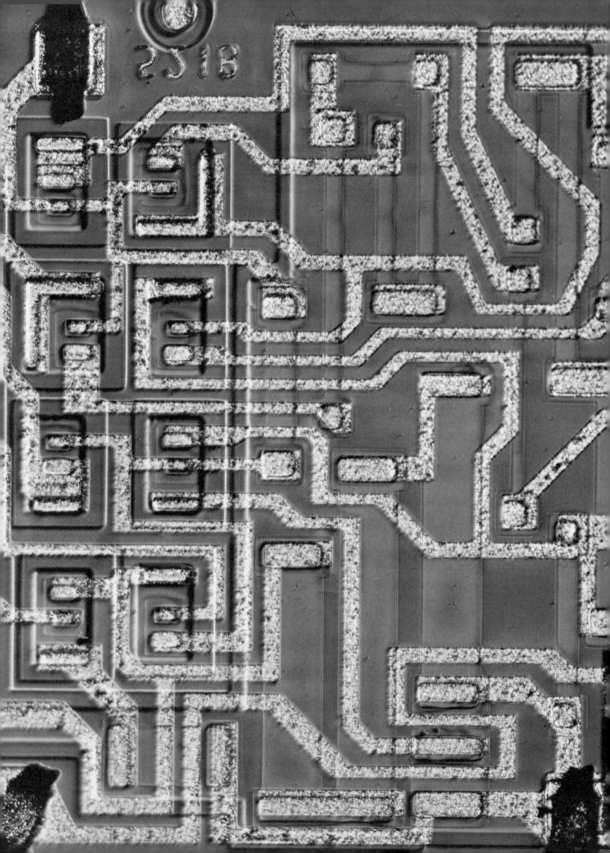

13

ELECTRICAL PROPERTIES OF MATERIALS

THIS illustration shows a linear integrated circuit used in a power amplifier. The grayish background is a "chip" of silicon 0.1 mm across alloyed with a small amount of material to make it an n-type material, as described in the text. By a combination of diffusion, etching, insulation with SiO_2, and further diffusion, a system of transistors and connections between transistors is built up into a complete circuit. There are eight transistors in a group in the left-hand side of the photograph.

In this chapter we will consider conduction first in the high-conductivity materials such as metals, then in the insulators and semiconductors. We will then take up other important materials such as dielectrics and piezoelectrics and their uses.

13.1 General

We will now take up some new and exciting fields for materials—their uses in devices for transmitting electrons, heat, and light. These are the rapidly growing fields of the semiconductor, magnetic tape, computer memory units and switches, television tube, and laser. Even in these phenomenal developments we shall see that the properties depend on the structure, and therefore a knowledge of the structure helps us to apply the materials to maximum advantage.

In this chapter we will discuss the theory of electrical conduction in metals, semiconductors, and insulators. The applications are to be found in metallic conductors, superconductors, resistors, and semiconductor devices. Then we will take up other electrical effects—dielectrics, thermionic emission, and piezoelectricity. Here the applications include capacitors, photocells, thermocouples, and crystal oscillators.

In Chap. 14 we will study magnetic properties, and then in Chap. 15 optical and thermal properties.

Electrical Conductivity

13.2 Conduction and carriers

Let us consider conductivity first from the large-scale engineering point of view and then see how the same relations can be developed from a knowledge of elementary particles. If we have a wire and apply a potential E, the current I that flows will depend on the circuit resistance R, as given by the well-known Ohm's law: $I = E/R$. Now the resistance depends on the nature of the wire itself: a copper wire has a lower resistance than an iron wire of the same size (length and cross-section). We use the term *resistivity* (ρ) to characterize the inherent ability of the wire to affect current flow, and multiply this by l/A to give the resistance:

$$R = \rho \frac{l}{A} \quad \text{or} \quad \rho = R\frac{A}{l} \qquad \rho = \text{ohm} \frac{\text{cm}^2}{\text{cm}} = \text{ohm-cm}$$

where R = resistance
ρ = resistivity
l = length of wire
A = cross-sectional area of wire

EXAMPLE 13.1 A student wants to build a dc heating coil rated at 110 volts and 660 watts. She has some Chromel wire 0.1 in. (0.254 cm) in diameter with a resistivity of 107.9 micro-ohms-cm. What length should she use?

ANSWER

Power $= EI$

$$I = \frac{660}{110} = 6 \text{ amp}$$

$$R = \frac{I}{E} = \frac{110}{6} = 18.3 \text{ ohms}$$

$$R = \rho \frac{l}{A} \quad \text{or} \quad l = \frac{RA}{\rho}$$

$$l = \frac{18.3 \text{ ohms} \times (0.1 \text{ in.} \times 2.54 \text{ cm/in.})^2 \times (\pi/4)}{107.9 \times 10^{-6} \text{ ohm-cm}}$$

$$l = 8,600 \text{ cm} = 86 \text{ m}$$

It is more positive and simpler in the text ahead to think of the material as conducting rather than resisting the passage of current, so we use the well-known parameter *conductivity* instead of resistivity. This is simply the reciprocal of resistivity:

$$\sigma = \frac{1}{\rho} = (\text{ohm-cm})^{-1} = \frac{\text{mho}}{\text{cm}}$$

Now let us examine the structural factors that go into conductivity. If we have a cube of material, 1 cm on the edge, the conductivity between opposite faces will depend directly on three factors:

1. The number of charge carriers, n (carriers/cm³)
2. The charge per carrier, q (coulombs/carrier)
3. The mobility of each carrier, μ (cm/sec)/(volt/cm)

The conductivity will depend on the product of all three of these factors:

$$\sigma = nq\mu$$

or, checking the units,

$$\sigma = \frac{\text{carriers}}{\text{cm}^3} \times \frac{\text{coul}}{\text{carrier}} \times \frac{\text{cm}}{\text{sec}} \times \frac{\text{cm}}{\text{volt}}$$

Since coul = amp-sec and volts = amp-ohm,

$$\sigma = \frac{1}{\text{ohm-cm}} = (\text{ohm-cm})^{-1} = \frac{\text{mhos}}{\text{cm}}$$

as in the large-scale example.

This is important to realize because we will be studying how the large-scale electrical properties are due to the movement of the elementary carriers.

13.3 Types of carriers

We find that there are four different types of carriers which give the phenomenon of "current flow."

1. *The electron* (charge $= 1.6 \times 10^{-19}$ coul). We recall that an ampere is a coulomb per second. Therefore, a movement of 6.25×10^{18} electrons is the motion of a coulomb of charge, and if it occurs across our cell each second, we have an ampere of current flowing.

2. *The electron hole* (charge $= 1.6 \times 10^{-19}$ coul). In Chap. 7, we discussed briefly the concept of an electron hole in the $(Fe^{2+}, Fe^{3+})O^{2-}$ lattice. Associated with each Fe^{3+} ion there is an electron hole, a place where a traveling electron is attracted because the electrical field of O^{2-} ions around the Fe^{3+} is adjusted for an Fe^{2+}. We can visualize the motion of three electrons, each moving one unit distance to the left (Fig. 13.1). As a result of the movement of the first electron, the Fe^{2+} marked "A" becomes Fe^{3+} in Step 1, then the Fe^{2+} marked "B," and so forth. Since we have electron motion, we have conductivity. However, rather than keep track of small movements of many electrons, it is simpler to focus on the movement of the hole. Instead of saying that three electrons (negative charges) moved three unit distances to the left, we say that one positive charge moved the same total distance to the right (A to D). Thus we classify an electron hole as a charge carrier with the same magnitude of charge as an electron but opposite in sign.

3 and 4. *Positive and negative ions* (charge $= n \times 1.6 \times 10^{-19}$ coul, where $n =$ valence). Since the current is the net charge transferred per second, if a positive ion such as Ca^{2+} moves from left to right, the electrical effect will be the same as if two electrons moved from right to left. In this way

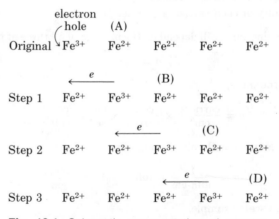

Fig. 13.1 *Schematic representation of counter-current electron and vacancy flow.*

ion movement can contribute to conductivity. Of course, the movement of an O^{2-} from right to left would be equivalent (in terms of charge) to the movement of two electrons to the left. In a perfect crystal the movement of ions would be difficult, but the presence of vacancies in real crystals makes ion movement possible. It is important to distinguish between electron hole movement that takes place by *electron* jumps from ion to ion and ion movement that takes place by *ion* jumps from lattice position to position. Naturally, we expect to find the mobility of ions much less than that of electrons because of their larger size.

13.4 Conductivity in metals, semiconductors, and insulators

Let us now discuss why there are vast differences in conductivity in different materials (Table 13.1).

In the metals we already have a partial picture of the metallic bond in which the valence electrons are contributed by the atom as it becomes an ion in the unit cell. We therefore expect to find a high mobility of these charge carriers. We have to refine this picture a little to explain why a metal with two electrons such as magnesium (2.5×10^5 mho/cm) does not have better conductivity than copper and silver with one valence electron and why silicon with four electrons has low conductivity (4×10^{-6} mho/cm).

To explain this and other important concepts we have to use the "band model," which runs contrary to classical thinking. Let us study the model and its explanation of conductivity.

First let us review the structure of the element sodium and the energy levels of its electrons (Fig. 13.2*a*). The electrons closer to the nucleus in the inner shells, such as $1s$, $2s$, $2p$, need not concern us here. It is the $3s$ electron which is furthest from the nucleus and therefore at a higher energy level that

Table 13.1 ELECTRICAL CONDUCTIVITY OF
SELECTED ENGINEERING MATERIALS

Material	Conductivity, mho/cm
Silver (commercial purity)	6.30×10^5
Copper (high conductivity)	5.85×10^5
Aluminum (commercial purity)	3.50×10^5
Ingot iron (commercial purity)	1.07×10^5
Stainless steel (301)	0.14×10^5
Graphite	1×10^3
Window glass	2×10^{-7}
Lucite	10^{-14} to 10^{-16}
Borosilicate glass	10^{-12} to 10^{-17}
Mica	10^{-13} to 10^{-17}
Polyethylene	10^{-17} to 10^{-19}

will leave the atom to form the electron gas. It is important to examine this electron gas more closely at this time. We recall that when we discussed the structure of an atom, we gave the Pauli exclusion principle, which holds that only two electrons in an atom can have the same energy level (and these will have opposite spins). When we put a group of atoms in a metal block, the same exclusion tendency exists. We find that the electrons that form the electron gas in sodium, for example, must have energies slightly different from one another. Therefore, instead of having the sharp energy level of the single $3s$ electron in an isolated sodium atom, a band of energy levels exists; hence the name "band theory" (Fig. 13.2b). It is postulated further that the number of energy levels in a band is equal to the number of electrons that can occupy the energy level times the number of atoms present in the block. This leads to the model of the conduction band of sodium having the number of "states" in the $3s$ band equal to twice the number of $3s$ electrons.

The energy levels or states at the top of the band are higher in energy. Also, since there is only one $3s$ electron per atom of sodium and the number of states equals twice the number of atoms, we will have half of the states unoccupied. Since the higher states have higher energy, we would expect that at absolute zero only the lower half of the states would be occupied, and this is indeed the case (Fig. 13.3a). As we warm up the metal, some electrons will

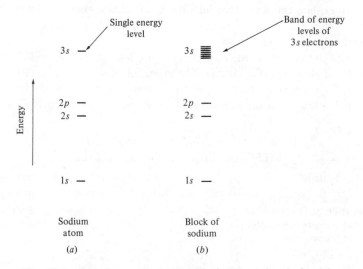

Fig. 13.2 *Development of a band of energy levels for (a) a single sodium atom, and (b) a block of sodium. The inner electrons (1s, 2p) are sufficiently shielded so that they form no bands.*

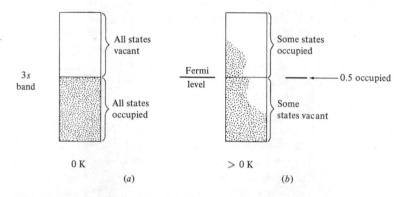

Fig. 13.3 *Enlarged (schematic) view of the 3s band of sodium at (a) 0 K and (b) higher temperatures.*

have higher energy and will leave lower states unoccupied (Fig. 13.3*b*). It is useful for later discussion to define the Fermi level (E_F), where the probability of occupancy by a conduction electron of a state is 0.5. In other words, at the Fermi level half the states are occupied, and there are as many occupied states above this level as there are unoccupied states below it.

It should be noted that the number of available states at different energy levels is a complex function. This is important later in analyzing the action of a thermocouple.

Returning now to sodium, we explain conductivity by saying that the electrons in the 3s band can move readily to empty states in the band. There-fore, when an electrical potential is applied, the electron is accelerated and acts as a charge carrier.

The important point is that in a monovalent metal we have an energy band that is half-filled, and the electrons in the upper part of the band can be energized and accelerated easily because there are open higher-level states nearby.

How do we explain the conductivity of magnesium with two valence electrons? Using the Pauli exclusion principle, we could calculate that all the spots in the 3s band, for example, would be filled because we have two valence electrons per atom; and since there are no empty adjacent spots, the element would be an insulator.

The saving factor is that the energy levels of the 3p band overlap the 3s band, and we have a continuous series of possible states (Fig. 13.4). Electrons can be accelerated into the upper levels of the combined band and serve as conductors. However, magnesium is not as good a conductor as sodium because of the complexity of electron motion accompanying the overlap of the 3s and 3p bands.

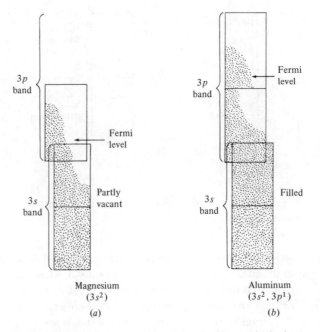

3p band

Fermi level

3s band

Partly vacant

Magnesium
$(3s^2)$

(a)

3p band

Fermi level

3s band

Filled

Aluminum
$(3s^2, 3p^1)$

(b)

Fig. 13.4 *The 3s and 3p bands in divalent and trivalent metals.*

13.5 Applications

Now let us review the conductivity of some common alloys (Fig. 13.5). We see that the monovalent elements silver, copper, and gold are best. Pure aluminum also has good conductivity because there are many unfilled p levels above its higher-energy electrons (i.e. the p level is only half-filled for a trivalent metal). Iron and the other transition metals have lower conductivity because of the complex energy levels in the region where s and p or d bands overlap.

The wide range of conductivity in any family of alloys deserves careful analysis. The conductivity of all elements decreases as a second element is added in solid solution. Typical data for copper are shown in Fig. 13.6. The more dissimilar the element from copper, the greater the change in resistivity. An example is the effect of phosphorus vs. the effect of silver. When two phases are present, the conductivity is determined by the volume fraction of each.

Because of the severe increase in resistivity accompanying alloying, it is often useful to improve the mechanical properties of a part by using cold-worked pure metal rather than by adding an alloy. Cold working increases the resistivity only a few percent because there are large blocks of pure undistorted metal available. On the other hand, the addition of even a small percentage of another element in solid solution produces irregularities in the lattice every

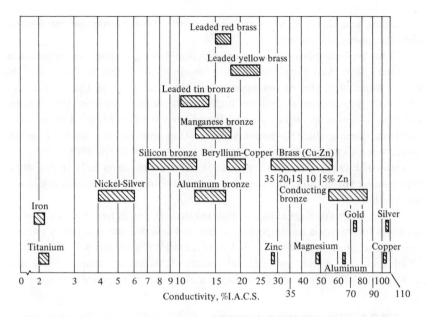

Fig. 13.5 *Conductivity of common metals and alloys in terms of percentage of the international copper standard (which is not of optimum purity).*

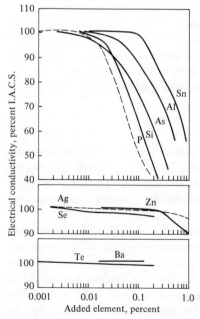

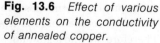

Fig. 13.6 *Effect of various elements on the conductivity of annealed copper.*

few atoms, and therefore has a pronounced effect in impeding electron motion, thereby raising the resistivity.

As an example of conductor selection let us review the competition between copper and aluminum. The conductivity of copper is higher, but the *conductance* of a wire of the same weight per foot is lower than aluminum. Therefore, aluminum is used for high-tension wires. However, the distance between towers is an important cost factor, so a steel core is used in the cable for strengthening. Once the power is close to the consumer, it is more convenient to use less bulky, more fatigue-resistant copper wires with the added advantage of easier soldering. For high frequencies the current flows only in the outer regions near the surface, and therefore copper tubing finds extensive use in this application.

Resistors are of great importance in circuitry and in heating. In industrial furnaces resistivity is not important compared to oxidation resistance. For this reason the heat-resistant nickel-chromium alloys such as 80% Ni, 20% Cr (Nichrome) are widely used. In laboratory furnaces, for temperatures over 1800°F (982°C) silicon carbide resistors and even platinum are employed.

13.6   Conducting glasses

Although glass is normally an insulator, there are a number of applications in which slight conductivity is useful. In high-voltage glass devices such as x-ray tubes it is desirable to avoid a gradual build-up of charge. In this case a glass containing lead oxide is heated in hydrogen, producing a thin layer of metallic lead. This is then grounded. In another case a transparent layer of tin oxide, SnO_2, is deposited on a glass surface. When a voltage is applied, a small current can flow through the high resistance, and the heating prevents fogging. Bulk conductivity can be developed in glass by adding substances such as iron oxide. This leads to conduction by electron holes, as described in Chap. 7.

13.7   Superconductivity

Many years ago Kamerlingh Onnes found that when mercury was cooled below a critical temperature, the electrical resistance fell to zero. In one demonstration he showed that a current flowed indefinitely in a ring of mercury, and the phenomenon was called "superconductivity." The effect has since been discovered in a number of other elements and even in alloys.

To point out some of the features of this effect, we will review first normal conductivity and then some of the special characteristics of superconductors.

From elementary physics we learn that as the temperature decreases, the vibration frequency of the positive ions of the lattice decreases. The resistance decreases because there is less conflict with electron motion (less scattering). Therefore, we find a gradually decreasing resistance with decreasing temperature in normal metals as we approach absolute zero. However, to explain the *sudden* appearance of superconductivity at a *critical* temperature above zero, Bardeen used the concept of pair formation. Below the superconducting critical temperature electrons of opposite spin and the same energy form pairs. In this condition the electrons are exempt from scattering and therefore exhibit perfect conductivity.

The phenomenon is now the basis for a number of developments; for example, high-field solenoid magnets operated at low temperatures are in use in many laboratories.

13.8 Semiconductors; general

We come now to a fascinating group of materials which, although they are lower in electrical conductivity than the metals, are essential in a number of newly developed devices, from Dick Tracy wrist-size radios to tiny remote communication units in distant space satellites. The transistor, the solar battery, and the integrated miniature circuit are all dependent on the properties of semiconductors.

The heart of all these appliances is not the complex circuitry but the fact that materials have been developed that can marshal and direct electron motion in a tiny space with a precision never before attained. The key to these devices is a very small crystal of a semiconductor such as silicon, with controlled amounts of impurities in solid solution.

To understand the operation and vast potential of these materials, we will follow this course. First we will observe the differences in the band structures of these elements compared to the metals. Then we will see the effect of adding different types of impurities (called "dopants"), leading to *n*- and *p*-type semiconductors. Finally, we will take up the use of these materials in a few applications.

13.9 Semiconductors and insulators

In semiconductors we find that we have a filled band (Fig. 13.7*b*) and that the next energy level into which we can accelerate electrons is *separated* by a small energy gap E_g in contrast to the metals, where open states exist. For example, in the series carbon, silicon, germanium, tin (four valence electrons) the next bands are separated by the following values:

Material	Energy gap, E_g (at 20°C), eV
Diamond	5.30 (insulator)
Silicon	1.06 (semiconductor)
Germanium	0.67 (semiconductor)
Tin	0.08 (conductor)

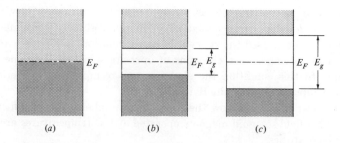

Fig. 13.7 *Energy band models of (a) a metal, (b) a semi-conductor, and (c) an insulator. E_F is the Fermi level (0.5 occupied).*

We express the width of the energy gap in electronvolts. The electronvolt is the energy required to move an electron through a field of 1 volt. To give a feel for the magnitude of this quantity, the thermal energy of an electron at room temperature is only about 0.03 eV on the average, so we would not expect many carriers to be produced from this source except in tin. Diamond, therefore, is a good insulator, while the small number of conduction band electrons in silicon and germanium leads us to call them semiconductors.

At this point we have a difficult question to answer about the existence of an energy gap, since there are four electrons in the outer shell in each of these cases. Why should the gap exist if we have filled only two of the six $2p$ states in carbon, for example? The explanation is rather complex and depends principally on the fact that while eight electrons can be accommodated in the two $2s$ plus six $2p$ states, an effect called "hybridization" takes place in a diamond. Instead of the four valence electrons falling into two $2s$ and two $2p$ levels, a *hybrid* group of four electrons is formed, giving rise to the four equal tetrahedral covalent bonds. This acts like a filled band, and there is an energy gap between this band and the conduction band above it.

Now we have the problem of why in silicon, for example, we encounter any conductivity *at all* if the band is filled and a gap exists to the next higher band. The answer lies in the fact that there is a variation in the energies of the outer shell electrons, and as we raise the temperature above 0 K, there will be an increasing, though small, number of electrons with enough energy to reach the upper band (Fig. 13.8a). Furthermore, the shifting of these electrons to the upper band creates other carriers by forming electron holes in the lower band (Fig. 13.8b).

The band above the valence band is called by convention the "conduction band," although conduction also occurs in the valence band because of the electron holes.

In addition to the group IV elements, semiconductors have been produced by combining elements with three outer shell electrons with elements with five

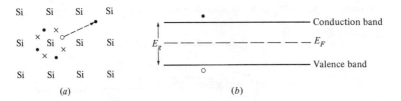

Fig. 13.8 (a) *Formation of a conduction electron and an electron hole in pure silicon (intrinsic semiconductor). Through the acquisition of thermal energy, an electron attains a high enough energy to leave its valence position. (b) The process shown in (a) has led to the presence of an electron in the conduction band and a hole in the valence band. The Fermi level is halfway between the bands (50 percent occupation).*

outer shell electrons to give an average of four, as for example in AlP and InAs. These are called III-V compounds. The band structure is similar to that of group IV, and a range of energy gaps is obtained for different combinations.

13.10 Extrinsic vs. intrinsic semiconductors

Up to this point we have discussed only semiconductors with an average of four electrons per atom or *intrinsic* semiconductors. If this balance is varied by "doping" with impurities, we obtain *extrinsic* semiconductors of two types, *n* and *p*.

To understand these important materials, let us consider the effect of adding the different impurities or dopants to an intrinsic semiconductor such as silicon. If we add a small quantity of phosphorus, these atoms will form a substitutional solid solution with silicon, giving the structure shown in Fig. 13.9.

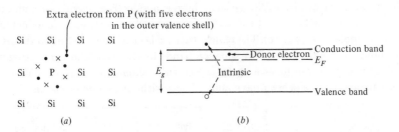

Fig. 13.9 (a) *Structure of silicon with a phosphorus atom as dopant, giving rise to an extra electron, thus making an n-type semiconductor (extrinsic). (b) The electron from the phosphorus shown in (a) is found just below the conduction band. The Fermi level E_F is higher than in Fig. 13.8 because of the addition of an electron above the old Fermi level and no increase in the electron holes below (schematically shown).*

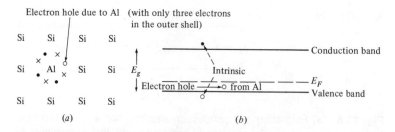

Fig. 13.10 (a) *Structure of silicon with an aluminum atom added as a dopant, giving rise to an electron hole, thus making a p-type semiconductor (extrinsic). (b) The electron hole produced by the aluminum shown in (a) is found just above the valence band. The Fermi level E_F is lower than in Fig. 13.8 because of the addition of a hole below the old Fermi level and no increase in the number of electrons above (schematically shown).*

We have shown the valence electrons contributed by the four silicon atoms adjacent to a phosphorus atom by the symbols ×. However, phosphorus has five outer shell electrons, shown by •, so we have an extra electron, as shown in the figure. The phosphorus, therefore, is an electron donor, and the structure is called an "n-type extrinsic semiconductor" (n refers to the *negative* extra electron). The electrons from the phosphorus are not in exactly the same positions as those that would be produced by an electron in the valence band, but they lie just below it in energy. This is because the electrical field of the phosphorus is not identical with that of a silicon atom. The Fermi level E_F is close to the conduction band. We have added electrons to the structure without adding holes, so E_F shifts upward.

Now let us consider the effect of adding an element with less than enough electrons to satisfy the covalent bonds of silicon. Aluminum is such an atom, as shown in Fig. 13.10. Again the silicon electrons are shown by × and the aluminum electrons by •. We see that there is an electron missing. Aluminum is an acceptor element, and the structure is called a "p-type semiconductor" (electron holes are *positive*). The band structure is also shown in Fig. 13.10, with the holes just adjacent to the valence band. In this case the conduction is mainly from electron holes and the 50 percent occupied level (E_F) is closer to the valence band, since we have added holes without adding electrons.

EXAMPLE 13.2 Calculate the conductivity of the *intrinsic* semiconductor germanium from the following characteristics (300 K):

Hole density = electron density = 2.4×10^{13} carriers/cm³
Electron mobility = 3,900 cm²/volt-sec
Hole mobility = 1,900 cm²/volt-sec
Charge/electron = charge/hole = 1.6×10^{-19} coul/carrier

ANSWER

$$\sigma = nq\mu \quad \text{(from Sec. 13.2)}$$
$$= nq_n\mu_n + pq_p\mu_p$$

where n = number of electrons/cm^3

$\quad q_n$ = charge/electron

$\quad \mu_n$ = mobility of electron carrier

$\quad p$ = number of holes/cm^3

$\quad q_p$ = charge/hole

$\quad \mu_p$ = mobility of hole carrier

Since in an intrinsic semiconductor

$$n = p \quad \text{and} \quad q_n = q_p$$

we have

$$\sigma = nq(\mu_n + \mu_p)$$
$$= (2.4 \times 10^{13} \text{ carriers/cm}^3) \times (1.6 \times 10^{-19} \text{ coul/carrier})$$
$$\times [(3,900 + 1,900) \text{ cm}^2/\text{volt-sec}]$$
$$= 0.0222 \frac{\text{coul}}{\text{volt-sec-cm}} = 0.0222 \frac{\text{amp}}{\text{volt-cm}}$$
$$= 0.0222 \frac{1}{\text{ohm-cm}} = 0.0222 \frac{\text{mho}}{\text{cm}}$$

13.11 *P-n* junctions, rectification

One of the important uses of semiconductors is in rectifying alternating to direct current. For this a junction is made between n- and p-type material. This can be done by changing the impurity (dopant) from a p type such as aluminum to an n type such as phosphorus during crystallization. Let us examine the relative amounts of carriers (holes and electrons).

Assume two electrically neutral blocks of n- and p-type material are brought into contact (Fig. 13.11a). Some movement of holes and electrons will occur. The important point is the large current flow if the n side is made negative and the p side positive (Fig. 13.11b). Electrons will be attracted across the junction to the p material, and holes will travel in an opposite direction, giving substantial current. (Remember a hole moving in a direction opposite to an electron is equivalent to a second electron moving in the direction of the first electron.) However, if the electric field is reversed, the holes and electrons will be attracted away from the junction, and since no carriers move across the junction, there is no current (Fig. 13.11c). The current, therefore, flows in pulses in only one direction.

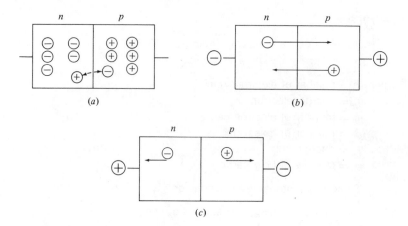

Fig. 13.11 *Simplified sketch of the action of a p-n junction used as a rectifier for alternating current. (a) No applied voltage; some local movement of electrons and holes to attain neutrality. (b) Voltage applied as shown. Considerable current flows across the n-p junction. Note that the effect of the motion of a — to the left and a + to the right equals 2— to the left. (c) Voltage applied in the opposite direction; no current flows across the interface.*

Here we may ask how is it possible to distinguish between holes and electrons since both contribute to conductivity. This is done by a very important measurement in the field of semiconductor development which depends on the *Hall effect.*

If we apply a voltage at the ends of a semiconductor crystal, a current will flow because of electron or hole movement or both. Now if we impose a magnetic field *across* the conductor, the electrons and holes will be affected differently. (Recall that an electrical charge moving in a magnetic field is diverted by the field.) Electrons are turned to one side of the bar and holes are turned to the other. Therefore, a charge will build up across the bar and will be predominantly in one direction if it is due to holes and in the other if it is due to electrons.

13.12 Temperature effects on electrical conductivity

One of the salient features of a metal is the decrease in conductivity with increasing temperature. The reasons for this were discussed briefly in the section on superconductivity. In contrast, the conductivity of semiconductors and insulators increases with temperature. In the case of heating an intrinsic semiconductor more electrons are pumped up to energies where they can enter

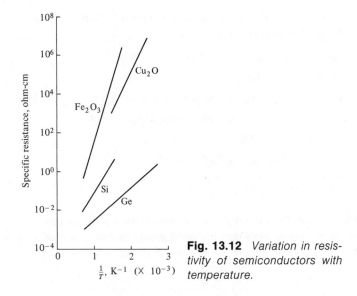

Fig. 13.12 *Variation in resistivity of semiconductors with temperature.*

the conduction band, leaving electron holes. This effect is disadvantageous in a circuit that depends on a *p-n* junction because the number of natural carriers can obscure the effects of the carriers added by the dopant.

On the other hand, because of the sensitivity of the semiconductor to temperature, the change in resistance can be used to indicate temperature accurately from 1 to 723 K (-273 to 450°C). These devices are called "thermistors" (Fig. 13.12).

EXAMPLE 13.3 The general formula for the number of electrons and holes in an *intrinsic* semiconductor is

$$n = p = AT^3 e^{-11,600 E_g / T}$$

where $n = p$ = number of carriers/cm^3
$\quad\quad A$ = constant
$\quad\quad T$ = absolute temperature in K (K = 273 + °C)
$\quad\quad E_g$ = energy gap (eV)
$\quad\quad e$ = 2.718

If we wish to construct a thermistor from silicon, what is the change in the number of carriers per cubic centimeter in going from 27 to 47°C? For silicon we are given that $n = p = 1.5 \times 10^{10}$ carriers/cm^3 at 300 K (27°C) and also $E_g = 1.06$ eV and remains constant over the temperature range in question.

ANSWER First we must arrive at the general equation, which requires calculation of A from the known conditions:

$$1.5 \times 10^{10} \text{ carriers/cm}^3 = A(300 \text{ K})^3 e^{(-11,600 \times 1.06)/300}$$

$$5.56 \times 10^2 = Ae^{-40.987}$$

$$A = 5.56 \times 10^2 \times 6.31 \times 10^{17} = 3.51 \times 10^{20}$$

At 47°C
$$n = p = 3.51 \times 10^{20}(320)^3 e^{(-11,600 \times 1.06)/320}$$

$$= 1.15 \times 10^{28} e^{-38.425}$$

$$= 2.36 \times 10^{11}$$

or
$$\frac{2.36 \times 10^{11}}{1.5 \times 10^{10}} = 15.7$$

Therefore, for the 20°C change the number of carriers increases by 15.7 times, which of course increases the conductivity by this factor plus a factor for the corresponding increase in mobility of both the holes and electrons.

The conductivity of ionic materials also increases with temperature. This is shown quantitatively by the effect on mobility:

$$\mu = \frac{qD}{kT}$$

where μ = mobility (cm^2/volt-sec)
q = charge on carrier (coul/carrier)
D = diffusion coefficient (cm^2/sec)
T = absolute temperature (K)
k = Boltzmann's constant (1.3×10^{-16} erg/ K)

The mobility of the ions in a solid is orders of magnitude lower than that of electrons or holes because of their large size. Movement occurs by jumps, using vacancies. In liquid salts the conductivity is higher because of easier ion movement in the liquid state.

Other Electrical Properties

13.13 Dielectric properties

A dielectric generally has two functions: (1) as an *insulator* and (2) to add to the total *capacitance* of a condenser compared to an air gap between the plates. The use as an insulator involves two principal factors: the breakdown strength and the useful temperature range. Cotton, silk, and many plastics are used below 90°C. Inorganic fillers such as mica and asbestos are used with plastics up to 130°C. The range can be extended with silicone to 180°C. Above this range inorganic materials such as mica, porcelain, and glass are necessary.

Typical values of breakdown voltage [the voltage across 1 mil (0.001 in. = 0.00254 cm) to cause considerable discharge], called the "dielectric strength," are given in Table 13.2. It is evident that for ordinary operating temperatures only mica exceeds the plastics. However, the plastics can absorb water, as explained earlier, which lowers the breakdown voltage, and this should be taken into account in their application. Therefore, the breakdown voltage is governed by a complex number of internal and environmental factors.

13.14   Dielectric constant

The relative permittivity (dielectric constant) ε_r is the quantity used to evaluate the charge-storing capacity of a dielectric in a capacitor:

$$\varepsilon_r = \frac{\varepsilon}{\varepsilon_0}$$

where ε is the permittivity of the dielectric and ε_0 that of a vacuum. Typical values of the dielectric constant are given in Table 13.2.

In absolute values, for a parallel-plate capacitor, the capacitance is:

$$C = \frac{0.224\varepsilon_r A}{10^6 d}\,(n-1)$$

where C = microfarads (μF)
   A = area in square inches of one plate
   d = distance between plates in inches
   n = number of plates
   ε_r = dielectric constant

The value of the dielectric constant depends on the ability of the material to react and orient itself to the field. The greater the reaction, the greater the energy stored, and hence the higher the dielectric constant. The behavior of the dielectric can be made up of the following effects.

1. *Electronic polarization.* This is present in all dielectrics. The electron positions around atoms will be affected by the field, and this takes place very rapidly.
2. *Ionic polarization.* Ions of opposite sign move elastically because of the effect of the field. This is also rapid and takes place only in ionic solids.
3. *Orientation of molecules.* When asymmetric (polar) molecules are present, their orientation is changed by the field.
4. *Space charge.* This is the development of charge at the interface of phases.

Of these effects the orientation of molecules contributes heavily to the differences in the dielectric constant. As an example, a greater charge can be

Table 13.2 DIELECTRIC STRENGTH AND DIELECTRIC CONSTANT FOR A NUMBER OF ENGINEERING MATERIALS

Dielectric Strength of Nonmetallics, volts/mil*

Material	High	Low	Material	High	Low	Material	High	Low
Micas, natural and synthetic	2,000	1,000	Nylon, glass-filled	500	400	Silicones (molded)	400	250
Polymethylstyrene	1,950	890	Nylons 6 and 11	500	420	Ureas	400	300
Polyvinylchloride	1,400	24	Polyesters (cast), allyl type	500	330	Melamines, shock-resistant	370	130
Acetal copolymer	1,200		TFE fluorocarbons	500	400	Phenolics (molded), very high shock resistance	370	200
Polyvinyl formal	1,000	860	Polyethylenes	480		Alkyds	350	300
Plastic laminates, high pressure	1,000	70	Polycarbonate, filled	475		Polycrystalline glass	350	250
Polypropylene	800	520	Nylons 6/6 and 610	470	385	Phenolics, heat-resistant	350	100
Plastic laminates, low pressure	800	100	Epoxies (molded)	468	334	Melamines, GP	330	310
Phenolics (cast), GP	800		Cellulose propionate	450	300	ABS resins, extra high impact	312	
Modified polystyrenes	650	300	Diallyl phthalate	450	275	Beryllia	300	250
Polyallomer	650		Phenolics (cast), GP	450	300	Alumina ceramics	300	200
Cellulose acetate	600	250	Melamines, electrical	430	350	Standard electrical ceramics	300	55
Cellulose nitrate	600	300	Phenolics (molded), GP	425	200	Zircon	290	60
CFE fluorocarbons	600	530	Polystyrenes, glass-filled	425	340	Steatite	280	145
Hard rubber	600	344	ABS resins, high impact	416	350	Forsterite	250	
Mica, glass-bonded	600	270	Cellulose acetate butyrate	400	250	Phenolics (cast), GP transparent	250	75
Polyesters (cast), rigid	570	340	Chlorinated polyether	400		Cordierite	230	140
Epoxies (cast)	550	350	Melamines, cellulose, electrical grades	400	350	Polyethylene foam, flexible	220	
Acrylics	530	400	Phenolics (cast), mechanical and chemical grades	400				
Polystyrenes, GP†	>500		Polycarbonate	400	350			
Acetal	500		Polyesters (cast), nonrigid	400	220			
Ethyl cellulose	500	350	Polyvinyl butyral	400	350			

Dielectric Constant of Nonmetallics‡

Material	High	Low
Mica, glass-bonded	40.0	6.9
Phenolics (cast)	11.0	4.0
Alumina ceramics	9.6	8.2
Lead silicate glass	9.5	6.6
Zircon	9.2	5.3
Polyvinylchloride	9.1	2.3
Micas, natural and synthetic	8.7	5.4
Phenolics (molded)	8.0	4.0
Soda-lime glass	7.4	7.2
Melamines	7.2	4.7
Beryllia	7.0	6.4
Cellulose acetate	7.0	3.2
Standard electrical ceramics	7.0	5.4
Ureas	6.9	6.4
Plastic laminates, high pressure	6.8	3.3
Forsterite	6.5	6.2
Steatite	6.5	5.5
Aluminum silicate glass	6.3	
Cellulose acetate butyrate	6.2	3.2
Cordierite	6.2	4.0
Polyesters (cast), nonrigid	6.1	3.7
Plastic laminates, low pressure	5.6	3.4
Polycrystalline glass	5.6	
Borosilicate glass	5.1	4.0
Silicones (molded)	5.1	3.6
Alkyds, GP†	5.0	4.8
Rubber phenolics	5.0	
Vinylidene chloride	5.0	3.0
Hard rubber	4.95	2.90
Polyesters (cast), allyl type	4.8	3.3
Alkyds, electrical and impact	4.5	4.2
Diallyl phthalate	4.5	3.3
Nylons 6 and 11	4.5	3.5
Epoxies (cast)	4.4	2.6
Epoxies, GP	4.4	3.4
Boron nitride	4.2	
ABS resins	4.1	2.8
Epoxies, heat-resistant	4.0	3.5
Modified polystyrenes	4.0	2.5
Polyesters (cast), rigid	4.0	2.8
Nylon, glass-filled	3.9	3.4
Phenoxy	3.8	3.7
Silica glass	3.8	3.4
Acetal	3.7	
Cellulose propionate	3.6	3.4
Ethyl cellulose	3.6	2.8
Nylons 6/6 and 610	3.6	3.4
Polycarbonate, filled	3.50	
Polystyrenes, glass-filled	3.41	2.74
Modified polystyrenes, extra high impact	3.3	
Polyvinyl butyral	3.3	1.9
Acrylics	3.2	2.7
Polyvinyl formal	3.0	
Polycarbonate	2.96	
Chlorinated polyether	2.92	
Methylstyrene-acrylonitrile	2.81	
Epoxies, resilient	2.8	2.6
Polystyrenes, GP	2.65	2.45
Polymethylstyrene	2.48	
CFE fluorocarbons	2.37	2.30
Polyethylenes	2.3	
Propylene-ethylene polyallomer	2.29	
Polypropylene	2.1	2.0
TFE fluorocarbons	2.0	
Prefoamed epoxy, rigid	1.55	1.19
Polyethylene foam, flexible	1.49	
Urethane rubber foamed-in-place, rigid	1.40	1.05
Silicone foams, rigid	1.26	1.23
Polystyrene foamed-in-place, rigid	1.19	
Prefoamed cellulose acetate, rigid	1.12	
Prefoamed polystyrene, rigid	<1.07	1.10

° Values represent high and low sides of a range of typical values. To obtain volts/cm, multiply by 393.7.

† GP = general-purpose grades.

‡ Values represent high and low sides of a range of typical values at 10^6 cycles.

Source: "Materials Selector," Reinhold Publishing Corp., New York, 1973.

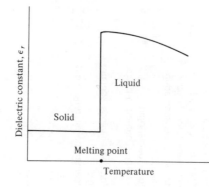

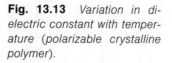

Fig. 13.13 *Variation in dielectric constant with temperature (polarizable crystalline polymer).*

built up on the plates of a capacitor if a material with a strong dipole is placed between the plates. In other words, the material with larger dipoles has a higher dielectric constant. Similar reasoning suggests that liquids will have higher dielectric constants than solids since polarization or dipole orientation is easier. This effect is shown schematically in Fig. 13.13. After the change due to the melting, the gradual fall-off with increasing temperature of the liquid reflects the higher atomic or molecular mobility which decreases polarization.

Polymers have the same general characteristics. The amorphous polymers are not as tightly bonded as the crystalline polymers. Therefore, amorphous polymer dipoles are more easily aligned, giving a higher dielectric constant. The crystalline polymers show a large increase in the dielectric constant at their melting point, as shown in Fig. 13.13.

The effect of frequency is different for the various materials. The small physical movements encountered with electronic and ionic polarization suggest that these changes take place over a broad range of frequencies. This is not the case, however, for dipole motion of molecules. Generally, the dielectric constant decreases as the frequency increases, since it becomes difficult for the dipole to shift at high frequencies.

As shown in Fig. 13.14, the dipoles can rotate when the polarity of the capacitor is charged. However, it takes time to shift from (*a*) to (*c*), and as

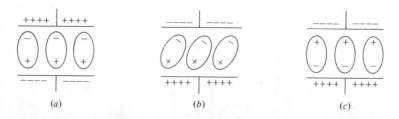

Fig. 13.14 *Dipole motion caused by a change in polarity. When the polarity reverses, (a) → (b), it takes a finite time to make the complete shift to (c).*

the complexity of the molecule increases, a longer time is required to shift the dipole.

If we impose a frequency that causes a slight shift each time, we can have dielectric heating because of the energy loss in each cycle. This phenomenon has been used in industry in such applications as the setting of glues in furniture manufacture and now it has also found use in the home. Since most of our foods are made up of organic molecules with dipoles (e.g. proteins and water), it is possible to rapidly "cook them from within" by exposure to suitable frequencies, delivered by the "microwave" oven.

13.15 Barium titanate type dielectrics (ferroelectrics)

In trying to reach objectives such as pocket-size television sets, there has been great demand for materials with higher dielectric constants to reduce the size of the capacitors. Since it was known that the mineral rutile, TiO_2, had a value of ε_r about 10, the search led to the titanates. Finally, it was found that barium titanate showed values of ε_r over 1,000, several orders of magnitude better than any known material. In addition, it was noticed that a permanent charge was developed. This is called "ferroelectric behavior."

It is interesting to examine how these effects result from the structure of $BaTiO_3$. In Chap. 7 we discussed the fact that the Ti^{4+} ion was larger than the octahedral site formed by the six oxygen ions (radius $Ti^{4+} = 0.64$ Å vs. 0.625 Å for the site) (Fig. 13.15). As a result, the titanium ion is located to one side of the center. Thus in each unit cell one side of the center is positive and the other negative or, in other words, a dipole develops.

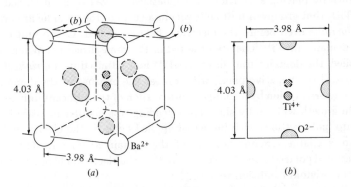

Fig. 13.15 (a) Structure of barium titanate. (b) The source of the dipole is due to two possible positions for the Ti^{4+} ion.

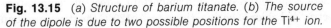

(L. H. Van Vlack, "Elements of Materials Science," 2d ed., Addison-Wesley Publishing Company, Inc., Reading, Mass., 1964, Fig. 8-26, p. 222.)

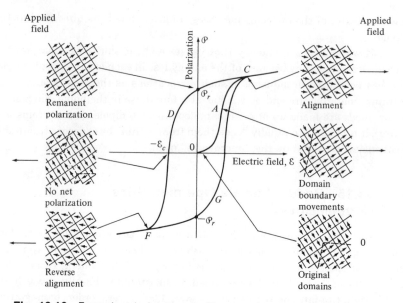

Fig. 13.16 *Ferroelectric hysteresis. The spontaneous polarization can be reversed; however, energy is consumed with each cycle.*

(L. H. Van Vlack, "Materials Science for Engineers," Addison-Wesley Publishing Company, Inc., Reading, Mass., 1970, Fig. 13.24, p. 273.)

When an electric potential is applied across the capacitor plates, the Ti^{4+} ions are attracted to the negative side. This leads to a high charge storage in the plates and, therefore, a higher dielectric constant.

It is now important to examine the hysteresis effect (Fig. 13.16). It can be shown by placing a solution of magnetized powder over an etched sample of $BaTiO_3$ that groups of unit cells within a grain of the ceramic are organized into *domains* even before the electrical field is applied. In the unit cells of a given domain the Ti^{4+} ions are oriented in the same direction. When the field is applied, the domains that have the Ti^{4+} ions situated in the *right* direction (i.e. toward the negative plate) will grow at the expense of the domains with the *wrong* orientation by a change in orientation of the adjacent unit cells. This domain growth will finally slow down, leading to a smaller increase in charge build-up (polarization) with increased applied fields. If the electric field is cut to zero, a charge remains because of the domain alignment. This is called the "residual polarization" $\mathcal{P}_r$. To remove this, the electric field must be reversed from the original condition to $-\mathcal{E}_c$.

There is an interesting and important effect of temperature on this "ferroelectric" behavior of barium titanate. At 120°C the structure changes from tetragonal to cubic and the effect disappears. This is called the "Curie temperature," in honor of Pierre Curie. This general term is used for specifying the temperature at which a change in electrical or magnetic properties takes place, and the change may or may not be accompanied by a phase change. To obtain ferroelectric materials for use at higher temperatures, other compounds with similar structures were investigated with the result that lead titanate, with a Curie point of 480°C, and lead metaniobate, with a Curie point of 570°C, were found. The most common material used at present is PZT [$Pb(Zr,Ti)O_3$].

EXAMPLE 13.4 Show that lead titanate ($PbTiO_3$) would be expected to have a structure equivalent to $BaTiO_3$.

ANSWER

	Ba^{2+}	Ti^{4+}	O^{2-}	
Ionic radius, Å:	1.43	0.64	1.32	(From Table 7.1)

	Pb^{2+}	Ti^{4+}	O^{2-}
Ionic radius, Å:	1.32	0.64	1.32

In general, in the perovskite structure (Figs. 7.12 and 13.15), the ions in the unit cell corners will have an average valence of 2 and radii close to Ba^{2+}. It is of additional interest that to obtain extensive solid solution the divalent radii should be ±15 percent or 1.21 to 1.55 Å. Ions in the $\frac{1}{2}$, $\frac{1}{2}$, $\frac{1}{2}$ sites must average 4+ (3+ in conjunction with Nb^{5+} in niobates, for example) and be about the radius of Ti^{4+}. See the problems for further data.

13.16 Interrelated electrical-mechanical effects (electromechanical coupling)

Every radio amateur has heard of the piezoelectric effect of a quartz crystal. If pressure is applied to a quartz crystal, the ends become charged or, conversely, if an electric field is applied, the crystal changes length. If an alternating voltage is applied, the crystal oscillates, giving out a sound wave of constant frequency.

The reason for the generation of a sound wave is shown in Fig. 13.17. Let us assume we have a single crystal with the dipoles of the unit cells aligned as shown. If we apply an electric potential as shown, the crystal will lengthen because of the attractions of the ions to the pole plates. If we use an ac voltage, the crystal will alternately expand and contract, sending out a wave into the surrounding medium, whether air or water. On the other hand, if we supply the mechanical motion, we will receive an electric charge by the reverse procedure. This phenomenon is called "piezoelectricity," from the Greek word meaning "to press."

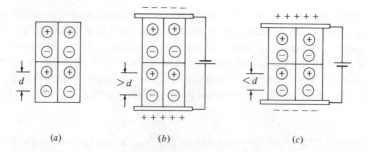

Fig. 13.17 (a) *Piezoelectric material.* (b) *An electric field induces dimensional expansion.* (c) *Reverse polarity causes a corresponding contraction. The procedure can be inverted by applying a pressure and obtaining a change in the voltage.*

(L. H. Van Vlack, "Elements of Materials Science," 2d ed., Addison-Wesley Publishing Company, Inc., Reading, Mass., 1964.)

Stronger piezoelectric effects can be obtained with ferroelectric materials because domain movement occurs. In quartz, rotation of unit cells and domain growth are absent.

(In one of the problems at the end of the chapter, we will take up the use of the ferroelectrics in such applications as sonar and phonographic pickups.)

13.17 Thermocouples, thermoelectric power

If we produce a temperature difference in a rod, the energy levels of the electrons will be different, giving more high-energy electrons in the hotter region (Fig. 13.18). The high-energy electrons will flow toward the cold end, producing a difference in charge.

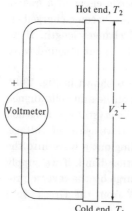

Fig. 13.18 *Development of a voltage in a thermocouple.*

(R. M. Rose, L. A. Shepard, and J. Wulff, "The Structure and Properties of Materials," vol. 4: Electronic Properties, John Wiley & Sons, Inc., New York, 1965.)

If we now attach a voltmeter with contact wires of the same material, there will be no difference at the meter. If the wires are of a different material than the rod, there will be a different voltage induced and the net voltage will appear on the meter. This voltage is the Seebeck potential S used in thermocouples. Examples are:

Thermocouple	Maximum useful temperature, °C
Copper vs. constantan (60% Cu, 40% Ni)	315
Iron vs. constantan	950
Chromel (90% Ni, 10% Cr) vs. Alumel (94% Ni, 2% Al, 3% Mn, 1% Si)	1200
Platinum vs. platinum– 10 to 13% rhodium	1500
Tungsten vs. tungsten-rhenium, also tungsten vs. molybdenum	>1500

There is a basic reason for the potential difference of a thermocouple which goes back to the concept of the Fermi level. In different metals the Fermi level changes at a different rate with increased temperature. Therefore, in a pair of metals the element with the greatest temperature coefficient of change will have more energetic electrons which flow into the element with the lower coefficient.

If we survey the metals used for thermocouples, we see that many are transition metals. This is because the distribution of electrons at different energy levels shows the greatest difference with temperature in metals with unfilled $3d$ and $4f$ shells.

SUMMARY

The electrical properties of materials depend principally on the number and mobility of the charge carriers in the structure: electrons, electron holes, and ions.

The flow of current is related to the conductivity, $\sigma =$ (number of carriers/cm^3) $\times$ (charge/carrier) $\times$ mobility. The number of carriers per cubic centimeter is relatively great in a metal because the electrons donated by the atoms exist in a band of energies which has many vacant states to which the electrons can be accelerated and thereby carry current. By contrast, in an insulator the valence electrons are bound in a filled energy band, and to raise an electron to an energy level at which it can conduct, the conduction band, requires considerable energy. The intrinsic semiconductors occupy an intermediate position. The number and type of carriers in a semiconductor can be increased by "doping" with elements that are electron donors or acceptors. By setting up junctions of these "extrinsic" semiconductors, the *p-n* junctions are formed. Certain metals and alloys exhibit superconductivity below critical temperatures.

Dielectrics of greatly improved dielectric constant can be made from ferroelectric materials such as barium titanate. In these materials a permanent dipole or charge imbalance exists in the unit cell. These dipoles result in the formation of electrical domains or regions of similar dipole alignment. This configuration leads to a high capacitance when the material is used in a capacitor. Electromechanical coupling from these materials or materials with fixed dipoles leads to valuable devices for conversion of electrical impulses to sound. Thermocouples are important for temperature measurement and operate because of the development of an emf by pairs of metals subjected to two different temperatures.

DEFINITIONS

Conductivity, σ   A measure of the ease of passage of an electric current. $\sigma = 1/\text{resistivity}$.

Resistivity $\rho = R(A/l)$, where A is the area of a conductor, l is its length, and R is its resistance.

Resistance $R = E/I$, where E is the voltage and I the current. The resistance of a conductor to the passage of a current increases with length and decreases with increasing cross-section. Note that this applies to direct current not to high-frequency alternating current, in which case most of the current flows in the surface layers.

Resistor   An electrical component designed to provide a desired resistance. Examples are a tiny component in a radio or the heating element in a toaster.

Charge carrier   A tiny particle with an electric charge, specifically electrons, holes, and ions.

Electron   A negative charge carrier having a charge of 1.6×10^{-19} coul.

Electron hole   A positive charge carrier having a charge of 1.6×10^{-19} coul.

Metal   A material with an unfilled valence band, leading to a very high concentration of carriers.

Insulator   A material with a filled valence band and a large energy gap between the valence band and the conduction band.

Semiconductor   A material with a filled valence band and a small energy gap between it and the conduction band.

Band model   A model in which the outer electrons of the atoms are combined as existing in a band of energy levels. If there are empty levels within a band, the material is a good conductor, because the upper-level electrons can easily be raised to a higher level of energy and therefore move to conduct current.

Energy gap, E_g   The energy required to move an electron from the valence band to the conduction band, usually expressed in electronvolts.

Pauli exclusion principle   No more than two electrons in an atom or block of material can occupy the same energy level.

Fermi level, E_F   The energy level at which the probability of occupancy of an energy state is 0.5. In other words, at this level the possible states that an electron or electron hole can occupy are half-filled.

Superconductivity   The phenomenon of zero resistivity which occurs in some metals and alloys below a critical temperature which is different for each material.

Intrinsic semiconductor   A semiconductor material that is essentially pure and for which conductivity is a function of the temperature and the energy gap of the material.

Extrinsic semiconductor   A semiconductor material that has been doped with an *n-type element* (such as phosphorus) which donates electrons that have energies close to the conduction band, or with a *p-type element* (such as aluminium) which provides electron holes close to the valence band level.

Mobility   A measure of the ease of motion of the carrier.

p-n junction   The boundary between *n*- and *p*-type materials.

Rectification   The conversion of alternating to direct current.

Hall effect   The development of a voltage by applying a magnetic field perpendicular to the direction in which a current is flowing. Electron carriers are forced to one side and electron holes to the other, giving a voltage which is perpendicular to both the field and the current.

Thermistor   A device used to measure temperature using the change in resistivity with temperature.

Dielectric constant   The ratio of the capacitance obtained using the material between the plates of a capacitor compared to that obtained when the material is replaced by a vacuum.

Capacitance   The charge-storing ability of a capacitor. Capacitance is a function of the geometry of the device and the dielectric constant of the material between the plates. The unit of capacitance is the farad (**F**).

Electronic polarization   The movement of the electrons in a dielectric toward the positively charged plate.

Ionic polarization   The displacement of ions in the material toward oppositely charged plates.

Dipole   An atomic or ionic grouping such as a molecule or unit cell in which the positive and negative centers of charge do not coincide. The *dipole moment* is the product of the charge and the distance between the charges.

Ferroelectric A material such as barium titanate in which the permanent dipoles are aligned in domains.

Electromechanical coupling A material that develops a charge at its ends when compressed or, conversely, changes dimensions when an electric potential is applied.

Piezoelectric material A material that shows electromechanical coupling. All ferroelectric materials are piezoelectric, but only materials with electrical domain structures are ferroelectric.

Thermocouple A pair of joined materials that develops an emf when one junction is at a different temperature than the other.

PROBLEMS

13.1 Suppose that in Example 13.1 the wire length is considered too awkward and it is decided to use a 10-ft (3.05-m) length. What diameter Chromel wire should the student buy? Suppose she decides to use only a 1-in. (2.54-cm) length. What diameter should she use? What practical difficulties might she encounter by using such short wires?

13.2 Sketch the electron occupation of the valence and conduction bands for the following materials at 0 and 273 K: potassium, magnesium, aluminum, pure silicon, silicon doped with aluminum, germanium doped with phosphorus. Indicate the approximate Fermi level E_F in each case.

13.3 It is known that chromium is insoluble in copper at room temperature yet will dissolve to over 1 percent at elevated temperatures. It can form a precipitation-hardening system. Compare the relative usefulness of this alloy with a solid solution of copper or work-hardened copper in improving the yield strength when maximum electrical conductivity is desired.

13.4 The following table summarizes the characteristics of the two intrinsic semiconductors, silicon and germanium, at 300 K.

Characteristic	Silicon	Germanium
Electrical resistivity, micro-ohm-cm	2.3×10^{11}	4.5×10^7
Energy gap, eV	1.06	0.67
Density of carriers, (electrons plus holes)/cm³	1.5×10^{10}	2.4×10^{13}
Lattice drift mobility, cm²/volt-sec		
Electrons	1,350	3,900
Electron holes	480	1,900

The conductivity of germanium was calculated in Example 13.2. Calculate the conductivity of silicon and check how it corresponds to the resistivity given in the table.

13.5 Assume we were to use germanium rather than silicon, as proposed in Example 13.3. Show by calculations whether the germanium thermistor would be more or less sensitive within this temperature range (27 to 47°C).

13.6 In a doped or extrinsic semiconductor the carriers produced by the dopant usually overpower the instrinsic carriers. Assume that we add phosphorous to silicon and find a conductivity of 1 mho/cm and a mobility of 1,700 cm^2/volt-sec. What is the concentration of conduction electrons per cubic centimeter? (*Hint:* Determine what type of carriers the phosphorus contributes.)

13.7 Iron oxide is made up with a ratio of Fe^{3+} to Fe^{2+} of 0.1. What is the mobility of the electron holes if the oxide has a conductivity of 1 mho/cm and 99 percent of the charge is carried by the electron holes. $a_0 = 4.3$ Å (*Hint:* FeO has an NaCl type lattice; for every two Fe^{3+} there must be one vacancy at an Fe^{2+} position to balance the charge. If we take 100 Fe^{2+}, there are 10 Fe^{3+} and 5 vacancies. Accompanying these atoms will be 115 oxygen atoms. Since the oxygen ion lattice is perfect, calculate the number of unit cells formed by 115 oxygen atoms at 4 oxygen per unit cell. Since each Fe^{3+} has one electron hole, the density of carriers can be calculated, then the mobility.)

13.8 Indicate how we might conduct an experiment on a crystal of LiCl whereby we could determine the diffusion coefficient of Li^+ from a measurement of the electrical conductivity. Be as specific as possible as to what measurements must be made and what assumptions would be necessary. In most cases the major portion of the ionic conductivity is due to the cation. For this problem assume that 100 percent is due to this effect.

13.9 A capacitor consists of two plates 2 by 3 in. (5.08 by 7.62 cm), spaced 0.01 in. (0.0254 cm) apart and filled with mica. Calculate the capacitance.

13.10 Which dielectric would you use for the smallest two-plate capacitors for the following application, assuming constant plate thickness?

Capacitance: 0.001 μF
Operating temperature: 300°F (149°C)
Voltage: 1,000 volts

Specify plate size and the distance between plates, using a safety factor of 5 against puncturing due to breakdown. (*Hint:* Compare several materials in Table 13.2.)

13.11 Indicate which material of each pair would have the higher dielectric constant and explain why for the following comparisons:

    a. Polyvinylchloride (PVC) and polytetrafluoroethylene (PTF)
    b. PVC at 25°C and 100°C
    c. PVC at 10^2 Hz (hertz) and PVC at 10^{10} Hz

13.12 Why should the cooling of barium titanate ceramics through the Curie temperature in the presence of a strong electric field improve their piezoelectric properties?

13.13 Explain how the characteristics of the barium titanate structure lead to its use in the following devices:

    a. Ultrasonic applications such as emulsification of liquids, mixing of powders and paints, homogenization of milk
    b. Microphones and phonograph pickups
    c. Accelerometers
    d. Strain gages
    e. Sonar devices

13.14 As mentioned in the text, a variety of alloys are used for thermocouples. Suppose that you can read a given potentiometer (a device for measuring emf) to 0.01 millivolt at 400°F (204°C). What will be the precision of your temperature measurement using a Chromel vs. alumel, iron vs. constantan, copper vs. constantan, or a platinum vs. platinum–10 percent rhodium thermocouple. There are many places where charts can be found, such as the "Handbook of Chemistry and Physics" and instruction manuals accompanying the potentiometer.

13.15 From the calculations of Prob. 13.1 explain why the manufacturer's catalog of Chromel wire gives recommendations for the wire diameters for different currents. Why is the allowable current lower for wires embedded in refractory than for wires in air when the thermal conductivity of air is lower than that of the refractory?

13.16 Several years ago a rash of failures due to brittle failure at contacts of transistors developed. The fractures showed a purple color and the effect was promptly called the "purple plague." From the following facts deduce the cause and suggest a remedy:

    1. The transistor is made of silicon doped with aluminum.
    2. The contact wire is gold.
    3. Gold and aluminum form a compound, $AuAl_2$, which is purple in color.

13.17 Show that the following complex materials meet the criteria for the perovskite structure and therefore might react in a similar way to barium titanate:

a. $KLaTi_2O_6$
b. Sr_2CrTaO_6
c. $BaKNbTiO_6$
d. $Pb(Zr, Ti)O_3$

14

MAGNETIC PROPERTIES OF MATERIALS

THIS is an x-ray diffraction pattern from an iron–3 percent silicon single crystal. The interaction of the x-rays with the magnetic domains in the iron caused the features shown. The "magnetic fish" shows no structure because the orientation of this region does not satisfy the Bragg law, discussed in Chap. 2, for diffraction. The square region to the right of the tail of the "fish" is a ferromagnetic domain. The chevron markings show other domains of different orientation.

In this chapter we will take up the relation of magnetic properties to structure in two groups of materials—the metallic and the ceramic. We will see that in each group important recent growth has occurred to meet the demands of the electronics industry.

14.1 General

It is not generally recognized that the development of materials with new high levels of magnetic properties is responsible for the growth of many new devices such as the tape recorder, ferrite magnets in television sets, memory cores in computers, permanent magnets in motor controls, and particle accelerators used in basic research.

We will see how these new developments involve two important groups of materials:

1. The metallic materials in which the advances have taken place in recent years, from the conventional iron-base alloys, to Alnico, and, more recently, to samarium-cobalt magnets.
2. The ceramic materials called "ferrites," which are spinels and related structures. The growth of this field has been comparable to that of the semiconductors in the electrical materials.

In order to understand the specification of magnetic materials and how their structure is related to magnetic properties, we need to review first a few definitions of the properties of greatest importance. These are relatively simple and are all related to the magnetic flux that is produced in a material as a result of a magnetic field. These properties, such as permeability and remanent induction, are given quite simply by the graph of the flux density B as a function of the field strength H, called a "hysteresis loop" or "magnetic hysteresigraph."

Following this review we will take up important metallic magnetic materials, encountering magnetic domains and domain movement, just as we found domains in ferroelectric materials. We will see that the structure required for the soft or low-hysteresis magnets is quite different from that for the hard or permanent magnets. Finally, we will discuss the magnetic properties available in the ceramic magnets or ferrites and their relation to structure.

14.2 The magnetic circuit and important magnetic properties

First let us recall that in an electric circuit (Fig. 14.1) if we apply a voltage E, the current I that flows in the conductor (load) is proportional to the conductivity σ of the material. For a sample 1 cm long with a 1-cm^2 cross-sectional area,

$$\sigma = \frac{I}{E}$$

In an analogous way, if we apply a magnetic field across a gap, the flux in the gap is proportional to the permeability μ of the material in the gap. If we

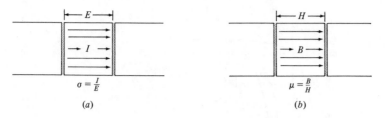

Fig. 14.1 *Analogy between (a) electric and (b) magnetic circuits.*

represent the field strength by H and the flux density by B, then

$$\mu = \frac{B}{H}$$

The usual meter-kilogram-second (mks) units for the quantities in this equation are derived in physics as

$$H = 1\,\text{amp/m}$$
$$B = 1\,\text{weber/m}^2 = 1\,\text{tesla} = 1\,\text{volt-sec/m}^2$$

For a vacuum,

$$\mu_0 = 4\pi \times 10^{-7}\,\text{henry/m}$$

where $\qquad\qquad$ 1 henry = 1 ohm-sec

Often centimeter-gram-second (cgs) units are still used, and the following conversions apply:

$$H\text{: } 1\,\text{amp/m} = 4\pi \times 10^{-3}\,\text{oersted}$$
$$B\text{: } 1\,\text{tesla} = 10^4\,\text{gauss}$$
$$\mu_0\text{: } 4\pi \times 10^{-7}\,\text{henry/m} = 1\,\text{gauss/oersted}$$

The important difference between the electric and magnetic circuits is that while the conductivity in an electric circuit is constant (independent of the values of E and I), the permeability μ of a magnetic material changes with the applied field strength H. The ratio B/H is not constant. Also, it is possible to have the flux persist after the field is removed, and this is called "remanent magnetization," as in a permanent magnet. Clearly, we need to examine these important effects in a magnetic circuit and to relate them to the structure of the material if we are to understand and construct magnetic devices.

Let us first discuss the B-H curve, which is as important in expressing magnetic properties as the stress-strain curve is for understanding mechanical properties. We place a sample in a gap in which we provide a controlled magnetic field strength H (as, for example, by changing the current in an electromagnet

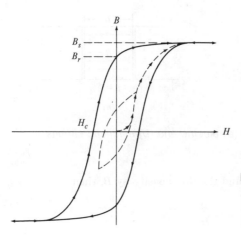

Fig. 14.2 B *(flux density) vs. H (field strength) hysteresis loop for a ferromagnetic material.*

with its poles at the sides of the gap), and we measure the magnetic flux density B and plot the B-H curve (Fig. 14.2). In a typical virgin curve the value of B rises rapidly and then practically levels off at B_s, called the "saturation induction." (The curve obtained the first time the material is magnetized—the virgin curve—is the dashed curve.) As the field strength is reduced to zero, there is still a magnetic flux density B_r, called the "remanent induction" or "residual induction." To lower the flux to zero, we need a definite field strength H_c in the opposite direction to the original, and this is called the "coercive magnetic force" (or the "coercive field"). Since the graph never·retraces the virgin curve but traces out a loop, it is called a "hysteresis loop" and is an index of the energy lost in a complete cycle of magnetization.

The shape of the B-H curve varies greatly with different magnetic materials, and various types of curves have different uses. To illustrate this, consider the B-H curves for a transformer core, a permanent magnet, and a computer memory unit, as shown in Fig. 14.3. In the curve for a transformer core there is only a small area enclosed by the hysteresis loop (Fig. 14.3a). This represents work done per cycle. It is related to power loss and is kept as small as possible. By contrast, in a memory unit we wish to magnetize easily to a sizable value of B and then retain most of this flux density when the power is off, so that it can be used as "memory" to activate the controls when called on. For this a "square loop" is desired, so that even with a reversed field a strong flux is still obtained for "readout" (Fig. 14.3b). In the permanent magnet the "power" of the magnet is related to the BH product.† We note that in this graph the x axis, the field strength, is many times more compressed than in

---

†The maximum BH product in the second quadrant of the B-H curve is used.

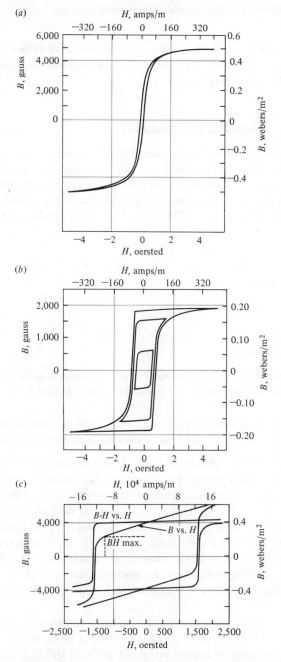

Fig. 14.3 (a) *Hysteresis curve of a soft ferrite.*
(b) *Hysteresis curves of a square-loop ferrite.* (c)
Hysteresis curve of a permanent magnet ferrite.

(E. C. Henry, "Electronic Ceramics," Doubleday & Company, Inc., Garden City, N.Y., 1969.)

the other two graphs, which reflects the large value of H needed to produce a permanent magnet with a large B_r. This large value of remanent magnetization should be noted (Fig. 14.3c).

14.3 Magnetic permeability

As stated earlier, the permeability of a vacuum is $\mu_0 = 4\pi \times 10^{-7}$ henry/m. Now if we insert a material into the same space, we obtain a new value for μ because the same applied field will produce a different flux density B. Instead of discussing the absolute value, it is useful to consider the relative value vs. that of the vacuum. We define the *relative permeability* as

$$\mu_r = \frac{\mu}{\mu_0}$$

and for a vacuum

$$\mu_r = \mu_0 = 1$$

After determining μ_r for a wide variety of materials, we find that they fall into three general classes:

1. If μ_r is less than 1 by about 0.00005, the material is called "diamagnetic."
2. If μ_r is slightly greater than 1 up to 1.01, the material is called "paramagnetic."
3. If μ_r is much greater than 1 up to 10^6 times μ_0, the material is called "ferromagnetic," or in the case of ceramic magnets, "ferrimagnetic."

All three of these phenomena are related to the structure of the material. Diamagnetism, which is negligible from an engineering standpoint, can be explained as follows: When a magnetic field is applied to a material, the motions of all the electrons change. This produces a local field around each electron. By a law of physics known as Lenz's law this field is in the opposite direction to the applied field and gives a small decrease in the total magnetism. Paramagnetism is also a small effect noted in all atoms and molecules possessing an odd number of electrons. Here the electrons line up with the applied field, thus reinforcing the field and giving a value of permeability slightly greater than 1.

Ferromagnetism is of the greatest engineering importance because a high permeability leads to a great flux density B for a given magnetizing force H compared to a vacuum or normal materials. Without this high flux density and the ability to change or reverse it (as in soft magnetic materials such as electromagnets) or maintain it after the field is removed (as in hard or permanent magnets), most electrical devices such as transformers, motors, tape recorders, and computers would not work.

Ferromagnetism is encountered in only a few elements: those with partially filled $3d$ shells, such as iron, cobalt, and nickel, and those with partially filled $4f$ shells, such as gadolinium. Although only the four elements named exhibit ferromagnetism at room temperature, alloys or ceramic compounds of the other elements with partially filled $3d$ or $4f$ shells, such as manganese, have important magnetic properties.

Now let us examine the structural factors responsible for these magnetic properties. We will see that, as with mechanical properties, both the atomic structure and the microstructure govern the magnetic properties. We will first discuss the effects of atomic structure on permeability and remanent induction in general terms and then take up metallic and ceramic magnets.

14.4 Atomic structure of ferromagnetic metals

We should recall from Chap. 2 that, in addition to its other quantum numbers, each electron revolving about the nucleus has the quantum number m_s, which describes the way an electron is spinning on its axis. Also, when $m_s = +\frac{1}{2}$, we think of the electron as spinning in one direction, and when $m_s = -\frac{1}{2}$, in the other. A spinning charge generates a magnetic field, and we can say by definition that $+\frac{1}{2}$ means that the axis of the field is up and $-\frac{1}{2}$ means it is down. Furthermore, we find that in an atom of manganese, for example, the five $3d$ electrons all have their spins in the same direction rather than alternating in direction. This is an example of Hund's rule, which states that as electrons are added in progressing to higher elements in the Periodic Table, there is a strong tendency to attain the maximum number of spins in one direction, such as five for the $3d$ level, before electrons of opposite spin are added. This leads to a magnetic moment for the atom as a whole. Each unpaired electron, i.e. electron not balanced by one of opposite spin, contributes a magnetic moment that is given the special name the *Bohr magneton*, and this has the value of 9.27×10^{-24} amp-m^2 in mks units or 0.927 erg/gauss in cgs units. The isolated manganese atom, therefore, has a magnetic moment of 5 Bohr magnetons. It is important to establish the magnetic moments of some other nearby elements (Fig. 14.4).

We note that iron has one less Bohr magneton than manganese because there are only four unpaired electrons and that the moments for cobalt and nickel are correspondingly reduced. The magnetic moment of copper is zero because the tendency to attain a second completed group of electrons of the same spin is stronger than that to add a second $4s$ electron. The $4s$ electrons do not contribute to the magnetic moment.

From this table we would predict that a block of manganese metal would have the greatest magnetic moment because its atoms have the highest moment. This is not the case because when the atoms are placed together in a block, complex alignment called "exchange interaction" occurs. In the case of *metallic*

Magnetic moment	Element	Number of electrons	Electronic structure 3d shell					4s electrons
1	Sc	21	[↑]	[ ]	[ ]	[ ]	[ ]	2
2	Ti	22	[↑]	[↑]	[ ]	[ ]	[ ]	2
3	V	23	[↑]	[↑]	[↑]	[ ]	[ ]	2
5	Cr	24	[↑]	[↑]	[↑]	[↑]	[↑]	1
5	Mn	25	[↑]	[↑]	[↑]	[↑]	[↑]	2
4	Fe	26	[↑↓]	[↑]	[↑]	[↑]	[↑]	2
3	Co	27	[↑↓]	[↑↓]	[↑]	[↑]	[↑]	2
2	Ni	28	[↑↓]	[↑↓]	[↑↓]	[↑]	[↑]	2
0	Cu	29	[↑↓]	[↑↓]	[↑↓]	[↑↓]	[↑↓]	1

Fig. 14.4 *Magnetic moments (un-ionized) of the transition elements.*

manganese the atoms line up so that the directions of the magnetic fields in the individual atoms are antiparallel and cancel out each other. This is called "antiferromagnetism." (We will find later that this is not the case for manganese in ceramic magnets, where the ions are isolated.) However, in the case of iron, cobalt, and nickel a fraction of the atoms do line up to produce ferromagnetism.

To get a feel for the relation between the atomic-scale property of magnetic moment and the engineering-scale measurement of the gross effect called "saturation induction," discussed earlier, let us make a calculation to compare theory with experiment, just as we related atomic radius and density in Chap. 2.

First, we write again for a vacuum with a given field strength H

$$B = \mu_0 H$$

If we place a ferromagnetic material in the circuit, B will rise greatly, and we can satisfy the equation by changing μ_0 to a new μ to show the new permeability. However, it is simpler to express the increase in B over the original value by adding a term $\mu_0 M$, where M is called the "magnetization":

$$B = \mu_0 H + \mu_0 M$$

The term $\mu_0 M$, therefore, is the increase in field strength, which is the same as the increase in magnetic moment caused by the new material.

As a final step, when measurements are made for ferromagnetic materials, we can usually omit $\mu_0 H$, since it is so small compared to $\mu_0 M$. Let us now consider an example.

EXAMPLE 14.1 Calculate the saturation induction B_s you would expect from iron in webers per square meter (teslas).

ANSWER $B = \mu_0 H + \mu_0 M$, but since $\mu_0 M$ is always many times greater than $\mu_0 H$ for a ferromagnetic material,

$$B \simeq \mu_0 M$$

The magnetic moment of an iron atom is 4 Bohr magnetons. Let us assume that when $B = B_s$, all magnetic moments are aligned. Converting to mks units, we have

$$\frac{\text{Magnetic moment}}{\text{Atom}} = \frac{4 \text{ Bohr magnetons}}{\text{Fe atom}} \left(9.27 \times 10^{-24} \frac{\text{amp-m}^2}{\text{Bohr magneton}} \right)$$

Since the density of iron is 7.87 g/cm³, there are n atoms per cubic centimeter:

$$n = \frac{(7.87 \times 10^6 \text{ g/m}^3)(0.602 \times 10^{24} \text{ atoms/at. wt.})}{55.85 \text{ g/at. wt.}}$$

$$= 8.5 \times 10^{28} \text{ Fe atoms/m}^3$$

The magnetization will be the magnetic moment per atom times n, or

$$M = \frac{4 \text{ Bohr magnetons}}{\text{Fe atoms}} \left(9.27 \times 10^{-24} \frac{\text{amp-m}^2}{\text{Bohr magneton}} \right)$$

$$\times \left(8.5 \times 10^{28} \frac{\text{Fe atoms}}{\text{m}^3} \right)$$

$$= 3.15 \times 10^6 \frac{\text{amp}}{\text{m}}$$

$$B_s = \left(4\pi \times 10^{-7} \frac{\text{volt-sec}}{\text{amp-m}} \right) \left(3.15 \times 10^6 \frac{\text{amp}}{\text{m}} \right)$$

$$= 3.96 \frac{\text{volt-sec}}{\text{m}^2} \quad (3.96 \text{ teslas})$$

We find experimentally that the actual B_s is less, 2.1 teslas. This is because the actual interaction between iron atoms in the metallic state does not lead to the sum of the magnetic moments, or in other words, not all the magnetic moments are aligned. We shall see that only when the atoms are isolated and ionized as in a ceramic will the full magnetic moment be realized.

At this point the question arises: If we have the atoms (and therefore their electrons) lined up in a block of iron, why is not the block spontaneously a magnet without the necessity of applying a magnetic field? This leads us to the topics of magnetic domains and Bloch walls.

14.5 Magnetic domains

For many years physicists speculated that a block of iron contained many tiny magnets that would line up under the influence of a magnetic field. In 1912 Weiss went further and postulated that there were *domains* (regions) where the atoms were aligned to produce a magnetic field in one direction and that these were balanced by domains of opposite alignment, so that no magnetic flux was present outside the specimen. Finally, in 1931 Francis Bitter decided to look for these domains under the microscope. He sprinkled magnetic powder on the polished sample and found something like grain boundaries but within the individual grains of the material. These were indeed the domains of Weiss, and the structure is shown in Fig. 14.5.

In the unmagnetized sample the domains are unaligned and assume balanced orientation, with a resultant magnetic moment of zero. However, if a magnetic field is applied, those domains with their magnetic directions parallel to the field will grow at the expense of those opposed. If the microstructure is such that the new domain structure is retained, we have a permanent magnet; if the material returns to the original state, we have a soft magnet.

As one domain grows at the expense of another, a change in alignment of the individual atomic magnets takes place. The domain boundary is called a "Bloch wall" and is highly important because it helps us understand the reasons for the magnitude of the magnetic field required to magnetize a material. First we should explain that the ease of magnetization varies with direction in a crystal, just as the ease of straining elastically varies in a single crystal, as shown by the modulus of elasticity (Chap. 3). Typical magnetization data for iron and nickel are shown in Fig. 14.6. Note that the preferred orientation for iron is [100] (higher induced magnetization for low magnetic field strengths). We would expect to encounter a domain polarized in the [001] direction on one side of a Bloch wall and another in the [00$\bar{1}$] direction on the other side. In the wall itself (about 1000 Å wide) we find intermediate polarizations and pass through the directions of difficult magnetization (Fig. 14.7). These domains are shown very clearly in Fig. 14.8 by another technique, using the interaction

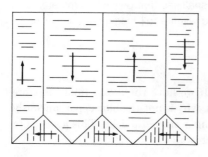

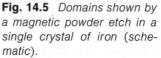

Fig. 14.5 *Domains shown by a magnetic powder etch in a single crystal of iron (schematic).*

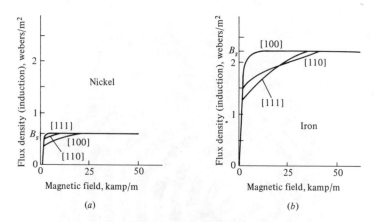

Fig. 14.6 *Magnetic anisotropy. (a) Nickel. (b) Iron.*

[After S. Kaya et al., *Sci. Reports* (*Tohoku Imp. Univ.*), (I)**15**: 721 (1926); (I)**17**: 1,157 (1928).]

of the domains with polarized light. The effect of a magnetic probe in growing domains is shown.

14.6   Effect of temperature on magnetization

Every ferromagnetic material will lose its magnetic properties in some temperature range upon heating, and the ferromagnetism will finally disappear at a temperature called the "Curie temperature." For example, the Curie temperature for iron is 770°C, similar to Fig. 14.9. At this temperature the thermal oscillations overcome the orientation due to exchange interaction and a random grouping of the atomic magnets results. Domains are re-formed on cooling, so the application of a magnetic field from the Curie temperature to room temperature can be helpful in producing a permanent magnet.

Let us now, for review, consider the characteristics of the *B-H* curve in terms of domain theory.

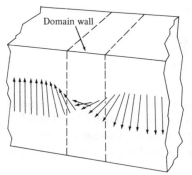

Fig. 14.7 *Domain wall showing the gradual change in direction of field between adjacent domains.*

Fig. 14.8 *Magnetic domains in a thin (0.1-mm) single crystal of a magnetic ceramic, samarium terbium ortho-ferrite (perovskite structure). A beam of polarized light passes through the crystal, then through an analyzer. The domains then show on the screen because of the effect of their magnetic field on the light beam (Faraday effect). The magnetic needle (white finger) has caused growth of the domains in the crystal which have the same magnetic direction as the domains in the needle (black areas).*

(Courtesy of P. C. Michaelis, Bell Telephone Laboratories.)

14.7 Magnetic saturation

If we start with a small field strength H and then increase it, the flux density B will rise rapidly and then attain *saturation induction* (Fig. 14.2). After this point the flux will only increase slightly with H. This phenomenon can be readily explained by the effect of increasing field strength on domains, as

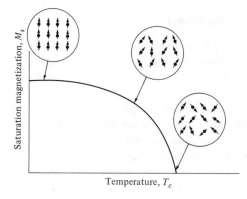
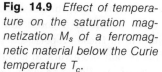

Fig. 14.9 *Effect of temperature on the saturation magnetization M_s of a ferromagnetic material below the Curie temperature T_c.*

Saturation magnetization, M_s

Temperature, T_c

discussed earlier. To begin with, as the field is increased, the domains in line with the field grow and the permeability is high. As the flux approaches saturation, the atomic magnets in the domains are practically all aligned and contributing to the high flux density. Beyond this point of saturation induction further increase in field strength only produces the small increases in flux density that would be obtained without the core of magnetic material.

14.8 Remanent induction

When the field strength H that was applied originally is decreased to zero, a magnetic flux B_r will persist in a magnetic material (Fig. 14.2). In a permanent magnet this will have a high value, while in a soft magnetic material it will be low. The reverse field H_c needed to demagnetize the bar and produce zero flux is called the "coercive field." The extreme differences between soft and hard magnets are shown in Fig. 14.3.

To explain this phenomenon structurally, it is necessary to observe the effects of the metal structure on retaining the domain structure that is present at high imposed field strengths. In a bar of pure iron, a soft magnetic material, the domains will revert to the original balance when the field is removed. However, in a magnetic material with obstacles to uniform domain movement, such as inclusions, highly cold-worked material, or strained regions, a considerable portion of the magnetism will be retained (B_r). For example, a quenched or cold-worked bar of high-carbon steel will make a good permanent magnet. Other methods of observing these effects include the isolation of domains by using powders suspended in an insulator so that each particle is a domain or, as in Alnico magnets, the development of a fine precipitate in which the domains are isolated from each other.

14.9 Metallic magnetic materials

Let us now compare the properties of some typical engineering materials with their structures. There are two principal groups: soft magnetic materials and hard or permanent magnetic materials.

Soft Magnetic Materials. In these materials, such as those used in a transformer core, the most sought-after properties are minimum hysteresis loss (area of the *B-H* loop) and maximum saturation magnetization. The attainment of these properties depends on the ease of domain movement, and for this reason cold-work and the presence of second phases are avoided. In the silicon-iron alloys the grain orientation in the sheet is also controlled. We find that the ease of magnetization varies with direction in a crystal, being easiest in the ⟨100⟩ directions. The sheet for transformers is prepared with the (100) plane parallel to the plane of the sheet and the [001] direction parallel to the rolling direction. This is the so-called "cubic" texture. The advantage of this is that the greater the ease of magnetization, the smaller the hysteresis loop and, consequently, the lower the power loss in the transformer core.

The special properties of "permalloy" (very low hysteresis) are obtained by controlled cooling to avoid the ordered structure that occurs in these iron-nickel alloys. This alloy is especially sensitive to work-hardening, and for this reason the shape or wire is formed and then annealed.

Hard Magnetic Materials. As we would expect, the characteristics desired in hard or permanent magnetic materials are quite the opposite of those wanted in soft materials. When the field is removed, we want a high value of remanent magnetization B_r and of coercive field H_c. It has been found, furthermore, that the lifting force of a magnet is related to the area of the largest rectangle that can be drawn in the second quadrant of the hysteresis curve, called the "*BH* product," and this energy product is taken as an index of the "power" of a magnet (Fig. 14.3c).

As discussed earlier, the key to retaining magnetism is to prevent relapse of the domains into a balanced orientation, giving no external magnetic moment. This can be prevented by two principal methods:

1. Use of a strained metal structure
2. Division of the magnetic part of the structure into small volumes that reorient with difficulty

In the first method high-carbon or alloy steels are quenched to produce martensite, which has a high internal stress.

In the second method alloys such as Alnico (Al, Ni, Co, and Fe) and Cunife (Cu, Ni, Fe) are used. The key to the Alnico alloys is that a single-phase BCC structure, stable at 1300°C, decomposes into two different BCC phases,

Table 14.1 MAXIMUM *BH* PRODUCTS
FOR SEVERAL MAGNETIC MATERIALS

Material	Maximum BH Product, amp-weber/m^3
Samarium-cobalt	120,000
Platinum-cobalt	70,000
Alnico	36,000
Ferroxdur*	12,000
Iron-cobalt sinter	7,000
Carbon steel	1,450

*A synthetic ceramic magnet ($BaFe_{12}O_{19}$) described in the following section.

α and α', at 800°C. The α', rich in iron and cobalt, has a higher magnetization and precipitates as fine particles. It takes a substantial field to magnetize these particles because each is essentially a single domain and must be rotated. However, once this is done, the B_r is high and the coercive force to demagnetize is also high. Many interesting techniques to improve the characteristics of hard magnets are used, such as controlled solidification to develop preferred orientation and heat treatment in a magnetic field.

More recently, samarium-cobalt and platinum-cobalt permanent magnets have been developed with very high maximum energy (*BH*) products, as shown in Fig. 14.3 and Table 14.1.

In the platinum-cobalt magnets a phase transformation (order to disorder) results in isolated domains, as it does in Alnico. This material is machinable but expensive—about $2,000/lb (1973 prices).

The samarium-cobalt magnet and other materials involving rare earths such as praseodymium and lanthanum are prepared as intermetallic compounds such as Co_5Sm (37 percent by weight samarium). Then the material is ground to fine powder, so that each particle is domain size. These tiny grains (of hexagonal structure) have great differences in ease of magnetization, depending on the direction in the crystal—easy along the c axis and difficult in the a direction. The powder particles are aligned by a strong magnetic field in the die, then pressed together and sintered. Care is taken to avoid grain growth during sintering. One use of these Co_5Sm magnets is in step motors in electronic wristwatches. The name of these motors arises from the fact that each time the motor receives a pulse from the vibration of a quartz crystal it moves the hands a step. The present cost of the magnet is $200/lb (1973 prices), but important decreases are expected.

14.10 Ceramic magnetic materials

For many years there was a search for magnetic materials with high electrical resistance. As an example of one important use of such a material, we know that a transformer heats up because the alternating electrical field

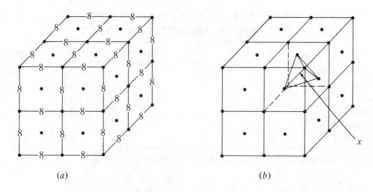

Fig. 14.10 *Octahedral and tetrahedral sites in a spinel. Oxygen atoms are indicated by • ; (a) possible octahedral sites are indicated by 8; and (b) location of the tetrahedral site is indicated by x.*

generates a stray current in the core, which leads to heating (eddy currents). In high-frequency equipment this heating is very severe, so a high-resistance magnetic material was sought. It was recognized that the mineral magnetite, Fe_3O_4, was a natural high-resistance ceramic magnet, so the search led to synthetic ceramic magnets of similar structure, generally called "ferrites" although many do not contain iron. (It is important to avoid confusion in the use of the word "ferrite" for these magnetic oxides and the same name for α iron!)

As a starting point in this highly important and rapidly growing field, we should look at the unit cell structure of magnetite itself. If we write the formula $Fe^{2+}Fe_2^{3+}O_4^{2-}$, we see that it is analogous to the spinel $Mg^{2+}Al_2^{3+}O_4^{2-}$ discussed in Chap. 7 as a refractory, and we find it has a similar unit cell. We have waited until this point to describe the unit cell (Fig. 14.10), because it is most important here to understand the magnetic properties. To explain the important ion positions, it is necessary to take a multiple of 8 times the chemical formula written in general terms, with the divalent and trivalent metal ions written as M^{2+} and M^{3+}: $[M^{2+}M_2^{3+}O_4^{2-}]_8$.

The 32 O^{2-} ions are arranged to form eight FCC cells. We recall from Chap. 7 that there are twice as many tetrahedral sites as FCC atoms in a unit cell, while the number of octahedral sites is equal to the number of FCC atoms (i.e. four per unit cell). Examples of these sites are shown in Fig. 14.10. In the model there are 64 tetrahedral and 32 octahedral sites, but only a portion of them are occupied in spinel. In the ferrites the iron ions are distributed as follows in the so-called "inverse" spinel structures:

Iron ions	Tetrahedral sites	Octahedral sites
16 Fe^{3+}	8	8
8 Fe^{2+}	0	8

Table 14.2 MAGNETIC MOMENTS FOR IONS FOUND IN SPINELS

Ion	Total Electrons	3d Electrons	Bohr Magnetons
Fe^{3+}	23	$3d^5$	5
Mn^{2+}	23	$3d^5$	5
Fe^{2+}	24	$3d^6$	4
Co^{2+}	25	$3d^7$	3
Ni^{2+}	26	$3d^8$	2
Cu^{2+}	27	$3d^9$	1
Zn^{2+}	28	$3d^{10}$	0

This is called *inverse* because in *normal* spinel, $MgAl_2O_4$ for example, the M^{2+} ions are in the tetrahedral sites and the M^{3+} are all in the octahedral sites.

Now, just as in ferromagnetic iron, the magnetism of a ceramic magnet is due to the magnetic moments of the ions. These are calculated in the same way as the magnetic moments of the atoms, but they are not the same if $3d$ electrons are removed in ionizing. For example, Fe^{2+} is formed by removing the two $4s$ electrons, so that it has the same moment as Fe^0. However, in forming Fe^{3+} we develop a magnetic moment of 5 Bohr magnetons because one electron of opposite spin is removed. Common magnetic moments are given in Table 14.2.

The eight Fe^{3+} ions in the tetrahedral sites are always opposite in magnetic alignment to the eight Fe^{3+} ions in the octahedral sites, and therefore no net magnetic moment results from the Fe^{3+} ions. This is called "antiferromagnetism," as in metallic manganese. However, the moments of the eight Fe^{2+} ions that also occupy octahedral sites are all aligned in the same direction in the unit cell and are responsible for the net magnetic moment (ferr*i*magnetic condition) (Fig. 14.11). Let us check this arrangement with a calculation.

EXAMPLE 14.2 Calculate the saturation induction of $[NiFe_2O_4]_8$. The unit cell is cubic, $a_0 = 8.34$ Å.

ANSWER The magnetic moment is due entirely to nickel ions, since spins for the 16 Fe^{3+} ions cancel out each other. Therefore, we have

$8 \times 2 = 16$ Bohr magnetons/unit cell contributed by the Ni^{2+} ions

$$M = \left(\frac{16 \text{ Bohr magnetons/unit cell}}{(8.34 \times 10^{-10})^3 \text{ m}^3/\text{unit cell}}\right)\left(9.27 \times 10^{-24}\frac{\text{amp-m}^2}{\text{Bohr magneton}}\right)$$

$$= 2.56 \times 10^5 \frac{\text{amp}}{\text{m}}$$

$B_s = \mu_0 M$ \quad (neglecting the $\mu_0 H$ term)

Assuming all magnetic moments are aligned at B_s,

$$B_s = \left(4\pi \times 10^{-7}\frac{\text{volt-sec}}{\text{amp-m}}\right)\left(2.56 \times 10^5\frac{\text{amp}}{\text{m}}\right)$$

$$= 0.32\,\frac{\text{volt-sec}}{\text{m}^2}\quad(0.32\text{ tesla})$$

The experimental value 0.37 tesla is even greater than the calculated value because, due to cation vacancies in the crystal structure, some Fe^{3+} is present with Ni^{2+} and contributes a higher magnetic moment per atom (see Table 14.2).

However, note that although the saturation induction is greater than the theoretical value, it is less than pure iron in Example 14.1. The two reasons for this are the lower magnetic moment per atom and the lower density of atoms in the ceramic which contribute to the magnetic moment.

Returning to the development of ceramic magnets in general, we find that in most cases the Fe^{2+} ions are replaced partly or completely with other ions. Since the ions in these positions are responsible for the magnetization, a partial replacement with an ion with lower moment will reduce the overall magnetization. Another technique is to add nonmagnetic ions which will replace some of the Fe^{3+} ions and force them into the Fe^{2+} sites.

For *soft* magnets for electronic and radio communications the following substitutions for Fe^{2+} ions are used: $Mn^{2+} + Zn^{2+}$, $Ni^{2+} + Zn^{2+}$, $Ni^{2+} + Cu^{2+} + Zn^{2+}$, giving a hysteresis loop similar to that in Fig. 14.3a. The Mn^{2+}, Ni^{2+}, and Cu^{2+} ions have magnetic moments, but the role of the Zn^{2+} ions is interesting because they have no magnetic moment themselves but do raise the overall moment in the following way. We mentioned that in the normal spinel structure the divalent ions occupy the tetrahedral sites. The Zn^{2+} ions

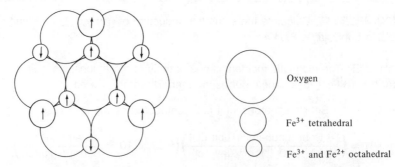

Oxygen

Fe^{3+} tetrahedral

Fe^{3+} and Fe^{2+} octahedral

Fig. 14.11 *Spin directions in Fe_3O_4. Fe^{3+} ions in tetrahedral sites and in octahedral sites have opposite spin directions. The net magnetization is due to the Fe^{2+} ions in the octahedral sites (schematic).*

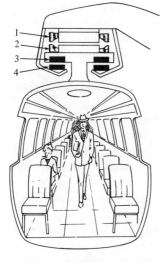

Fig. 14.12 *An artist's conception of a magnetic roadway.* Electric drive: (*1*) *portion of linear motor attached to roadway and serving the same function as the stator of a conventional electric motor;* (*2*) *portion of linear motor attached to the car, functioning as a rotor.* Magnetic suspension: (*3*) *ceramic (ferrite) permanent magnets attached to the car;* (*4*) *ferrite magnets attached to the roadway.*

(Courtesy of Westinghouse Corporation.)

continue to follow this tendency when added to the inverse spinel and thereby force some Fe^{3+} from their "unproductive" tetrahedral sites to the octahedral sites, where they align to increase the magnetic moment. It is important to realize that, just as we added alloying elements in solid solution to change mechanical properties, in this case we are adding ions in solid solution to modify magnetic properties.

To provide magnets with square loops (Fig. 14.3*b*), the substitutional ions are $Mn^{2+} + Mg^{2+}$ and Co^{2+}. These are important for computer operations.

The field of ceramic *permanent* or *hard* magnets is especially exciting. Already these materials have found a variety of uses, such as the magnetic inserts in refrigerator doors and in small motors. Formerly an electromagnet was needed in these motors, but with the development of small powerful permanent magnets to provide a field, the design is greatly simplified. These operate windshield wipers, heater blowers, air conditioners, and window raisers in modern automobiles. In the 30-billion-eV accelerator at the Brookhaven National Laboratory over 7 tons of ferrites are used. As an illustration of possible future use, a drawing of an elevated magnetic roadway in which the car would be floated by magnetic repulsion is shown in Fig. 14.12.

The crystal structure of these permanent ceramic magnets is slightly different from the spinel structure (see the problems), but since it follows the same principles, the materials are still called "ferrites." Just as the ferrites we have discussed have the structure of a naturally occurring mineral (spinel), so these hard ferrites have the hexagonal structure of a mineral called "magnetoplumbite," $Pb(Fe,Mn)_{12}O_{19}$. The source of the magnetism is similar: the alignment of ions such as Fe^{2+} and Mn^{2+} with magnetic moments and the formation of domains. In the synthetic ceramic magnets simpler formulas are used, such as $PbFe_{12}O_{19}$ or $BaFe_{12}O_{19}$. One of the ways of obtaining a strong initial alignment of domains is to heat the material above its Curie temperature and then cool it under the influence of an externally imposed magnetic field. This is analogous to polarizing a ferroelectric material by cooling it under an electric field.

In addition to the soft and hard types, there are two subtypes of great importance.

The square-loop materials retain their full remanent magnetization as the field is reversed until a sharp cutoff is reached (Fig. 14.3b). Therefore, in using the magnet as an information-giving device, we can get full strength without weakening until the cutoff is reached. If we want to clear out the information, this is done cleanly and the device is ready to remagnetize. For these magnets manganese-magnesium-iron oxides and cobalt-iron oxides are used.

For microwave communication it is important to have low losses per cycle because of the high frequencies. In iron-core transformers we can tolerate the hysteresis losses at 60 Hz but at 10^6 Hz they would be prohibitive. Here the high resistance and very narrow loops of certain ferrites are important. Typical oxides are of the $(Al,Ni,Zn)O$ type.

Another family of magnetic ceramics is the iron garnet group. The general formula is $3M_2O_3 \cdot 5Fe_2O_3$. The unit cell is more complex than spinel and the chief advantage is that elements with high magnetic moment such as samarium and yttrium can be accommodated as M in the formula.

SUMMARY

In the electrical materials the most important concept was the effect of structure on the ease of movement of charge carriers. In an analogous way, in magnetic materials the important concept is the effect of structure on the value of magnetic flux. There are two important differences to be established immediately.

The value of magnetic flux density B obtained by the application of a magnetic field of strength H is a function of the permeability μ, but at some value of H very little increase in B is encountered. This is the saturation magnetization B_s.

Second, the flux density does not fall to zero upon removal of the field, but only drops to the residual magnetization B_r, which is low in a soft magnetic material and high in a permanent or hard magnetic material.

The features of hard and soft metallic magnetic materials are reviewed first. It is established that the key feature is the magnetic domain, a region of structure in which the magnetic moments of the atoms are all aligned in one direction. When a field is applied, the domains with moments aligned parallel to the field grow at the expense of others. In a soft magnetic material, domain wall movement is made easy by simple grain structure, avoidance of cold-work, etc. Conversely, in hard magnetic materials the desired structure consists of tiny (300 Å) particles, each with only one domain and aligned to provide a strong magnetic field.

In ceramic magnets the inverse spinel structure is used to produce soft magnets of the desired characteristics. For hard magnets fine domain-sized particles of the magnetoplumbite structure are used.

DEFINITIONS

Permeability, μ The ratio of the flux density B developed when a magnetic field of strength H is applied; $\mu = B/H$. This is analogous to the concept that the conductivity σ is equal to the ratio of the current I produced by a voltage E in a wire (for unit length and area of the conductor).

Relative permeability, μ_r The ratio of permeability of a material to the permeability of a vacuum; $\mu_r = \mu/\mu_0$.

Diamagnetism, paramagnetism   Very small deviations from $\mu_r = 1$.

Ferromagnetism   Large values of μ_r produced by alignment of certain atoms such as iron, cobalt, and nickel into domains.

Domain   A region of a metal or ceramic structure in which the magnetic moments due to the individual atoms are aligned in the same direction.

Domain wall, Bloch wall   The boundary between two domains. On either side of the wall the magnetic moments are oriented at 90 or 180° to each other.

Antiferromagnetic   Material with ferromagnetic atoms that are aligned with magnetic moments in opposing directions, leading to no net magnetic moment.

Ferrimagnetic   Material with ferromagnetic atoms that are aligned in opposing directions in the unit cell but with a net moment.

Bohr magneton   The magnetic moment produced in a ferromagnetic atom by one unpaired electron. 1 Bohr magneton = 9.27×10^{-24} amp/m^2 (0.927 erg/gauss).

Saturation induction, B_s The maximum magnetic flux obtainable in a material. For a ferromagnetic material, B_s may be taken as $\mu_0 M$.

Magnetization, M The magnetization times the permeability of a vacuum gives the increase in flux due to insertion of a given material into a field of strength H. Since μ_0 is a constant ($4\pi \times 10^{-4}$ volt-sec/amp-m), the magnetization is a measure of the increased flux density produced by the material.

Remanent induction, B_r The flux density remaining when magnetic field is removed.

Coercive field, H_c The field needed to reduce flux density to zero after initial application of a magnetic field to obtain B_s.

Soft magnetic material   An easily magnetized and demagnetized material, with a small area in the *B-H* loop.

Hard magnetic material   Material that is magnetized and demagnetized with difficulty and that has a large area in the *B-H* loop.

Normal spinel   A ceramic of the general formula $M^{2+}M_2^{3+}O_4$. There are divalent ions in tetrahedral sites and trivalent ions in octahedral sites.

Inverse spinel   Material with the same general formula as a normal spinel but with the trivalent ions occupying tetrahedral and octahedral sites and the divalent ions occupying only octahedral sites.

Ferrite   A magnetic inverse spinel. Do not confuse this term with the same word used earlier to mean α iron.

PROBLEMS

14.1   An important characteristic in a power-transformer core is the power loss which leads to heating and lowered efficiency. The power loss is made up of stray currents induced in the core and the hysteresis loss which is related to the area inside the *B-H* loop for a complete cycle.

    a. Describe the type of metallic material you would select for a transformer and explain how it would be fabricated (laminations, preferred orientation, etc.).

    b. Describe the type of ceramic material you would select and the ideal *B-H* curve you would specify.

14.2   Ceramic materials with square *B-H* curves are used for memory units in computers. Explain the differences in *B-H* curves you would specify compared to ceramic *transformer* materials.

14.3   Explain how the following characteristics of a ferromagnetic material could affect its *B-H* curve:

    a. Grain size
    b. Inclusions
    c. Residual stresses
    d. Heat treatment (specify)

14.4   After an external magnetic field is removed, the domains tend to revert to their original size and distribution. Explain. What structural conditions inhibit this motion, i.e. tend to retain the induced magnetism?

14.5   Estimate the magnetic moment per unit cell of the following substances:

    a. $[Fe_3O_4]_8$, considered as $[Fe^{2+}Fe_2^{3+}O_4]_8$
    b. $[MgAl_2O_4]_8$
    c. $[CoFe_2O_4]_8$

14.6 Estimate the saturization magnetization of gadolinium (atomic weight = 157.26 g/mole, density = 7.8 g/cm^3), assuming the magnetic moment per atom is equal to the number of unpaired $4f$ electrons and that the atom follows Hund's rule. In this case Hund's rule states that the first seven $4f$ electrons will have the same spin direction and the next seven the opposite spin direction.

14.7 Calculate the saturation induction B_s of $[Li_{0.5}Fe_{2.5}O_4]_8$, assuming $a_0 =$ 8.37 Å in an inverse spinel.

14.8 In the manufacture of Alnico magnets the alloy is solidified and cooled in the magnetic field. What is the advantage?

14.9 In making magnetic sealer sticks for refrigerators the magnetic material is placed in a thermoplastic in the liquid state, and a magnetic field is maintained while the plastic sets. Why?

14.10 The terms "ferromagnetic," "antiferromagnetic," and "ferrimagnetic" are used to describe the different types of alignment of electron spins in iron, manganese, and magnetite (Fe_3O_4, spinel), respectively. For each case sketch a group of atoms illustrating the difference in spin alignment. In the ferrimagnetic structure mark which group of ions is antiferromagnetic and which group is aligned to produce magnetization.

14.11 Show how the net magnetic moment of $[(Zn_{0.5}Ni_{0.5})Fe_2O_4]_8$ is greater than that of $[NiFe_2O_4]_8$. (*Hint:* The zinc ions displace Fe^{3+} ions from tetrahedral sites.)

14.12 In the magnetoplumbite structure of hard magnets the general formula is $AO \cdot 6B_2O_3$, where A is barium, lead, or strontium; O is oxygen; and B is aluminum, chromium, gadolinium, or iron.

    a. The valences of A and B are constant but different. What are they?

    b. Using these valences, what is the percentage variation in ionic radius for A metals and B metals?

14.13 A magnetic ferrite for an oscilloscope component is to have a final rectangular shape measuring 0.621 by 0.356 by 0.109 in. (1.58 by 0.905 by 0.277 cm). An investigator finds that the specific gravity of a similar test piece changes from 3.57 green (before sintering) to 5.02 after sintering. Design and give the dimensions of a die for producing the green shapes that will sinter to the desired dimensions. Assume linear shrinkage is the same percentage in all directions. Also, note that the volumetric shrinkage is not 3 times the linear shrinkage for appreciable amounts of shrinkage.

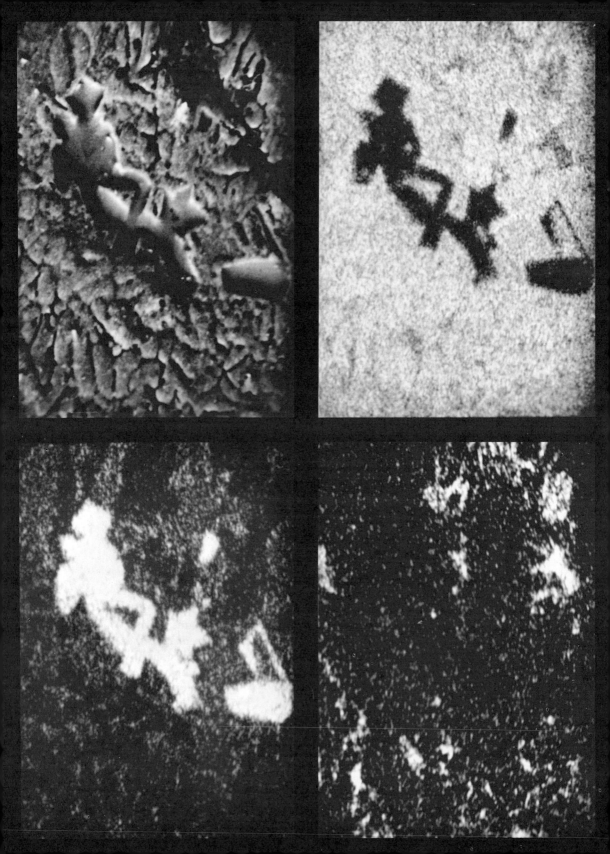

15.1 Introduction

In this final chapter we will take up some of the optical and thermal properties of materials, and at this point it should hardly be necessary to anticipate that these properties, like all the others we have studied, are dependent on the structures of the materials. We will begin with the optical properties because they are closely related to the electrical properties recently discussed in Chap. 13.

Optical Properties

By "optical properties" we mean those properties governing the emission, absorption, transmission, reflection, and refraction of light. This field covers not only the well-known properties such as emission from a tungsten filament, the sources of color, transparency, opacity, and index of refraction, but also the newer areas of explosive growth such as fluorescence in lighting, phosphors for television tubes, and emissions from lasers.

15.2 Emission

There are two principal types of emitted light: that with a continuous spectrum, such as "white" light from a tungsten light bulb, and light confined to definite wavelengths, such as that from a sodium-vapor lamp. We shall concentrate on the case of emission of definite wavelengths because of its great importance in new devices.

From physics we learn that when light is emitted from any source, we need first to think of it as traveling in discrete units or little energy packets called "photons" (Fig. 15.1). Second, we must recognize that light has a definite wavelength which is related to the energy of the photon. We will need to use both these aspects in explaining the interaction of light with materials, especially with the electrons of materials.

Since visible light can originate from radiation of nonvisible wavelengths, let us review a broad portion of the spectrum. On one side of the visible spectrum we find the infrared and longer-wavelength photons, and on the other, the ultraviolet and x-ray wavelengths (Fig. 15.2).

From physics we have the relation between energy and the wavelength of a photon:

$$E = \frac{hc}{\lambda}$$

15

OPTICAL AND THERMAL PROPERTIES OF MATERIALS

THIS illustration shows the use of the microprobe to find the chemical composition of the different phases in the aluminum silicon copper engine block alloy discussed in Chap. 4. The overall microstructure is shown in the upper left and consists of a silicon-rich phase in relief in a matrix of aluminum-rich alloy. This structure is then exposed to a sweeping beam of electrons which excites x-rays from the surface. The wavelength of the x-rays depends on the atoms emitting them.

To find the location of silicon atoms for example, the x-rays are filtered so that only those characteristic of silicon are passed through to the TV screen where the image of the structure is shown. Therefore bright areas show the silicon-rich phase, lower left. In the upper right, the x-rays from aluminum atoms were admitted, showing the aluminum-rich matrix. Finally, in the lower right, the locations of the copper-rich phase are indicated.

In this chapter first we will take up the optical properties of materials, relating the emission, absorption, etc., of light to the structure. Then we will consider the thermal properties which have not yet been described.

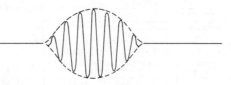

Fig. 15.1 *Schematic drawing showing both particle and wave characteristics of a photon.*

where E = energy
h = Planck's constant = 6.62×10^{-27} erg-sec
c = velocity of light = 3×10^{10} cm/sec
λ = wavelength in centimeters

If we express this relation in simple terms of electronvolts, which we used in Chap. 13, we have

$$E = \frac{1.24 \times 10^{-4}}{\lambda} \text{ eV}$$

EXAMPLE 15.1 Calculate the energy of photons of infrared ($\lambda = 11{,}500$ Å), yellow ($\lambda = 5800$ Å), and ultraviolet ($\lambda = 2000$ Å) light.

ANSWER

Infrared $\qquad E = \dfrac{1.24 \times 10^{-4}}{11{,}500 \times 10^{-8}} = 1.08$ eV

Yellow $\qquad E = \dfrac{1.24 \times 10^{-4}}{5800 \times 10^{-8}} = 2.14$ eV

Ultraviolet $\qquad E = \dfrac{1.24 \times 10^{-4}}{2000 \times 10^{-8}} = 6.20$ eV

This interrelation of energy and wavelength is important in understanding both emission and absorption of light, and we will consider emission first.

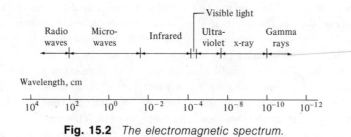

Fig. 15.2 *The electromagnetic spectrum.*

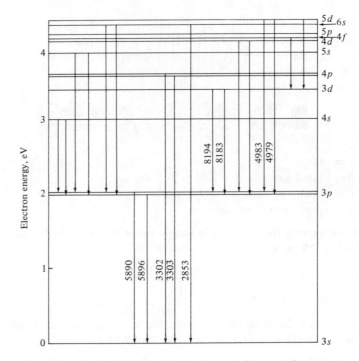

Fig. 15.3 *Electron energy level diagram for a sodium atom. The photon wavelengths associated with some of the electronic transitions from excited states are given in units of 10^{-8} cm.*

We are all familiar with the emission of yellow light that occurs when we throw salt into a fire. Let us examine the reason for this phenomenon. We recall from Chaps. 2 and 13 that the electrons in an atom may occupy different energy levels. As an example, in Fig. 15.3 the energy levels above the 3s level in sodium are shown. If an electron in the 3s level is raised to the 3p level and then falls back to the 3s level, a photon of about $\lambda = 5890\,\text{Å}$ or yellow light is emitted. The steps involving energy absorption (raising the electron to an upper level) and the fall need not be the same. For example, we could supply the energy to raise the electron to the 6s level, and it could jump down to the 3p level and then to the 3s. An analogy is to consider playing a pinball machine. The ball is sent to the top in one stroke but it can come down by a series of jumps, some of which may give a blink of light, others just noise or heat.

At this point we should also consider the companion problem of the emission of shorter wavelengths (x-rays) when an electron falls to a still lower level. If we bombard sodium atoms (or ions) with sufficiently energetic electrons, we can knock out an electron from the first (K) shell. This creates a hole into which an L-shell electron can fall. For example, the transition $2p \rightarrow 1s$ gives a photon of wavelength 11.91 Å or an x-ray wavelength. Just as the color of visible light emitted from a given jump depends on the particular levels involved, so the wavelength of the x-ray depends on the specific case. For instance, it takes much more energy to knock out an electron from the inner ring (K shell) of an atom of high atomic number because of the greater charge on the nucleus, and we expect to find a shorter-wavelength x-ray emitted when an L-shell electron falls into the K shell. Typical wavelengths illustrating this effect for the K_α x-ray are manganese: $\lambda = 2.10578$ Å; iron: $\lambda = 1.939980$ Å; tungsten: $\lambda = 0.213828$ Å.

The scanning electron microprobe that we mentioned in Chap. 4 depends on the emission of these characteristic x-rays. A scanning electron beam is swept across a given microstructure. From each tiny region of the structure x-rays are produced with wavelengths which depend on the elements present. For example, if we want an indication of the amount of silicon in a given region, we put in a filter that allows only the silicon wavelength rays from this region to pass through. Then the picture screen showing the microstructure is irradiated by these x-rays and shows bright spots only at regions of high silicon concentration (Chap. 15 frontispiece). Similarly, the regions of high aluminum concentration are found by allowing only the x-rays that are characteristic of aluminum to pass through in another picture.

We should discuss one more phenomenon, called "luminescence," which is quite important in the new light-emitting materials. A familiar sight in natural history museums is a case of fluorescent and phosphorescent minerals. After exposure to ultraviolet light the specimens glow different colors, and some continue to glow after the ultraviolet light source is turned off. This conversion of ultraviolet light to visible light is given the general term "luminescence." If this glow ceases when the ultraviolet stimulus is removed, it is called "fluorescence." If the glow persists for over 10^{-8} sec after the ultraviolet light is turned off, the light is called "phosphorescence." In either case we have to explain why light of a visible wavelength results from the exposure to ultraviolet light or electron bombardment.

In a luminescent material a solid, such as zinc sulfide, contains a small amount (one part per million) of an impurity, such as copper. This impurity is the key to luminescence. Again the process is a little like a sophisticated pinball machine in which an electron is raised to a high level and may fall into traps from which it is later released (Fig. 15.4). There are essentially three steps in this process:

1. A photon or electron is shot at the solid, such as zinc sulfide, and its energy is absorbed. This enables an electron of the solid to rise to the conduction band, which is separated from the valence band by a definite energy gap. A hole is created in the valence band as well.
2. The electron wanders through the structure until it falls into a trap or luminescent center. These traps are produced by the impurity, in this case

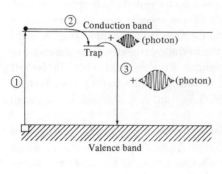

Fig. 15.4 *Development of luminescence. Step 1: The electron is excited by ultraviolet radiation into the conduction band. Step 2: The electron falls into the trap, emitting an infrared photon (the trap is an intermediate energy level due to an impurity). Step 3: The electron falls into a lower level, emitting a photon of visible light. Note: The hole will also be attracted to the trap and will emit light.*

the copper ions. The "energy level" traps lie between the more elevated conduction band of the solid and its valence band.

3. After a time the electron acquires enough energy to leave the trap and fall to the valence band, giving up a photon of definite wavelength (in the visible spectrum) related to the ion producing the trap, in this case copper. Phosphorescence occurs when materials have deep traps. Therefore, it takes more time for the trapped electrons to receive enough energy to escape. As a consequence, they are liberated over a longer period by continued electron movement. It is possible to produce approximately "white" light by using a mixture of impurities or luminescent centers such as silver which gives a peak at 4300 Å, copper at 5100 Å, and manganese at 5900 Å.

Applications. These variables in color and in duration of luminescence are of great practical importance. In the dots of a color television tube, for example, phosphors that give pure primary colors and do not have long decay times are desired. The development of materials with these characteristics provides an example of the extremes to which manufacturers will go for color quality. It was found that europium, a rare earth, provided good red-color centers. As a result, a new mining and chemical-separation industry developed, and the other rare earths that are found with europium are now available as low-cost byproducts. Another application of these effects is in electroluminescent panels in which the initial radiation step is eliminated. Here the application of a voltage across a strip provides directly the energy required for electrons of the material to move to the phosphor centers, and the jumps that occur later give off light. The color of the panel is a function of the active ions chosen.

Lasers. Light amplification by stimulated emission of radiation (*laser*) is an even more sophisticated effect. In the ruby laser, for example, a single crystal rod of alumina (Al_2O_3) contains a small quantity of Cr^{3+}

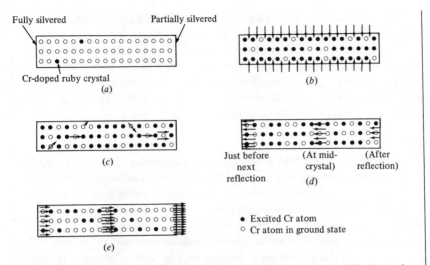

Fig. 15.5 *Schematic drawing of stimulated emission and amplification of a light beam in the ruby laser. (a) At equilibrium. (b) Excitation by xenon light flash. (c) A few spontaneously emitted photons start the stimulated emission chain reaction. (d) Reflected back, the photons continue to stimulate emission as they travel. (e) Increasing in power, the beam is finally emitted.*

(R. M. Rose, L. A. Shepard, and J. Wulff, "The Structure and Properties of Materials," vol. 4: Electronic Properties, John Wiley & Sons, Inc., New York, 1966.)

ions as impurities (Fig. 15.5). The rod is ground with flat surfaces at each end, and one end is heavily silvered to be opaque while the other is lightly silvered for partial reflection. Shining on the rod is a tube containing xenon gas, which emits constant-wavelength light (5600 Å) when energized. The sequence is as follows: At first only a few electrons in the chromium ions are above the ground state, but after the xenon lamp is flashed, the electrons are energized. These electrons either fall to the ground state or take up a higher-energy metastable state for an average of 3 millisec. A high percentage are pumped up to this metastable state. Next a few electrons fall to the ground state, emitting photons of a characteristic wavelength. A percentage of these emitted photons are reflected from the ends and stimulate emission throughout the structure in a short period, $\frac{6}{1,000}$ sec.† A pulse of coherent wavelength emission thus travels through the lightly silvered end. (By "coherent" we mean that the photons are in phase with each other.) It is a beam with very low divergence because photons emitted through the sides of the rod are lost.

---

†There is a critical relation between the length of the rod and the wavelength of the emission, so that a "standing wave" of photons is built up within the rod.

Many uses have developed for lasers, including surgical and metal-lurgical heating, surveying, and precision measurement. Various materials, including gas, can be used in the laser. In a semiconductor laser with a *p-n* junction an electrical impulse energizes the carriers, so that as they fall through an energy gap, a stimulated emission of photons of longer wave-length (6500 to 8400 Å) is obtained. Also, lasers with continuous emission of light (rather than pulsed) have been developed.

Emission from Solid Masses. As distinguished from the emission of light by separated ions or electrons, the radiation from heated solids such as a tungsten wire filament in a conventional light bulb is important. Here we obtain white light or a continuous spectrum, because in the solid tungsten metal we have a *band* of electron energies and a band of sites into which excited electrons can fall. Another case of emission involves the "boiling off" of electrons from the incandescent filament of a radio tube to be accelerated through space by the potential between the electrodes.

15.3 Absorption

It is important to distinguish between color produced by emission and color produced by absorption. We see a yellow color *emitted* when salt is thrown in a fire because of the characteristic wavelength of the electron jumps in the sodium atom. However, if we look through a yellow glass filter at a white light, we see yellow light transmitted because the filter has absorbed the other colors and has merely allowed the already existing photons of yellow wavelength to pass through. The reaction, therefore, is between the electrons of the material and the photons of nonyellow wavelengths. The range of colors produced this way is called the "absorption spectrum."

For example, glass is colored only by ions of the transition elements. Copper is considered a transition element in the Cu^{2+} state because it has an unfilled $3d$ shell. Typical colors are Cr^{2+}, blue; Cr^{3+}, green; Cu^{2+}, blue-green; Cu^+, colorless; Mn^{2+}, orange; Fe^{2+}, blue-green; U^{6+}, yellow (unfilled $5f$ shell). These colors result from the interaction of the inner-shell electrons with the photons of the various wavelengths of white light, and the characteristic color that emerges is the unabsorbed part of the spectrum. Because the ions do not act independently of the glass matrix, the color is not a sharp wavelength, as it is in the emission spectrum, where the ions in the vapor have no such interaction with a matrix. The rare earths also lead to colored glasses because of the electron interaction in the $4f$ shell.

Another example of absorption is the photoelectric effect in which the absorption of a photon results in the emission of an electron.

When a photon of light of energy $h\nu$ (ν = frequency = c/λ) strikes an electron in a block of metal and the electron is sufficiently energized to leave the block, we say that the energy of the electron is made up of two components:

1. The energy required to leave the block ($e\phi$)
2. The kinetic energy (K.E.) of the electron outside the block

Therefore, $h\nu = e\phi + \text{K.E.}$

The value $e\phi$ is a constant for the material at a given temperature and is called the "work function." Multiplying the work function by the charge on the electron gives the energy barrier that is surmounted as the electron leaves the material.

It would be expected that the alkali metals would make good emitters because of the ease with which their electrons leave. This is found to be the case, and the yield of electrons varies with the wavelength of the light and the element. As the atomic number increases from sodium to cesium, the maximum yield occurs at larger wavelengths. In other words, a photon of higher energy is required to knock out an outer-shell electron in sodium compared with cesium because the outer electrons are further from the nucleus and are more loosely held in the latter case.

EXAMPLE 15.2 The work function $e\phi$ for tungsten is 4.52 eV. What is the maximum wavelength of light that will give photoemission of electrons from this element?

ANSWER The energy of the photons will just equal the value they require to escape; in other words, kinetic energy is equal to zero. Therefore,

$$h\nu = e\phi = 4.52 \text{ eV}$$

$$(6.62 \times 10^{-27} \text{ erg-sec})\left(10^{-7}\,\frac{J}{\text{erg}}\right)\nu = 4.52 \text{ eV}\left(1.6 \times 10^{-19}\,\frac{J}{\text{eV}}\right)$$

or

$$\nu = 1.09 \times 10^{15} \text{ sec}^{-1}$$

$$\lambda = \frac{c}{\nu} = \frac{(3 \times 10^{10} \text{ cm/sec})(10^8 \text{ Å/cm})}{1.09 \times 10^{15} \text{ sec}^{-1}}$$

$$\lambda = 2750 \text{ Å}$$

15.4   Reflection

Reflection can be an apparently simple phenomenon or a rather complex one. In the simple case we merely shine a beam of white light at a surface and receive a reflected beam of almost the same intensity and "spectrum," with the angle of incidence equaling the angle of reflection. On the other hand, we find a red color in the reflected beam from copper, a yellow color from gold, and similar but less noticeable variations from other metals. In all cases the photons interact with the electrons, since a photon can be considered an electromagnetic wave. However, the interaction is not the same for all wavelengths of light and depends on the energies of the different electron levels of the material. In metallic reflection, significant interaction can occur between the photons and the outer-shell electrons of the metal. Thus, in the case of copper the photons corresponding to blue wavelengths are absorbed by $3d$ electrons, leaving a reddish color to be reflected.

15.5 Transmission

Now let us consider why metals are opaque in all but very thin sections and why ionic and covalent solids are transparent to visible light. At the same time we can discuss the more interesting subject of why materials we consider opaque, such as silicon and germanium, are transparent to infrared.

Metals. We recall that the valence electrons of a metal can be thought of as having a band of energies with many unfilled energy states at the upper energy levels. When a photon of light strikes an electron, the electron can be raised to an unoccupied higher level, and the radiation is absorbed rather than transmitted. We can transmit visible light in metals only in very special cases, such as in a very thin sheet of gold, in which the probability of collision during transmission is reduced.

Ionic and Covalent Solids. In this case the electrons are bound to the atoms. It takes a high-energy photon to break an electron bond and accelerate the electron to the conduction band. (Recall that there is a gap between the valence and conduction bands in these materials.) For this reason photons of visible and infrared wavelengths pass through ionic solids with ease. However, very high-energy radiation such as ultraviolet is absorbed because the energy of these photons can be used to accelerate the electrons to the conduction band. If atoms of an impurity are present, absorption can occur. Such is the case with chromium ions in ruby, which absorb the photons of blue and green wavelengths. The remaining photons (or other wavelengths) in white light pass through, giving an overall red spectrum.

Semiconductors. These materials provide a very interesting special case of transmission. Here we recall that only relatively little energy is required to raise the carriers to the conduction band and therefore absorb radiation. It happens that the energy of the photons in infrared is not sufficient (long wavelength and, therefore, low energy), so that semiconductors are transparent to infrared radiation and can be used as windows or lenses for efficient transmission. However, the energy of visible light is sufficient to raise the carriers to the conduction band; the semiconductors appear opaque in visible light.

EXAMPLE 15.3 Explain the following transmission effects of white light on the given materials.

Material	Energy gap, eV	Color
Diamond	5.6	Colorless
Sulfur	2.2	Yellow
Silicon	1.1	Opaque

ANSWER Referring to Example 15.1, we see that none of the photons of the visible spectrum is energetic enough to react with the electrons of diamond and raise them to the conduction band. Therefore, the light is transmitted unchanged. In sulfur the green and blue wavelengths are absorbed, leaving yellow (and red). In silicon all the visible light is absorbed because only 1.1 eV is required for photon-electron interaction, and this is available from the light.

The use of light to activate electrons of a semiconductor to the conduction band is called "photoconductivity." It is possible to make a semiconductor with an energy gap corresponding to the energy of photons of light of a certain wavelength. If the wavelength is longer than the given wavelength, the photons are not energetic enough to activate electrons to the conduction band. If the wavelength is much shorter, the light is so heavily absorbed that only the surface is affected; since most of the crystal is inactive, the overall photoconductivity is small. At just the critical wavelength the response is high, as shown in Fig. 15.6. By this technique a missile or aircraft, both rich infrared emitters, can be detected from hundreds of miles.

Color Centers. The characteristics of transmission may be altered by impurities, and this is especially evident in the colors developed in transparent crystals such as ruby and sapphire. This is due to the development of color centers. The most striking evidence of this effect is found by heating a transparent uncolored crystal of sodium chloride in sodium vapor and observing the development of a yellow color in the material. The mechanism is as follows: Sodium atoms condense on the surface, and chloride ions migrate from the inner regions to combine with the sodium. This leaves ionic vacancies. To balance the charges in the crystal, these vacancies attract electrons. These electrons are not held as rigidly as the normal valence electrons and can therefore be raised to the conduction band by lower-energy photons, such as those in the

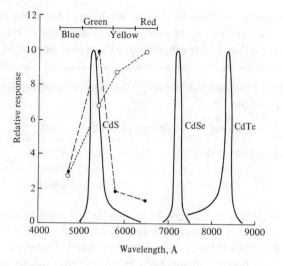

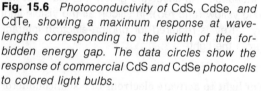

Fig. 15.6 *Photoconductivity of CdS, CdSe, and CdTe, showing a maximum response at wavelengths corresponding to the width of the forbidden energy gap. The data circles show the response of commercial CdS and CdSe photocells to colored light bulbs.*

(J. J. Brophy, "Semiconductor Devices," McGraw-Hill Book Company, New York, 1964.)

visible spectrum. Thus, when white light passes through the crystal, certain wavelengths are absorbed and the remaining beam is colored. Ions of impurities can cause the same effect by providing electrons that are easily raised to the conduction band and therefore photons of certain wavelengths are selectively absorbed.

A number of important new glasses are dependent on the interaction of light with impurities in the glass. For many years it was realized that the windowpanes in old buildings became purple with age, and it was found this was due to the reaction of ultraviolet light with trace amounts of manganese. Specifically, the Mn^{2+} ion is colorless, but reaction with a photon causes the loss of another electron to become Mn^{3+}, which absorbs most of the visible spectrum except violet, thereby giving the glass a purple hue. In this case the electron from the Mn^{2+} is probably trapped by iron impurities, so the process is not reversible. However, in the very popular photochromic eyeglasses and windows containing silver ions, a reversible process is present. When the light is bright, the photons reduce Ag^+ ions to metallic silver, causing a darkening. When the light source is removed, the process reverses and colorless Ag^+ ions are regenerated.

15.6 Refraction

Everyone is familiar with the fact that a stick immersed in water appears to be bent, starting at the water-air interface. A useful concept is that the velocity of light appears to be lower in water. The index of refraction n is defined as

$$n = \frac{c}{v}$$

where c = velocity of light in a vacuum
v = velocity of light in the material

Liquids or solids with dense atomic packing or elements of high atomic number generally have the higher indices of refraction (i.e. lower velocities of light). In some crystals the index of refraction is not the same in all directions, just as there are differences in magnetic properties in different directions. Crystals of lower symmetry, such as calcite and quartz, will split the incident beam into polarized beams. This is called "birefringence" or "double refraction." Not only is this characteristic useful in mineral identification, but it is used in the material Polaroid. In this case crystals of iodoquinone are encased in plastic, and these filters absorb the glare component of the light which is polarized.

Thermal Properties

15.7 Introduction

When a solid is heated, three important engineering effects are encountered. The solid itself absorbs heat, transmits heat, and expands. The first effect is described by the heat capacity C_p, which is the energy required to raise the temperature of a mole of the solid 1 K, or cal/mole-K. The second is defined by the thermal conductivity or the energy (calories) flowing per second from one face to another of a cube of 1-cm dimensions with 1 K temperature difference between the faces, that is, (cal/cm²)/(cm/sec-K) or cal/cm-sec-K. The third is the change in length per unit of length per K, that is, cm/cm-K or simply $(K)^{-1}$. [The interrelation of these units with British thermal units (BTU) and with the joule—the preferred energy unit—is given in the problems.] The interesting problem we have to face is to relate these characteristics to structure.

We can make a start by considering what happens as we heat a simple metal such as silver from 0 K. The energy will be taken up by the increasing amplitude of vibration of silver ions, and the electrons at the upper level of the conduction band will be accelerated. From this simple model we can explain all three thermal properties.

15.8   Heat capacity

For our purposes it is sufficient to discuss C_p, the heat capacity at constant pressure. It is simpler to begin with the heat capacity at say 20°C. We find that most simple solids show a value of approximately 6 cal/mole-K (25.1 J/mole-K). Boltzmann showed that this could be rationalized by considering the energy of the vibrating atoms. From elementary chemistry we know that the internal energy of a mole of perfect gas is $\frac{3}{2}RT$ where R is the gas constant of 1.987 cal/mole-K (8.3 J/mole-K) and T is the absolute temperature (K). For bound atoms vibrating in a solid there is an added potential-energy term of the same $\frac{3}{2}RT$, giving the total energy as $3RT$ or approximately $6T$. Since the energy equals C_pT (from 0 to T K), then $C_p = 6$ cal/mole-K (25.1 J/mole-K).

This simple relation does not apply at low temperatures because of effects discussed in quantum mechanics or to complex solids in which the rotational energy of molecules is involved. The contribution of the electrons to specific heat is also low except at low temperatures.

We should note in passing that although the heat capacity per *mole* is constant for many simple substances, it is important to realize in engineering calculations that there is a large difference in heat capacity per *gram*. For example, the value for iron is 0.107 cal/g (0.448 J/g) while that for lead is 0.031 cal/g (0.1296 J/g). Also, for light elements the heat capacity per mole is lower until elevated temperatures are reached.

15.9   Thermal conductivity

The carriers of thermal energy are electrons in metals and phonons (quanta of energy) in other solids. The velocity of a phonon is the speed of sound, but since many collisions occur, the net thermal conductivity is low compared to that of metals (Table 15.1).

Since electrons serve as carriers in metals, we would expect that good electrical conductivity would be an indication of good thermal conductivity. This is shown by the Wiedemann-Franz relationship:

$$\frac{\text{Thermal conductivity}}{\text{Electrical conductivity} \times T} = L \text{ (practically a constant)}$$

where the Lorenz number $L = 1.6$ to 2.5 (volts/K)$^2 \times 10^{-8}$ for most metals at $T = 20°C$ (293 K).

As alloys are added the thermal conductivity, like the electrical conductivity, falls off rapidly, often to a value only one-tenth of the pure metal. An example is in the construction of stainless-steel cooking utensils. For better conductivity a sandwich of stainless steel with a low-carbon iron center is used

Table 15.1 THERMAL CONDUCTIVITY OF
VARIOUS MATERIALS AT 300 K

Material	Conductivity, cal/cm-sec-K*
Aluminum	0.53
Copper	0.94
Iron	0.18
Silver	1.00
Carbon (diamond)	1.50
Germanium	0.14
70% Ni, 30% Cr solid solution	0.034
70% Cu, 30% Zn solid solution	0.24
Steel	0.12
Sodium chloride	0.017
Potassium chloride	0.017
Silver chloride	0.0026
Glass	0.0019

*Multiply by 4.1868 to obtain J/cm-sec-K.
Source: R. M. Rose, L. A. Shepard, and J. Wulff, "The Structure and Properties of Alloys," vol. 4: Electronic Properties, John Wiley & Sons, Inc., New York, 1966.

so that the influence of the practically pure iron will level out temperature gradients. An alternative is the use of a copper base bonded to the stainless steel.

15.10 Thermal expansion

If we use a classical model of atom or ion vibrations back and forth about fixed centers as a sample is heated, we cannot explain expansion. However, let us review the attractive and repulsive forces between two ions which determine the interionic distance (Fig. 15.7).

As we bring two ions together, we see that the attractive forces dominate until we reach the bottom of the trough at radius r_0. To bring the atoms closer requires additional energy. Now the atoms will show radial spacing r_0 only at 0 K. At higher temperatures the atoms will vibrate with increasing amplitude, as at T_1 and T_2. Note that the average distance of separation will be r_1 and r_2, which do not lie above r_0 because of the different effects of temperature on the attractive and repulsive forces. The quantity $\Delta r/\Delta T$ is the familiar coefficient of thermal expansion, which is characteristically different for different structures.

The ceramics have low coefficients because the attractive forces are almost as great as the repulsive forces as the temperature increases. Values range from 1.2×10^{-6} (K)$^{-1}$ for diamond to 6.7×10^{-6} (K)$^{-1}$ for alumina (Table 15.2). Great variation is encountered in the glasses. In pure fused silica the

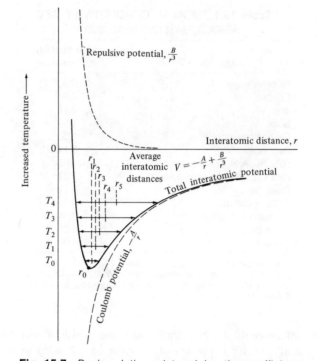

Fig. 15.7 *Basic relations determining the coefficient of thermal expansion. The change in interatomic distance with temperature is determined as follows. At 0 K the balance between the attractive forces of oppositely charged ions (the Coulomb potential) and repulsive forces as the electron clouds overlap leads to the interatomic distance r_0. As the material is heated, the average distance becomes r_1, r_2, etc. The change in distance per K is the coefficient of expansion.*

(R. M. Rose, L. A. Shepard, and J. Wulff, "The Structure and Properties of Materials," vol. 4: Electronic Properties, John Wiley & Sons, Inc., New York, 1966.)

coefficient is low, 0.05×10^{-6} (K)$^{-1}$, while in window glass the bonding is weaker and the coefficient is 7.2×10^{-6} (K)$^{-1}$. Correspondingly the melting and softening points of fused silica are higher than those of window glass.

In the metals there is a close correlation between the melting point, which is an index of bonding, and this coefficient. Aluminum, for example, has a coefficient of 24.1×10^{-6} (K)$^{-1}$, while that of tungsten is 3.95×10^{-6} (K)$^{-1}$. Of special interest is the difference between α iron with a coefficient of about 14×10^{-6} (K)$^{-1}$ and γ iron with 18×10^{-6} (K)$^{-1}$. As illustrated in the problems,

Table 15.2 THERMAL PROPERTIES OF SELECTED MATERIALS

Material	Specific Heat C_p (at 300 K), cal/g-K*	Atomic or Molecular Weight, g/mole	Linear Coefficient of Expansion [at 300 $(K)^{-1} \times 10^{-6}$]	Lorenz Number† $L = \sigma_T/\sigma_e T$, $(volts/K)^2 \times 10^{-8}$	Melting Point T_m, K
Aluminum	0.22	27	24.1	2.2	933
Carbon					
Graphite	0.18	12	2.3 to 2.8		3760
Diamond	0.12		1.2		
Copper	0.092	63.5	17.6	2.23	1356
Gold	0.031	197	13.8	2.35	1336
Iron	0.11	55.9	14.0	2.47	1810
Lead	0.32	207.2	28	2.47	600
Molybdenum	0.065	96	5.55	2.61	2880
Nickel	0.105	58.7	13.3	2.2	1725
Niobium	0.074	92.9	7.4		2740
Platinum	0.031	195.1	8.8	2.51	2040
Silver	0.056	107.9	19.5	2.31	1333
Tantalum	0.036	181	6.7		3270
Tin	0.54	118.7	23.5	2.52	505
Tungsten	0.034	183.9	3.95	3.04	3680
Type 304 stainless steel (austenitic)	0.14		17.3		~1690
Invar	0.12		1.26		~1700
Alumina ceramic	0.18	101.9	6.7		2470
Boron nitride		24.8	7.72		1950
Magnesia (MgO)	0.21	40.3	14		3020
Fused silica (or fused quartz)	0.19	60.1	0.05		1950
Glass	0.2		7.2		
Mica	0.12 to 0.25		40		
Phenolic resins	~0.4		15 to 45		
Polymethylmethacrylate	0.35		81		
Polytetrafluoroethylene	0.25		100		
Low-density polyethylene	0.5		180		
Natural rubber			670		
Silicone rubber			1,200		

* Multiply by 4.186 to obtain J/g-K.   † At 20°C (293 K).
Source: R. M. Rose, L. A. Shepard, and J. Wulff, "The Structure and Properties of Alloys," vol. 4: Electronic Properties, John Wiley & Sons, Inc., New York, 1966.

the austenitic stainless steels exhibit the same level of expansion as γ iron, while other steels resemble α iron. This difference can cause severe stresses when structures containing different types of steel are heated. In more than one case, in gas turbines interference between mating parts made of α and γ structures has been encountered.

A bimetallic strip is an example of how differential thermal expansion can be used in temperature control relays. If we bond two metals together, as by hot rolling, we form a bimetallic strip. If a piece is held at one end and heated, it will bend if the metals have different coefficients of expansion, and the amount will depend on the temperature.

EXAMPLE 15.4 (This example helps to understand the action of a thermostat.)

Given strips of iron and copper 0.010 in. wide, 0.005 in. thick, and 1 in. long, bond them together lengthwise along the flat faces at 70°F. Then heat the bimetal to 170°F.

a. Which way will it curl?
b. If the bimetal is restrained from curling (as in a die), what will be the stress developed in the iron and in the copper portions?

Assume the following data:

	Linear coefficient of expansion, in./in.-°F	Modulus of elasticity, psi
Iron	7.8×10^{-6}	30×10^6
Copper	10.0×10^{-6}	16×10^6

ANSWER

a. The copper strip if unbonded would expand more than the iron strip. The bonded strip, therefore, curls into an arc with the copper on the outside and the iron on the inside.
b. If the bonded strip is restrained between flat plates, the copper is prevented from attaining the length it would have reached if not bonded to the iron. It is under compressive stress. The iron, on the other hand, is stretched by the tendency of the copper to expand a greater amount than the iron, and the iron is under tension. Since the cross-sectional areas are equal, the compressive stress in the copper is the same magnitude as the tensile stress in the iron (or movement would occur due to the unbalanced force):

$$\sigma_{Cu} = \sigma_{Fe}$$

Now for a moment consider the expansion which would have occurred in each metal if it were not bonded to the other during heating from 70 to 170°F.

The change in length in the iron

$$\Delta l = (170 - 70)°F \times 7.8 \times 10^{-6} \text{ in./in.-°F}$$
$$= 780 \text{ microinches (in an inch length)}$$

For the copper

$$\Delta l = (170 - 70)°F \times 10.0 \times 10^{-6} \text{ in./in.-°F}$$
$$= 1,000 \text{ microinches (in an inch length)}$$

Therefore, the free iron specimen would be 1.000780 in. long and the free copper 1.001000 in.

Returning to the bonded case, we find that the specimen would have a length between these values. Furthermore, the difference in elastic strain between the copper and the iron would equal 1,000 − 780 microinches/in., or 220 microinches/in. In other words, the copper attempts to expand 1,000 microinches/in. and the iron only 780. An accommodation is reached in which the copper side is under compression and the iron side in tension. The important point is that the elastic strain in the copper plus that in the iron will equal 220 microinches:

$$\varepsilon_{Cu} + \varepsilon_{Fe} = 220 \times 10^{-6}$$

Substituting the relation $\varepsilon = \sigma/E$ in each case,

$$\frac{\sigma_{Cu}}{E_{Cu}} + \frac{\sigma_{Fe}}{E_{Fe}} = 220 \times 10^{-6}$$

But we know $\qquad \sigma_{Cu} = \sigma_{Fe}$

or $\qquad E_{Cu}\varepsilon_{Cu} = E_{Fe}(220 \times 10^{-6} - \varepsilon_{Cu})$

Substituting the numerical values given for E_{Cu} and E_{Fe}, we obtain:

$$\varepsilon_{Cu} = 143 \times 10^{-6} \text{ in./in. } (\sigma_{Cu} = 2,300 \text{ psi compression})$$
$$\varepsilon_{Fe} = 77 \times 10^{-6} \text{ in./in. } (\sigma_{Fe} = 2,300 \text{ psi tension})$$

In the polymers the coefficients of expansion are orders of magnitude greater as a result of the low intermolecular bonding forces. However, where bonding is weakest, in the elastomers, the coefficients are 670 to 1,200 × 10^{-6} (K)$^{-1}$, and in Teflon the coefficient is 100 × 10^{-6} (K)$^{-1}$. These values must be kept in mind if cracking due to differences in expansion on heating is to be avoided in metal–brittle plastic assemblies.

The low coefficient of expansion of the iron-nickel alloy known as Invar *below* 200°C is an interesting example of the interaction between thermal and magnetic effects. Increasing the temperature tends to produce expansion, but because of the special strong magnetic attraction between atoms in this alloy, only a small net expansion takes place. However, above the Curie temperature (where ferromagnetism disappears) the alloy expands normally.

Other anomalous effects are the differences in coefficients of expansion in different crystal directions in cubic as well as other systems. These exist in both metals and ceramics. Finally, as a word of caution in design, the coefficient of expansion should never be taken as an average over a temperature range where phase transformation takes place, or catastrophic failure can occur as a result of the great dilation due to transformation alone.

SUMMARY

The optical properties of major importance are emission, absorption, reflection, transmission, and refraction. The basis for understanding each of these phenomena is the relation between electrons and photons, which may be thought of as little energy packets with definite energy and wavelength.

Emission of a photon occurs when an electron falls from a higher state or quantum level in an atom. For example, sodium atoms emit yellow light because the predominant electron jump is accompanied by the emission of a photon of about $\lambda = 5890\,\text{Å}$. In fluorescent and phosphorescent materials photons of higher energy such as x-rays are absorbed by electrons which fall back through a series of electron traps formed by atoms of impurities. The fall from an intermediate trap to the normal level is accompanied by the emission of visible light.

Absorption is the reverse of emission. A photon strikes an electron which can be raised to an existing higher state by the energy absorption. This explains why metals are opaque. There is such a selection of empty energy states for electrons which are continuous with valence bands that photons of practically any wavelength can be absorbed.

Reflection is the re-emission of light from a surface. If there is no selective absorption, i.e. absorption of photons of certain energy levels by electrons, the light is re-emitted with the color unchanged. In some cases such as copper and gold, photons in the blue and green portions of the spectrum are absorbed.

Transmission is closely related to absorption. A diamond, for instance, is transparent to visible light because photons of these wavelengths do not have enough energy to raise the electrons through the energy gap to the conduction band. Therefore, the light passes through unaffected. Color centers may lead to the change from white to colored light. These centers result from impurities or lattice defects that lead to regions in the structure with electrons that can absorb certain photons of the visible spectrum. The remaining transmitted light is colored.

Refraction is due to the slower speed of light in a material compared to a vacuum, and the index of refraction increases with density.

The thermal properties of major importance are heat capacity, thermal conductivity, and thermal expansion.

Heat capacity is the energy required to raise the temperature of a system 1 K and may be thought of as increasing the kinetic and potential energies of the atoms. Since a mole contains a definite number of atoms, we find for most solids the heat capacity per mole is constant [6 cal/mole-K (25.1 J/mole-K)].

Thermal conductivity is due to the motion of carriers such as free electrons and photons. Because of the relatively great quantity of free electrons in metals, the conductivity is great. Thermal expansion results from the fact that although atoms vibrate about an average position, the distances between average positions change with temperature. The greater the bonding forces, the lower the change in position. For this reason the covalent and some of the ionic solids have the lowest coefficients. The metals are intermediate, and finally the polymers with van der Waals bonds have the highest coefficients.

DEFINITIONS

Light emission   The emission of photons of wavelengths in the visible spectrum, caused by electron jumps from higher to lower energy levels.

Fluorescence   The emission of light in the visible spectrum by a material exposed to ultraviolet light.

Phosphorescence   The continued emission of light in the visible spectrum after removal of the ultraviolet light source.

Absorption   The preferential absorption of light of certain wavelengths.

Photoelectric effect   The absorption of photons of definite wavelengths leading to emission of electrons, i.e. photoelectric current.

Transparency   The state in which photons of transmitted light do not interact with electrons of the structure. The photon energy is insufficient to liberate bound electrons. An example of a transparent material is the diamond.

Opacity   The state in which photons interact with electrons and raise electrons to higher levels (no light is transmitted). Opacity occurs, for example, in the metals.

Color centers   The preferential absorption and transmission of photons of certain (colored) wavelengths because certain ions (transition elements) have inner shell electrons that interact with photons of certain wavelengths and do not interact with others.

Heat capacity, C_p   The heat required to raise the temperature of a system 1 K, measured in cal/mole-K.

Thermal conductivity The quantity of heat transmitted through a unit volume in unit time. Thermal conductivity = cal/cm-sec-K.

Thermal expansion The change in unit length per unit of temperature change; $\Delta l/\Delta T = $ cm/cm-K = $(K)^{-1}$.

PROBLEMS

15.1 The most valuable diamonds have a bluish tint. Given that the wavelength of blue light is 4500 Å and the energy gap of pure diamond is 6 eV, explain the source of the color.

15.2 Explain why the electrical conductivity of a semiconductor can be increased by shining a light of a given wavelength on the conductor. This is called "photoconductivity." How could you use this effect to activate the beam at a remote lighthouse?

15.3 Glass is more easily fabricated into laser rods than single crystals of ruby. Early attempts to make lasers of glass with Cr^{3+} ions as impurity failed because the Cr^{3+} interacted with the ions of the glass rather than acting independently, as when substituted for Al^{3+} in alumina. However, satisfactory glass lasers have been made using neodymium in silicate glass. Look up the electronic structure of neodymium and explain what characteristic would lead to less interaction between the adjacent ions of the glass compared to Cr^{3+}. (*Hint*: Determine which electrons are the color centers in the neodymium.)

15.4 An absorption-vs.-wavelength curve for a material shows a strong maximum at a wavelength equal to 1500 Å but little absorption at longer wavelengths.

  a. Calculate the energy gap.
  b. Would the material be transparent, colored, or opaque in transmitting white light?
  c. Would you expect the material to be metallic or nonmetallic?

15.5 Calculate whether barium with a work function of 2.50 eV would be more suitable than the tungsten in Example 15.2 for use as a photocell with visible light.

15.6 The steady-state heat flow in a material is directly proportional to the thermal conductivity. Why are bricks for furnaces deliberately made with voids? What are the characteristics of home insulation?

15.7 Calculate how close the heat capacity values C_p of the following materials come to the ideal value of 6 cal/mole-K (25.1 J/mole-K) at 298 K (room temperature).

  a. Copper: 0.092 cal/g-K (0.324 J/g-K)
  b. Nickel: 0.105 cal/g-K (0.44 J/g-K)

c. N_2: 0.24 cal/g-K (1.04 J/g-K)

d. Aluminum: 0.22 cal/g-K (0.92 J/g-K)

15.8 In the production of aluminum cylinder heads wear-resistant valve guides are needed because the Inconel "X" valve stems wear the aluminum block too rapidly. After much experimentation austenitic gray cast-iron (Ni-resist) valve-guide inserts were found successful. Hardened ferritic steel inserts also wore satisfactorily but loosened up from a shrink fit in the aluminum when exposed to service. (A shrink fit is obtained by machining the outside dimension of the valve guide slightly oversize, cooling the guide in dry ice to below the outside dimension of the hole, and inserting while cold.) The coefficients of expansion of the materials are:

Aluminum block: $12 \times 10^{-6}\,(°F)^{-1}\,[21.6 \times 10^{-6}\,(°C)^{-1}]$
Austenitic cast iron: $11 \times 10^{-6}\,(°F)^{-1}\,[19.8 \times 10^{-6}\,(°C)^{-1}]$
Hardened steel: $7 \times 10^{-6}\,(°F)^{-1}\,[12.6 \times 10^{-6}\,(°C)^{-1}]$

Why do the austenitic cast-iron inserts maintain the shrink fit better?

15.9 Calculate the stresses in the bimetallic strip of Example 15.4 if the iron strip is 0.002 in. (0.00508 cm) thick and the copper strip is 0.005 in. (0.0127 cm) thick.

15.10 Calculate the stresses per 100°F (55.6°C) temperature change in a bimetallic strip made of austenitic stainless steel and SAE 1020 steel. The coefficient of expansion of austenitic stainless steel is 9.6×10^{-6} in./in.-°F $(17.3 \times 10^{-6}$ m/m-°C) and that of SAE 1020 steel is 7.8×10^{-6} in./in.-°F $(14.0 \times 10^{-6}$ m/m-°C). Assume the individual metals have equal cross-sectional areas.

REFERENCES

Without attempting to be comprehensive, we have assembled a few references for the convenience of the students and instructor. These are divided into four groups: (1) Metals, (2) Ceramics, (3) Polymers, and (4) General. In each group, the first references are those which can be used readily by the students after completion of this course, in order to find detailed specifications for materials in their professional work. Then follows a list of texts for study beyond the scope of the present work.

Metals

American Society of Testing Materials Specifications (*see General*)

"Metals Handbook," 8th ed., American Society for Metals
    Metals Park, Ohio
    This is not one handbook, but is composed of a group of eight large volumes at present, beginning with volume 1, "Properties and Selection of Metals," and advancing through topics such as heat treatment, casting, welding, phase diagrams, and microstructures. This is a very authoritative source because the material is written under the supervision of excellent committees of technical experts.

"Metals Reference Book," 4th ed., C. J. Smithells
    Plenum Publishing Corp., New York, 1967
    This work in three volumes provides many worthwhile tables of properties such as diffusivity, thermochemical data, etc.

"The Structure of Metals," 3d ed., C. S. Barrett and T. B. Massalski
    McGraw-Hill Book Company, New York, 1966
    This text has two important features: (1) a thorough discussion of crystallographic methods such as x-ray diffraction techniques, and (2) excellent discussions of metallic structures and transformation mechanisms. A good deal of the material is at senior or graduate student level.

Textbooks on Metallurgy

"Atomic Theory for Students of Metallurgy," W. Hume-Rothery
    Institute of Metals, London, 1955
"Elements of Mechanical Metallurgy," W. J. M. Tegart
    Macmillan Publishing Co., New York, 1966
"Elements of Physical Metallurgy," A. G. Guy
    Addison-Wesley Publishing Company, Inc., Reading, Mass., 1959
"Elements of X-Ray Diffraction," B. D. Cullity
    Addison-Wesley Publishing Company, Inc., Reading, Mass., 1956

"High Strength Materials," V. F. Zackay
    John Wiley & Sons, Inc., New York, 1965
"Mechanical Metallurgy," G. E. Dieter
    McGraw-Hill Book Company, New York, 1961
"The Mechanical Properties of Matter," A. H. Cottrell
    John Wiley & Sons, Inc., New York, 1964
"Phase Diagrams in Metallurgy," F. N. Rhines
    McGraw-Hill Book Company, New York, 1956
"Physical Metallurgy," E. Birchenall
    McGraw-Hill Book Company, New York, 1959
"Physical Metallurgy," Bruce Chalmers
    John Wiley & Sons, Inc., New York, 1959
"Physical Metallurgy Principles," 2d ed., R. E. Reed-Hill
    Van Nostrand Reinhold Co., New York, 1973
"Science of Metals," N. H. Richman
    Blaisdell Publishing Company, Waltham, Mass., 1967
"Structure and Properties of Alloys," 3d ed., R. M. Brick, R. B. Gordon, and
    A. Phillips
    McGraw-Hill Book Company, New York, 1965

Ceramics

American Society of Testing Materials Specifications (*see General*)
"Phase Diagrams for Ceramists," E. M. Levin, C. R. Robbins and H. F.
    McMurdie
    American Ceramic Society, Columbus, Ohio, 1964

Textbooks on Ceramics

"Electronic Ceramics," E. C. Henry
    Doubleday & Company, Inc., Garden City, N.Y., 1969
"Elements of Ceramics," F. H. Norton
    Addison-Wesley Publishing Company, Inc., Reading, Mass., 1952
    Excellent discussions of technique as well as basic material.
"Glass-Ceramics," P. W. McMillan
    Academic Press, London, 1964
"Glass Science," R. H. Doremus
    John Wiley & Sons, Inc., New York, 1973
"Introduction to Ceramics," W. D. Kingery
    John Wiley & Sons, Inc., New York, 1960
    General coverage of ceramic structures and properties.
"Physical Ceramics for Engineers," L. H. Van Vlack
    Addison-Wesley Publishing Company, Inc., Reading, Mass., 1964
    Basic coverage with many illustrative problems.

Polymers

American Society for Testing Materials Specifications (*see General*)

Materials Selector
  Reinhold Publishing Co., Stanford, Conn., published yearly
Modern Plastics Encyclopedia
  McGraw-Hill Book Company, New York, published yearly

Textbooks on Polymers

"Mechanical Properties of Polymers," L. E. Nielsen
  Van Nostrand Reinhold Co., New York, 1962
"Organic Polymers," T. Alfrey and E. F. Gurnee
  Prentice-Hall, Inc., Englewood Cliffs, N.J., 1967
"Properties and Structure of Polymers," A. V. Tobolsky
  John Wiley & Sons, Inc., New York, 1960
"Textbook of Polymer Science," 2d ed., F. W. Billmeyer
  John Wiley & Sons, Inc., New York, 1971.

General

American Society for Testing Materials Specifications
  American Society for Testing Materials, Philadelphia
  This is a number of separate volumes issued triennially. These cover metals, ceramics, and polymers and are the most commonly used standard for specification.
"Crystal Structures," R. W. G. Wykoff
  John Wiley & Sons, Inc., New York, 1963
  Several volumes giving details of crystal structures.
"Materials Handbook," G. S. Brady
  McGraw-Hill Book Company, New York, 1951
Materials Selector
  Reinhold Publishing Co., Stanford, Conn., published yearly

Textbooks on Materials in General

"Corrosion Control," H. H. Uhlig
  John Wiley & Sons, Inc., New York, 1963
"Corrosion Engineering," M. B. Fontana and N. D. Green
  McGraw-Hill Book Company, New York, 1967
"Electronic and Magnetic Properties of Materials," A. Nussbaum
  Prentice-Hall, Inc., Englewood Cliffs, N.J., 1967
"Electronic Processes in Materials," L. V. Azaroff and J. J. Brophy
  McGraw-Hill Book Company, New York, 1963

"Introduction to Materials Science," A. L. Ruoff
Prentice-Hall, Inc., Englewood Cliffs, N.J., 1972
"Introduction to Properties of Materials," D. Rosenthal
D. Van Nostrand Company, Inc., Princeton, N.J., 1964
"Introduction to Solids," S. V. Azaroff
McGraw-Hill Book Company, New York, 1960
"Materials Science for Engineers," L. H. Van Vlack
Addison-Wesley Publishing Company, Inc., Reading, Mass., 1970
"Mechanical Behavior of Engineering Materials," J. Marin
Prentice-Hall, Inc., Englewood Cliffs, N.J., 1962
"Modern Composite Materials," L. J. Broutman and R. H. Krock, eds.
Addison-Wesley Publishing Company, Inc., Reading, Mass., 1967
"Physics of Solids," 2d ed., C. A. Wert and R. M. Thomson
McGraw-Hill Book Company, New York, 1970
"Principles of Engineering Materials," C. R. Barrett, W. D. Nix, and
A. S. Tetelman
Prentice-Hall, Inc., Englewood Cliffs, N.J., 1973
"The Structure and Properties of Materials," A. T. Di Benedetto
McGraw-Hill Book Company, New York, 1967
"The Structure and Properties of Materials," 4 vols., J. Wulff et al.
John Wiley & Sons, Inc., New York, 1965

INDEX

Italic numbers indicate where definitions may be found.

PHOTO CREDITS

PHYSICAL PROPERTIES OF SELECTED ELEMENTS

Element	Symbol	Atomic Number	Atomic Weight	MP (°C)	Density (g/cm³)	Crystal Structure	Atomic Radius (Å)	Ionic Radius (Å)	Most Common Valence
Aluminum	Al	13	26.98	660	2.699	FCC	1.43	0.57	+3
Argon	A	18	39.99	−189	1.78×10^{-3}	FCC	1.92	—	—
Barium	Ba	56	137.36	714	3.5	BCC	2.24	1.43	+2
Beryllium	Be	4	9.01	1277	1.85	HCP	1.13	0.54	+2
Boron	B	5	10.82	2030	2.34	Ortho.	0.97	0.2	+3
Bromine	Br	35	79.92	−7.2	3.12	Ortho.	1.19	1.96	−1
Cadmium	Cd	48	112.41	321	8.65	HCP	1.52	1.03	+2
Calcium	Ca	20	40.08	838	1.55	FCC	1.97	1.06	+2
Carbon[1]	C	6	12.01	3727	2.25	Hex.	0.71	<0.20	+4
Cerium	Ce	58	140.13	804	6.77	HCP	1.82	1.18	+3
Cesium	Cs	55	132.91	28.7	1.90	BCC	2.70	1.65	+1
Chlorine	Cl	17	35.46	−101	3.21×10^{-3}	Ortho.	1.07	1.81	−1
Chromium	Cr	24	52.01	1875	7.19	BCC	1.28	0.64	+3
Cobalt	Co	27	58.94	1495	8.85	HCP	1.25	0.82	+2
Copper	Cu	29	63.54	1083	8.96	FCC	1.28	0.96	+1
Fluorine	F	9	19.00	−220	1.70×10^{-3}	—		1.33	−1
Germanium	Ge	32	72.60	937	5.32	Dia.	1.39	0.44	+4
Gold	Au	79	197.00	1063	19.32	FCC	1.44	1.37	+1
Helium	He	2	4.00	−270	0.18×10^{-3}	HCP	1.79	—	—
Hydrogen	H	1	1.01	−259	0.09×10^{-3}	HCP	0.46	1.54	−1
Iodine	I	53	126.91	114	4.94	Ortho.	1.36	2.20	−1
Iron	Fe	26	55.85	1536	7.87	BCC	1.28	0.87	+2
Lead	Pb	82	207.21	327	11.36	FCC	1.75	1.32	+2
Lithium	Li	3	6.94	180	0.534	BCC	1.57	0.78	+1
Magnesium	Mg	12	24.32	650	1.74	HCP	1.60	0.78	+2
Manganese	Mn	25	54.94	1245	7.43	Cubic	1.12	0.91	+2
Mercury	Hg	80	200.61	−38.4	13.55	Rhomb.	1.55	1.12	+2
Molybdenum	Mo	42	95.95	2610	10.22	BCC	1.40	0.68	+4
Neon	Ne	10	20.18	−249	0.90×10^{-3}	FCC	1.60	—	—
Nickel	Ni	28	58.71	1453	8.90	FCC	1.25	0.78	+2
Niobium	Nb	41	92.91	2468	8.57	BCC	1.47	0.74	+4
Nitrogen	N	7	14.01	−210	1.25×10^{-3}	Cubic	0.71	0.1 to 0.2	+5
Oxygen	O	8	16.00	−219	1.43×10^{-3}	Ortho.	0.60	1.32	−2
Phosphorus[2]	P	15	30.98	44.3	1.83	Ortho.	1.09	0.3 to 0.4	+5
Platinum	Pt	78	195.09	1769	21.45	FCC	1.38	0.52	+2
Potassium	K	19	39.10	63.7	0.86	BCC	2.38	1.33	+1
Scandium	Sc	21	44.96	1539	2.99	FCC	1.60	0.83	+2
Silicon	Si	14	28.09	1410	2.33	Dia.	1.17	0.39	+4
Silver	Ag	47	107.88	961	10.49	FCC	1.44	1.13	+1
Sodium	Na	11	22.99	97.8	0.971	BCC	1.92	0.98	+1
Strontium	Sr	38	87.63	768	2.60	FCC	2.15	1.27	+2
Sulfur[3]	S	16	32.07	119	2.07	Ortho.	1.04	1.74	−2
Tin	Sn	50	118.70	232	7.30	Tetra.	—	0.74	+4
Titanium	Ti	22	47.90	1668	4.51	HCP	1.47	0.64	+4
Tungsten	W	74	183.86	3410	19.3	BCC	1.41	0.68	+4
Uranium	U	92	238.07	1132	19.07	Ortho.	1.38	1.05	+4
Vanadium	V	23	50.95	1900	6.1	BCC	1.36	0.61	+4
Zinc	Zn	30	65.38	419	7.13	HCP	1.37	0.83	+2
Zirconium	Zr	40	91.22	1852	6.49	HCP	1.60	0.87	+4

[1] Present as graphite—sublimes rather than melts.
[2] White phosphorus.
[3] Yellow sulfur.